Quantum Machine Learning and Optimisation in Finance

Drive financial innovation with quantum-powered algorithms and optimisation strategies

Antoine Jacquier

Oleksiy Kondratyev

Quantum Machine Learning and Optimisation in Finance

Assistant Group Product Manager: Kunal Sawant
Publishing Product Manager: Samriddhi Muraka
Project Manager: Prajakta Naik
Senior Editor: Rounak Kulkarni
Technical Editor: Vidhisha Patidar
Copy Editor: Safis Editing
Indexer: Pratik Shirodkar
Proofreader: Rounak Kulkarni
Production Designer: Alishon Mendonca
Business Development Executive: Jaron Richy

First published: October 2022
Second edition: December 2024

Production reference: 1021224

Published by Packt Publishing Ltd.
Grosvenor House
11 St Paul's Square
Birmingham
B3 1RB, UK.

ISBN 978-1-83620-961-4

www.packtpub.com

Foreword

As far as computational prospects (including hardware and software) and related developments are concerned, we live in the most exciting times. Every day brings new achievements, new promises, and, on occasion, new disappointments. Presently, some of the ideas discussed for decades are reaching the usable stage, the most exciting developments being thermonuclear fusion, all-purpose artificial intelligence, distributed ledger technology, and quantum computing. However, not surprisingly, breakthroughs in these fields are very hard to achieve, so, despite strenuous government and private efforts and lavish funding, their stated goals have not been reached yet.

Jacquier and Kondratyev, two of the strongest quants of their generation, have written a remarkable book about quantum computers and their applications in finance, emphasizing practical aspects of quantum machine learning and optimisation. To put the subject of this book in a proper context, let us briefly touch upon the history of computing devices. Without exaggeration, one can claim that the history of human progress is closely related to the history of computing devices. Original simple but beneficial instruments, such as abacuses, were invented at the dawn of civilization. Eventually, computing devices developed into potent tools, such as supercomputers, that define our day-to-day existence and future as species.

To start with, it is worth mentioning some milestones. Abacuses were used in Babylonia as early as c. 2700–2300 BC; eventually, they spread worldwide and became known as Roman abacuses. Computational devices were used together with memory devices, such as tally sticks and clay tablets. In classical antiquity, analog computers, such as the fabled

Antikythera, probably intended for astronomical calculations, were actively used. Subsequently, medieval Muslim astronomers and engineers brought such devices to the next level. As a result, they developed remarkable objects, such as sundials, planispheres, and astrolabes.

Eventually, the centre of progress moved to Europe and, later, to North America, where various calculating tools, such as the slide rule and mechanical calculator, were introduced in early modern times. However, the real breakthrough was achieved in 1804 when Jacquard created a loom programmable via punched cards. Babbage, "the father of the computer", invented the first mechanical computer programmable with punch cards in the United Kingdom. Unfortunately, his Difference Engine and Analytical Engine were never completed because their execution involved severe practical issues, including insurmountable engineering obstacles, lack of funding, and general ridicule (Babbage's story should serve as a fair warning to all the intrepid inventors who come with ideas centuries ahead of their time). Several decades later, the American inventor Hollerith was much more successful. He used punched cards to store data readable by a tabulator machine; IBM can trace its origins to these humble cards.

In the late 1800s to early 1900s, analog computers, such as tide-predicting machines, using physical phenomena to model the problem they were built to solve, became all the rage. Specifically, an analog computer's developers must find physical processes governed by the same or similar equations as the problem of actual interest. Since, by design, such computers use noisy continuous values subject to various errors, they produce approximate solutions. However, such solutions are highly beneficial in many applications, including warfare, navigation, and economics, to mention but a few. One of the more remarkable examples of analog computers is the Phillips Hydraulic Computer, using simple tools to control the water's flow to model the national economy of the United Kingdom with a 2% accuracy. Others are the battleship's fire-control systems and bomb sights. Shortly before and during World War II, digital computers evolved into formidable competitors of analog computers and, by the 1950s, replaced them altogether, except for some highly specialized applications. Stemming from theoretical insights due to Turing, who invented the celebrated eponymous

Turing Machine, and several other pioneers and practical engineering inventions, such as vacuum tubes and transistors, digital computers conquered the world. Initially, digital computers, such as the Ananasov-Berry computer, the Colossus, and the ENIAC, to mention but a few, were enormous machines built to serve military needs, including nuclear weapons design and cryptography. Eventually, computers found a wide plethora of commercial applications.

Modern digital computers store the necessary data in a magnetic memory in the form of 0-1 bits; they operate on this data via logical gates. The fact that the data is digitised has profound implications. On the one hand, while undeniably humongous, the amount of data a computer can store and the speed at which it can process this data are limited. On the other hand, inevitable errors inherent in any physical device are relatively easy to control.

Quantum computers were independently conceived in the early 1980s by Benioff (who proposed a quantum version of the Turing Machine), Feynman, and Manin. Feynman, with his usual eloquence, stated: "Nature isn't classical, dammit, and if you want to make a simulation of nature, you'd better make it quantum mechanical, and by golly, it's a wonderful problem because it doesn't look so easy". Quantum computers use quantum mechanical phenomena to solve computational problems that conventional computers cannot solve efficiently (or at all). To build a working quantum computer, one must overcome tremendous difficulties and achieve conditions such that quantum effects become dominant or, at least, noticeable. For example, such a computer must operate at low temperatures close to absolute zero. The primary distinction between quantum computers and their classical brethren is how they store and operate on the data. Instead of the 0-1 bits used by digital computers, quantum computers use the so-called qubits, capable of storing continuous information represented by a point on the so-called Bloch sphere; however, once measured, the data collapses to the classical 0-1 state. In theory, qubits should be entangled so that their quantum states depend on the state of all other qubits. The data, stored by a quantum computer, is processed via quantum gates. Since the data is continuous rather than discrete, the storage problem is not an issue per se; however, the continuity of the data makes quantum computers prone to inevitable errors. As a result, fault-tolerant quantum computing with suitably many qubits remains a very distant possibility. Furthermore, building an actual qubit is very difficult. Different possibilities have been tried, such as Josephson's junctions, trapped ions, and many others.

Quantum computing has numerous exciting prospective goals, although it is unclear if all (or any) of them can be reached. The best known is cryptography, more concretely, factorising huge integers into prime factors using Shor's algorithm, thus breaking conventional asymmetric encryption. It is worth mentioning that Shor's algorithm is probabilistic; hence, it is particularly well adapted to quantum computers. The promise of achieving the factorisation breakthrough galvanised governments and private companies into pouring billions of dollars into quantum computing and related fields. For instance, the US government launched the National Quantum Initiative to explore and promote quantum information science. In addition, the National Institute of Standards and Technology evaluates and standardizes quantum-resistant public-key cryptographic algorithms. However, the biggest number reliably factored by Shor's algorithm is 35, factored in 2021 using a computer with very few qubits. Another exciting application is to search problems, which can be efficiently handled with Grover's algorithm, although practical implementations are lacking. Several research groups, such as the Google-NASA collaboration, actively study machine learning applications. Nevertheless, large-scale applications are still far from being conquered.

This situation brings us back to the Jacquier and Kondratyev book. Given the relatively slow progress achieved in quantum computation, it is natural to go back to the early days of modern computing hardware and see whether one can use a quantum-like computer as a modern analog computer; this is precisely what Jacquier and Kondratyev do. Specifically, rather than waiting for quantum computers to reach their top form, they advocate using quantum annealing computers and similar analog machines to perform the required calculations. Such computers start with a simple Hamiltonian with a known ground state and slowly (adiabatically) evolve the original Hamiltonian into the Hamiltonian of interest while always preserving the ground state. Thus, when the process is completed, the measurement allows one to find the ground state for the actual problem rather than a simplified one. Jacquier and Kondratyev use analog quantum computers to solve several exciting mathematical finance problems with verve and panache. Specifically, they discuss quantum boosting and demonstrate how to apply it to predict credit card defaults and solve classification problems. Subsequently, Jacquier and Kondratyev turn their attention to quantum Boltzmann machines and explain how to use them for distribution sampling.

Analog quantum computers can be used to solve many financial engineering problems. For instance, a quantum annealer is a perfect tool for solving the multi-period integer portfolio optimisation problem, which is NP-Complete. Monte Carlo simulations, which are fundamental for derivatives pricing and other related tasks, are particularly natural to perform using quantum computers since they can be viewed as true random number generators.

When it comes to the digital gate model quantum computing, Jacquier and Kondratyev discuss quantum neural networks and their applications to machine learning. Later, the authors cover Quantum Circuit Born Machine, Variational Quantum Eigensolver, and the Quantum Approximate Optimisation Algorithm. Finally, Jacquier and Kondratyev discuss new quantum algorithms, such as quantum kernels, Bayesian Quantum Circuit, Quantum Fourier Transform, and Quantum Monte Carlo Simulation.

Jacquier and Kondratyev are excited about quantum algorithms and their potential applications; however, they are not starry-eyed and approach quantum algorithms cautiously. Specifically, they put much effort into showing how classical algorithms can solve problems they are interested in and when quantum algorithms outperform their classical brethren.

While building quantum computers originally envisioned by Feynman and Manin might still be decades away, more practical analog quantum computers already exist. Experience suggests that finance is one of the fields where breakthroughs in computing tend to be used in real time. Quants who want to use quantum algorithms in their day-to-day work could scarcely do better than starting their journey by studying this book.

Bon voyage!

Alexander Lipton and Marcos López de Prado,
Abu Dhabi Investment Authority

Contributors

About the authors

Antoine Jacquier graduated from ESSEC Business School before obtaining a PhD in mathematics from Imperial College London.

His research focuses on stochastic analysis, asymptotic methods in probability, volatility modelling, and algorithms in quantum computing. He has published about 50 papers and has co-written several books. He is also the director of the MSc in mathematics and finance at Imperial College and regularly works as a quantitative consultant for the finance industry.

He has a keen interest in running and whisky.

> *I would like to thank all the people who challenged and motivated me in applied mathematics, asking the right questions, pointing to interesting problems, and proposing diverse solutions from many angles. Chief among them are Jim Gatheral, Peter Friz, Mathieu Rosenbaum, Josef Teichmann, Mark Davis, Claude Martini, and Aleksandar Mijatović, whose constant and generous ideas led to numerous advances in my career.*
>
> *My journey through quantum computing, while not on the obvious path dictated by my mathematical background, started out of this scientific curiosity acquired over the years. I am indebted to Kostas Kardaras, Mugad Oumgari, Alexandros Pavlis, and Amine Assouel for crawling patiently with me through the dark meanders of quantum computing and quantum mechanics.*

Oleksiy Kondratyev obtained his PhD in mathematical physics from the Institute for Mathematics, National Academy of Sciences of Ukraine, where his research was focused on studying phase transitions in quantum lattice systems.

Oleksiy has over 20 years of quantitative finance experience, primarily in banking. He was recognised as Quant of the Year 2019 by Risk magazine. Oleksiy is a Visiting Professor at the Department of Mathematics, Imperial College London, and a Research Fellow at ADIA Lab.

Outside the world of finance and quantum computing, Oleksiy's passion is for sailing, in particular, offshore racing. Oleksiy holds the RYA Yachtmaster Ocean certificate of competence and is a member of the Royal Ocean Racing Club.

Writing a book is a long journey and a result of many years of intensive research. This book would not have been possible without the discussions, collaborations, exchange of ideas, and support of Majed Al Romaithi, Bhavesh Amin, David Bell, Michael Brett, Kasper Christoffersen, Brian Coyle, Michael Cuthbert, David Garvin, Tushar Gupta, Max Henderson, Mark Hodson, Blanka Horvath, Wendy Huang, Ray Johnson, Elham Kashefi, Geoff Kot, Rounak Kulkarni, Alexander Lipton, Charissa Liu, Marcos López de Prado, Aaron Lott, Alex Manson, Roger McKinley, Ashley Montanaro, Samriddhi Murarka, Prajakta Naik, Krzysztof Osiewalski, Marco Paini, Manos Papathanasiou, Amit Ramadas, Chad Rigetti, Safis Editing, Christian Schwarz, Elena Strbac, Robert Sutor, Davide Venturelli, Agnieszka Verlet, José Viñals, Colin Williams, Bill Winters, and Stefan Wörner.

About the reviewer

Gerhard Hellstern (Prof., Dr. rer. nat, graduate physicist, *1971) is a professor at the Faculty of Economics at the Baden-Württemberg Cooperative State University in Stuttgart. From 1990 to 1995, he studied physics at the University of Tübingen and the State University of New York at Stony Brook; in 1998, he graduated as Dr. rer. nat. From 1998 to 2018, he was employed by several commercial banks and then for 17 years at Deutsche Bundesbank. There, he was in charge of the banking audits division for many years. Gerhard Hellstern has been involved in the application of data science methods (data analytics as well as machine and deep learning) in finance for many years. These methods also include quantum computing and quantum machine learning. He is a Qiskit advocate at IBM and a member of the research network QuantumBW in Germany. His current research focuses on applications of quantum computing/quantum machine learning in the financial sector and beyond, and he has published several papers in this domain.

Table of Contents

Standard notations

$\mathbb{C}$	Complex numbers
$\mathcal{H}$	Hamiltonian
$\mathcal{I}$	Identity operator
$\log$	Natural logarithm
$\log_2$	Logarithm base 2
$\mathbb{N}$	Natural numbers
$\mathbb{P}$	Probability
$\mathcal{P}$	Projection operator (measurement operator)
q	Vector of binary variables, $q := (q_1, \ldots, q_N)$, $(q_i \in \{0, 1\})_{i=1,\ldots,N}$
$\mathbb{R}$	Real numbers
s	Vector of spin variables, $s := (s_1, \ldots, s_N)$, $(s_i \in \{-1, +1\})_{i=1,\ldots,N}$
σ_α^i	Pauli operator ($\alpha \in \{x, y, z\}$) acting on qubit i
u	Variable (scalar)
u	Column vector
$u^\top$	Transpose of a column vector
u^*	Complex conjugate of a column vector
$|u\rangle$	Ket: column vector in Dirac notation
$\langle u|$	Bra: complex conjugate transpose of column vector in Dirac notation
$\langle u|v\rangle$	Inner product
$|u\rangle\langle v|$	Outer product
$|u\rangle \otimes |v\rangle$	Tensor product
U	Matrix
U	Quantum logic gate
$\mathcal{U}$	Operator
$\mathcal{U}^\dagger$	Adjoint operator
$\langle \psi|\mathcal{U}|\psi\rangle$	Expectation value of operator $\mathcal{U}$ in state $|\psi\rangle$

Standard abbreviations

AI	Artificial Intelligence
ANN	Artificial Neural Network
AQC	Adiabatic Quantum Computing
BQC	Bayesian Quantum Circuit
CD	Contrastive Divergence
CMOS	Complementary Metal-Oxide-Semiconductor
CNN	Convolutional Neural Network
DBM	Deep Boltzmann Machine
DMC	Density Matrix Classifier
DNN	Deep Neural Network
FN	False Negative
FP	False Positive
GA	Genetic Algorithm
GAN	Generative Adversarial Network
HHL	Harrow-Hassidim-Lloyd
KS	Kolmogorov-Smirnov
LISP	List Processing
LSTM	Long Short Term Memory
ML	Machine Learning
MLP	Multi-Layer Perceptron
MMD	Maximum Mean Discrepancy
MPQC	Multilayer Parameterised Quantum Circuit
MPS	Matrix Product State
MPT	Modern Portfolio Theory
NISQ	Noisy Intermediate Scale Quantum
NP	Non-deterministic Polynomial
PDE	Partial Differential Equation

PDF	Probability Density Function
PQC	Parameterised Quantum Circuit
PSO	Particle Swarm Optimisation
QA	Quantum Annealing
QAML	Quantum Annealing for Machine Learning
QAO	Quantum Anharmonic Oscillator
QAOA	Quantum Approximate Optimisation Algorithm
QBM	Quantum Boltzmann Machine
QCBM	Quantum Circuit Born Machine
QCNN	Quantum Convolutional Neural Network
QFT	Quantum Fourier Transform
QGAN	Quantum Generative Adversarial Network
QHO	Quantum Harmonic Oscillator
QLS	Quantum Linear Solver
QLSTM	Quantum Long Short Term Memory
QMC	Quantum Monte Carlo
QML	Quantum Machine Learning
QNN	Quantum Neural Network
QPE	Quantum Phase Estimation
QPU	Quantum Processing Unit
QRAM	Quantum Random Access Memory
QSDK	Quantum Software Development Kit
QSDP	Quantum Semidefinite Programming
QUBO	Quadratic Unconstrained Binary Optimisation
QVA	Quantum Variational Autoencoder
RAM	Random Access Memory
RBM	Restricted Boltzmann Machine
ReLU	Rectified Linear Unit
RSA	Rivest-Shamir-Adleman

RT	Resistor-Transistor
SDP	Semidefinite Programming
SVD	Singular Value Decomposition
SVM	Support Vector Machine
TN	True Negative
TP	True Positive
TPQC	Tensor Network Parameterised Quantum Circuit
TTS	Time-to-Solution
VQE	Variational Quantum Eigensolver

Preface

> Quantum machine learning – the most
> **overhyped** and **underestimated**
> field at the same time
>
> ———————————————————
>
> Iordanis Kerenidis

Introduction

Why quantum computing?

Quantum computing and AI will revolutionise and disrupt our society in the same way that classical digital computing did in the second half of the 20th century and the internet did in the first two decades of the 21st century.

Quantum computing (or, more generally, quantum information theory) has been the subject of extensive research since the 1960s, but it was only in the last decade that progress on the hardware side has made it possible to test quantum computing algorithms, and it was only in the last several years that quantum computing's supremacy was finally claimed as an experimental fact (when a landmark experiment was conducted on Google's 53-qubit Sycamore quantum chip [16]).

The story of quantum computing is, in this respect, similar to the story of AI: AI was born in the 1950s but then experienced two "winters", when interest in AI and machine learning declined considerably (following the Lighthill report in the UK, the Speech Understanding Research debacle in the US in the 1970s, and the LISP collapse in the 1990s), before becoming widely used and adopted to the point that we can no longer imagine our lives without it.

Even though we cannot rule out a "quantum computing winter" before quantum computing technology becomes embedded in everyday life to the same extent as the internet, smartphones, and AI, the range of quantum computing breakthroughs we have witnessed in the last few years makes it somewhat unlikely.

With recent advances in the field, we have finally reached the era of Noisy Intermediate-Scale Quantum (NISQ) computing [259]. NISQ-era computers are powerful enough to test quantum computing algorithms and solve non-trivial real-world problems – and establish quantum speedup and quantum advantage over comparable classical hardware.

However, it is likely that the first real-world production-level business applications will be a hybrid quantum-classical protocol, where most of the computation and data processing is done classically, but the hardest problems are outsourced to the quantum chip. In finance, discrete portfolio optimisation problems, which are NP-hard, are such examples and clear objectives to tackle.

Why quantum machine learning?

It is a combination of quantum computing and AI that will likely generate the most exciting opportunities, including a whole range of possible applications in finance, but also in medicine, chemistry, physics, etc. We have already witnessed the first promising results achieved with parameterised quantum circuits trained as either generative models (such as a quantum circuit born machine, which can be used as a synthetic data generator) or discriminative models (such as a quantum neural network that can be trained as a classifier). The possible use cases include market generators, data anonymisers, credit scoring, and the generation of trading signals.

All the models and techniques mentioned so far rely on the existence of universal, gate model quantum computers. However, there is another type of quantum hardware – quantum annealers – which realise the principle of adiabatic quantum computing. Quantum annealers are analog quantum computers that are very well suited for solving complex optimisation problems that are NP-hard for classical computers. Optimisation problems form a large class of hard-to-solve financial problems, not to mention the fact that many supervised and reinforcement learning tools used in finance are trained via solving optimisation problems (minimisation of a cost function, maximisation of reward).

An example of discriminative machine learning problems solved using quantum annealers would be building a strong classifier from several weak ones – the quantum boosting algorithm. The strong classifier is highly resilient against overtraining and errors in the correlations of the physical observables in the training data. The quantum annealing-trained classifiers perform similarly to state-of-the-art classical machine learning methods. However, in contrast to these methods, the annealing-based classifiers are simple functions of directly interpretable experimental parameters with clear physical meaning and demonstrate some advantage over traditional machine learning methods for small training datasets.

Another application of quantum annealing is in generative learning. In deep learning, a well-known approach for training a deep neural network starts with training a generative deep Boltzmann machine, typically using the Contrastive Divergence (CD) algorithm, and then fine-tuning the weights using backpropagation or other discriminative techniques. However, generative training is often time consuming due to the slow mixing of Boltzmann (Gibbs) sampling. The quantum sampling-based training approach can achieve comparable or better accuracy with significantly fewer iterations of generative training than conventional CD-based training.

The main focus of this book is therefore on tackling practical real-world applications of Quantum Machine Learning (QML) algorithms executable on NISQ hardware rather than adopting the more traditional quantum computing textbook approach, diligently describing standard quantum computing algorithms (Shor's, Grover's, . . .), the quantum hardware demands of which are well beyond the capabilities of NISQ computers. The focus is also on the hybrid quantum-classical computational protocols that reflect the most productive way of harnessing the power of quantum computing – it is in tandem with classical computing that quantum computing solutions can provide maximum benefits to the users. In this book, we cover all major QML algorithms that have been the subject of intensive research by the industry and that have shown early signs of potential quantum advantage. We also provide a balanced view of both analog and digital quantum computers and do not try to make a call on which quantum computing technology (superconducting qubits, trapped ions, neutral atoms, etc.) will be the eventual winner. The material is presented in a hardware-agnostic way with a strong emphasis on the fundamental characteristics of the algorithms rather than their hardware realisations, although we do not ignore the question of algorithms' embedding and the practical limitations of the existing quantum computing hardware.

Why finance?

It is reasonable to expect that the incredibly fast rate of quantum hardware improvements we have witnessed over the last several years will lead to the widespread adoption of quantum computing techniques in finance. The finance industry is already investigating the potential of QML to solve classically hard practical problems and assist in achieving digital transformation. We might have moved past the point of quantum computing supremacy, but our quest to establish quantum computing advantage has just begun.

Quantitative finance is a discipline rich in interesting but computationally hard problems. Many such problems are interdisciplinary in nature and often require the transformation and adoption of mathematical and computational techniques developed in other fields. Here, we can mention, for example, the application of the theory of stochastic differential equations to option pricing [241], methods of optimal control theory to management science and economics [284], machine learning techniques to portfolio construction, and optimisation [208].

This is why we turn to finance when we are looking for a wide range of real-world use cases to test (and improve!) quantum computing algorithms. The book provides many examples of the quantum computing techniques and algorithms applied to solving practical financial problems such as portfolio optimisation, credit card default prediction, credit approvals, and the generation of synthetic market data. At the same time, the methods and techniques are formulated and presented in the most general form – we hope our readers will discover many new exciting quantum computing use cases in finance and beyond.

Who this book is for

The book is primarily aimed at three main groups: academic researchers and STEM students, finance professionals working in the field of quantitative finance and related areas, and computer scientists and ML/AI experts. At the same time, the book is organized in such a way as to be accessible and useful to a much wider audience.

The book does not require any prior knowledge of quantum mechanics and the complexity of the mathematical apparatus should not feel intimidating: although we do not sacrifice mathematical rigor, the emphasis is very much on the understanding of the fundamental properties of the models and algorithms.

What this book covers

The book is split into two parts reflecting the natural progression from analog to digital quantum computing, with an increasing depth in the analysis and understanding of algorithms. However, we start with a chapter that covers the basic principles of quantum mechanics and provides the motivation for the computational methods based on those principles.

Chapter 1, The Principles of Quantum Mechanics, covers the basic mathematical principles of quantum mechanics. It provides the necessary definitions and discusses the postulates of quantum mechanics and their relevance to quantum computing.

Part I: Analog Quantum Computing – Quantum Annealing

For a number of years, quantum annealers were the only large-scale quantum computing devices available for experiments in solving non-trivial NP-hard combinatorial optimisation problems. Although quantum annealing specifically targets solving classically hard optimisation problems, it can also be used for many different hybrid quantum-classical problems, such as samplers and classifiers. The book provides detailed coverage of these applications and illustrates them on specific financial use cases.

Chapter 2, Adiabatic Quantum Computing, introduces the concept of analog quantum computing. The chapter starts with the principles of adiabatic quantum computing and proceeds with the quantum adiabatic theorem. The physical realisation of adiabatic quantum computing is quantum annealing, which is explained alongside its classical counterpart – simulated annealing. The chapter also discusses the implementation, limitations, and universality of adiabatic quantum computing.

Chapter 3, Quadratic Unconstrained Binary Optimisation, describes the single most important application of quantum annealing: solving classically hard optimisation problems. A wide range of combinatorial optimisation problems can be formulated as Quadratic Unconstrained Binary Optimisation (QUBO) problems (or, equivalently, as Ising problems) solvable on a quantum annealer. The chapter provides in-depth coverage of the forward and reverse quantum annealing techniques and demonstrates the power of quantum annealing on a discrete portfolio optimisation use case.

Chapter 4, Quantum Boosting, extends the range of QUBO applications beyond combinatorial optimisation and outlines the quantum boosting algorithm designed to combine a large number of weak classical classifiers into a strong classifier. The algorithm is formulated as a QUBO problem executable on a quantum annealer and applied to the use case of building a strong predictor of credit card defaults from a large number of weak predictors.

Chapter 5, Quantum Boltzmann Machine, explores further machine learning applications of quantum annealing. The quantum Boltzmann machine can be used as a generative model for sampling from a learned probability distribution as well as an efficient method of pre-training deep feedforward neural networks.

Part II: Gate Model Quantum Computing

Gate model quantum computing hardware has seen enormous progress in recent years and is quickly approaching the quantum advantage threshold. The search for quantum advantage – the real-world productive application of a quantum computing solution that outperforms any viable classical alternative – is one of the strongest motivations for quantum computing research in finance and elsewhere. The book explores the main quantum computing algorithms implementable on existing NISQ devices and highlights a range of possible financial applications that may benefit from this new computing paradigm.

Chapter 6, Qubits and Quantum Logic Gates, introduces the paradigm of gate model quantum computing. We start with the basic concepts of classical digital computing and expand the computational logic to accommodate the new principles of superposition and entanglement. The chapter draws parallels between and contrasts classical and quantum logic gates and shows how to assemble quantum circuits from individual quantum logic gates.

Chapter 7, Parameterised Quantum Circuits and Data Encoding, proceeds with the construction of quantum algorithms covering both the theoretical and the practical aspects of building Parameterised Quantum Circuits (PQCs) and demonstrates how classical samples can be encoded into quantum states processed by the PQCs. The chapter provides a detailed description of specific data encoding techniques.

Chapter 8, Quantum Neural Network, considers parameterised quantum circuits trained as classifiers. Throughout this chapter, we show how differentiable and non-differentiable learning algorithms can be used to efficiently train quantum neural networks. The chapter also discusses the limitations of existing QPUs and how to design quantum circuits that extract maximum benefit from the available quantum computing hardware. We investigate QNN performance on a credit approval use case and benchmark it against several standard classical classifiers.

Chapter 9, Quantum Circuit Born Machine, introduces a quantum counterpart to classical generative models such as Boltzmann machines – the Quantum Circuit Born Machine (QCBM). The chapter starts with the definition of the QCBM and how it can be efficiently configured and run on available QPUs, continues with the differentiable and non-differentiable learning and training procedures, and concludes with the market generator use case benchmarked against the classical Restricted Boltzmann Machine.

Chapter 10, Variational Quantum Eigensolver, introduces the variational principle and formulates the Variational Quantum Eigensolver (VQE) approach to optimisation problems. The chapter discusses a hybrid quantum-classical approach to training the VQE and looks at the practical aspects of running it on NISQ devices.

Chapter 11, Quantum Approximate Optimisation Algorithm, describes the gate model quantum computing approach (inspired by quantum annealing) to solving QUBO-type problems, such as NP-hard Max-Cut optimisation problems.

Chapter 12, Quantum Kernels and Quantum Two-Sample Test, covers two cutting-edge QML algorithms based on quantum feature maps realised with the help of parameterised quantum circuits.

Chapter 13, The Power of Parameterised Quantum Circuits, investigates the main sources of quantum advantage we expect to demonstrate on practical applications of parameterised quantum circuits. The chapter focuses on two elements: strong regularisation provided by quantum neural networks and the expressive power of quantum generative models.

Chapter 14, Advanced QML Models, discusses new promising quantum algorithms and techniques such as quantum GAN, Bayesian quantum circuit, symmetric encryption, and quantum semi-definite programming.

Chapter 15, Beyond NISQ, looks beyond the capabilities of NISQ era computers. The chapter presents several algorithms that lie in the foundation of many envisaged future applications of quantum computers. The algorithms include quantum Fourier transform, quantum phase estimation, quantum Monte Carlo, and quantum linear solver.

To get the most out of this book

This book is intended as an in-depth introduction to the power of quantum computing techniques for quantitative finance problems. While it is designed to be self-contained, this book assumes that the reader has some familiarity with basic mathematical concepts in algebra, analysis, and computing. Knowledge of quantum mechanics is not required, and the main tools thereof shall be explained and made accessible to non-physicists.

Conventions used

There are a number of text conventions used throughout this book.

`CodeInText` indicates code words in the text, software packages, folder names, path names, and so on.

When we wish to draw your attention to a particular definition or notation, the relevant lines or items are set in either *italic* or **bold**.

Important remarks and conclusions are shown in boxes.

Get in touch

Feedback from our readers is always welcome.

General feedback: If you have questions about any aspect of this book, email us at customercare@packtpub.com, and mention the book's title in the subject of your message.

Errata: Although we have taken every care to ensure the accuracy of our content, mistakes do happen. If you have found a mistake in this book, we would be grateful if you could report this to us. Please visit http://www.packtpub.com/support/errata, and fill in the form.

Piracy: If you come across any illegal copies of our works in any form on the internet, we would be grateful if you would provide us with the location address or website name. Please contact us at copyright@packtpub.com with a link to the material.

If you are interested in becoming an author: If there is a topic that you have expertise in, and you are interested in either writing or contributing to a book, please visit http://authors.packtpub.com.

Share your thoughts

Once you've read *Quantum Machine Learning and Optimisation in Finance, Second Edition*, we'd love to hear your thoughts! Scan the QR code below to go straight to the Amazon review page for this book and share your feedback.

https://packt.link/r/1836209614

Your review is important to us and the tech community and will help us make sure we're delivering excellent quality content.

Download a free PDF copy of this book

Thanks for purchasing this book!

Do you like to read on the go but are unable to carry your print books everywhere? Is your eBook purchase not compatible with the device of your choice?

Don't worry, now with every Packt book you get a DRM-free PDF version of that book at no cost.

Read anywhere, any place, on any device. Search, copy, and paste code from your favorite technical books directly into your application.

The perks don't stop there, you can get exclusive access to discounts, newsletters, and great free content in your inbox daily.

Follow these simple steps to get the benefits:

1. Scan the QR code or visit the link below:

https://packt.link/free-ebook/9781836209614

2. Submit your proof of purchase
3. That's it! We'll send your free PDF and other benefits to your email directly

1

The Principles of Quantum Mechanics

Quantum mechanics is a framework for the development of physical theories but is not itself a physical theory. Actual physical theories are built upon a foundation of quantum mechanics. This is why quantum mechanics plays such an important role in all natural sciences. Information theory is no exception and also derives inspiration from the ideas and methods of quantum mechanics.

Understanding quantum computing requires some familiarity with the basic principles of quantum mechanics. This book does not assume any prior knowledge of quantum mechanics and provides all the necessary definitions and explanations when needed. At the same time, the reader is encouraged to learn more about this fascinating subject at the level of mathematical formalism that they are comfortable with. Out of the extensive universe of textbooks on quantum mechanics that provide an introduction to this discipline, it is necessary to mention the classical book by Landau and Lifshitz [197] as well as the equally classical book on quantum computing by Nielsen and Chuang [238], which covers the most relevant aspects of quantum mechanics from the quantum computing perspective. For someone taking their first steps in quantum computing who would like to get an overall picture and some historical perspective, the excellent book by Bernhardt [32] provides both without the heavy usage of complex mathematical apparatus. Readers looking for a more formal modern take on the subject of quantum mechanics may find it in the book by Robinett [272]. The practical aspects of quantum computing are covered in great detail in the book by Sutor [302], and anyone looking for a python quantum computing programming textbook will find it in the work by Loredo [210].

1.1   Linear Algebra for Quantum Mechanics

Quantum computing and quantum mechanics rely on a specific notational formalism, due to Dirac, and are supported by classical linear algebra, in particular Hermitian structures of matrices and tensor products. We provide here a self-contained review of these tools to facilitate the understanding of the rest of the book. We start with basic linear algebra principles before introducing Dirac notations and the quantum counterparts of linear algebra tools. Sections 1.1.1 to 1.1.4 concentrate on standard definitions of finite-dimensional Hilbert spaces and matrices, while Sections 1.1.5 to 1.1.7 review the key details and properties of complex matrices (decompositions, Hermitian property, and rotations). Sections 1.1.9 to 1.1.11 introduce Dirac's formalism and the essential aspects of quantum operators.

1.1.1   Basic definitions and notations

We let $\mathbb{F}$ denote either the real field $\mathbb{R}$ or the complex one $\mathbb{C}$. For a complex number $z = x + iy \in \mathbb{C}$, with $x, y \in \mathbb{R}$, we write the conjugate $z^* := x - iy$. We let $\mathcal{M}_{m,n}(\mathbb{F})$ denote the space of matrices of dimension $m \times n$ with entries in $\mathbb{F}$ and $\mathcal{M}_n(\mathbb{F})$ whenever $m = n$. For $A := (a_{ij})_{1 \leq i \leq m; \ 1 \leq j \leq n} \in \mathcal{M}_{m,n}(\mathbb{F})$, $A^* := (a_{ij}^*)_{1 \leq i \leq m; \ 1 \leq j \leq n}$ is the complex conjugate. If $A \in \mathcal{M}_n(\mathbb{F})$, we write $A^\top$ for its transpose and $A^\dagger := (A^*)^\top$ for its *Hermitian conjugate*. We finally denote $\mathbf{I}$ as the identity matrix and write $\mathbf{I}_n$ whenever we wish to emphasise the dimension, and $\mathbf{0}_{m,n}$ as the null matrix in $\mathcal{M}_{m,n}(\mathbb{F})$. Recall that a matrix $A \in \mathcal{M}_n(\mathbb{F})$ is invertible (or non-singular) if there exists $B \in \mathcal{M}_n(\mathbb{F})$ such that $AB = BA = \mathbf{I}_n$.

Given two matrices, $A \in \mathcal{M}_{p,m}(\mathbb{F})$ and $B \in \mathcal{M}_{q,n}(\mathbb{F})$, we define their tensor product as

$$A \otimes B := \begin{bmatrix} a_{11}B & \cdots & a_{1m}B \\ \vdots & \ddots & \vdots \\ a_{p1}B & \cdots & a_{pm}B \end{bmatrix} \in \mathcal{M}_{pq,mn}(\mathbb{F}).$$

Since a vector is a particular case of a matrix, for $u \in \mathbb{F}^m$ and $v \in \mathbb{F}^n$, we can write

$$u \otimes v = \begin{bmatrix} u_1 \\ \vdots \\ u_m \end{bmatrix} \otimes \begin{bmatrix} v_1 \\ \vdots \\ v_n \end{bmatrix} = \begin{bmatrix} u_1 v_1 \\ \vdots \\ u_1 v_n \\ u_2 v_1 \\ \vdots \\ u_m v_n \end{bmatrix} \in \mathbb{F}^{mn}.$$

1.1.2   Inner products

A vector space $\mathbf{V}$ over the field $\mathbb{F}$ is a set endowed with

- A commutative, associative addition operation.
- An operation of multiplication by a scalar.

The addition and multiplication by a scalar have the following properties (for scalars $\alpha, \beta \in \mathbb{F}$ and vectors $u, v \in \mathbf{V}$):

- $v + \mathbf{0} = v$;
- $v + (-v) = \mathbf{0}$;
- $\alpha(\beta v) = (\alpha\beta)v$;
- $(\alpha + \beta)v = \alpha v + \beta v$;
- $\alpha(u + v) = \alpha u + \alpha v$;
- $1 \cdot v = v$.

Armed with this, we can now define an inner product on $\mathbf{V}$:

Definition 1. A map $\langle \cdot, \cdot \rangle : \mathbf{V} \times \mathbf{V} \to \mathbb{F}$ is called an *inner product* if, for u, v, w $\in \mathbf{V}$ and $\alpha \in \mathbb{F}$,

- *(Positive definiteness)* $\langle u, u \rangle \geq 0$ and $\langle u, u \rangle = 0$ if and only if $u = 0$.
- *(Conjugate symmetry)* $\langle u, v \rangle = \langle v, u \rangle^*$.
- *(Linear in the first argument)* $\langle u + v, w \rangle = \langle u, w \rangle + \langle v, w \rangle$ and $\langle \alpha u, v \rangle = \alpha \langle u, v \rangle$.
- *(Antilinear in the second argument)* $\langle u, v + w \rangle = \langle u, v \rangle + \langle u, w \rangle$ and $\langle u, \alpha v \rangle = \alpha^* \langle u, v \rangle$.

The inner product is further called non-degenerate if $\langle u, v \rangle = 0$ for all $v \in \mathbf{V} \setminus \{0\}$ implies $u = 0$.

For example, the following spaces carry a natural inner product:

- The vector space $\mathbb{C}^n$ with the inner product $\langle u, v \rangle := u^\dagger v = \sum_{i=1}^n u_i^* v_i$.
- The space of complex-valued continuous functions on $[0, 1]$ with $\langle f, g \rangle := \int_0^1 f(t)^* g(t) dt$.
- If $X, Y \in \mathcal{M}_{m,n}(\mathbb{R})$, then $\langle X, Y \rangle := \mathrm{Tr}(X^\top Y) = \sum_{i=1}^m \sum_{j=1}^n X_{ij} Y_{ij}$ defines an inner product on the space of (real) matrices.

Projection matrices are particularly useful for geometric purposes:

Definition 2. A matrix $P \in \mathcal{M}_n(\mathbb{F})$ is called a (orthogonal) *projection* if $P^2 = P$.

In particular, if $\mathbf{W}$ is a vector subspace of $\mathbb{F}^n$ with some orthonormal basis $(w_1, \ldots, w_d)$, it is then easy to check that the map $\mathcal{P}_{\mathbf{W}} : \mathbb{F}^n \to \mathbb{F}^n$ onto $\mathbf{W}$ satisfying

$$\mathcal{P}_{\mathbf{W}}(v) := \sum_{i=1}^d \langle v, w_i \rangle w_i, \quad \text{for any } v \in \mathbb{F}^n,$$

defines an orthogonal projection.

1.1.3 From linear operators to matrices

Let $\mathbf{V}$ be a finite-dimensional vector space over $\mathbb{F}$ and $\langle \cdot, \cdot \rangle$ a non-degenerate inner product on $\mathbf{V}$. Given a linear operator $\mathcal{A} : \mathbf{V} \to \mathbf{V}$, then, according to the Riesz representation theorem [336, Section III-6], there exists a unique linear operator $\mathcal{A}^\dagger : \mathbf{V} \to \mathbf{V}$, called the adjoint operator, such that

$$\langle \mathcal{A}u, v \rangle = \langle u, \mathcal{A}^\dagger v \rangle, \quad \text{for all } u, v \in \mathbf{V}.$$

Indeed, for any $v \in \mathbf{V}$, the map $u \in \mathbf{V} \mapsto \langle \mathcal{A}u, v \rangle$ is a linear functional, hence an element of the dual space $\mathbf{V}^\dagger$ (the space of bounded linear functionals on $\mathbf{V}$). Therefore, for each $v \in \mathbf{V}$, there exists $v' \in \mathbf{V}$ such that $\langle \mathcal{A}u, v \rangle = \langle u, v' \rangle$. It is then easy to show that the map $v \mapsto v'$ is linear, proving that the adjoint operator is uniquely defined. In the particular case where $\mathcal{A} = \mathcal{A}^\dagger$, the operator $\mathcal{A}$ is called Hermitian, which is a key requirement in quantum mechanics:

Definition 3. The operator $\mathcal{A}$ is called *Hermitian*, or *self-adjoint*, if $\mathcal{A} = \mathcal{A}^\dagger$.

For a Hermitian operator $\mathcal{A}$, we then have, for any $u \in \mathbf{V}$,

$$\langle \mathcal{A}u, u \rangle = \langle u, \mathcal{A}^\dagger u \rangle = \langle u, \mathcal{A}u \rangle = \langle \mathcal{A}u, u \rangle^*$$

by conjugate symmetry (Definition 1), and therefore $\langle \mathcal{A}u, u \rangle$ is real. Conversely, if $\langle \mathcal{A}u, u \rangle$ is real, then

$$\langle \mathcal{A}u, u \rangle = \langle \mathcal{A}u, u \rangle^* = \langle u, \mathcal{A}u \rangle = \langle \mathcal{A}^\dagger u, u \rangle.$$

Therefore, $\langle (\mathcal{A} - \mathcal{A}^\dagger) u, u \rangle = 0$; since this is true for all $u \in \mathbf{V}$, then $\mathcal{A} = \mathcal{A}^\dagger$.

The following property of operators shall be useful to ensure that systems driven by operators preserve distances, or norms:

Definition 4. The linear operator $\mathcal{A} : \mathbf{V} \to \mathbf{V}$ is called *unitary* if it is surjective and

$$\langle \mathcal{A}u, \mathcal{A}v \rangle = \langle u, v \rangle, \quad \text{for all } u, v \in \mathbf{V}.$$

Recall that a linear operator between two finite-dimensional normed spaces is bounded, and therefore continuous. For any $u \in \mathbf{V}$, this implies that $\|\mathcal{A}u\| = \|u\|$, so that a unitary operator $\mathcal{A}$ preserves the norm. In that case, $\mathcal{A}$ is an isometry, therefore injective. Being also surjective, it is bijective and therefore its inverse exists. For a unitary operator $\mathcal{A}$ and any $u, v \in \mathbf{V}$, we have

$$\langle u, v \rangle = \langle \mathcal{A}u, \mathcal{A}v \rangle = \langle u, \mathcal{A}^\dagger \mathcal{A}v \rangle$$

by definition of the adjoint, implying that

$$\mathcal{A}^\dagger \mathcal{A} = \mathcal{I} = \mathcal{A}\mathcal{A}^\dagger,$$

where $\mathcal{I}$ is the identity operator.

Example (Real Matrices): If $\mathbf{V} = \mathbb{R}^n$ with inner product $\langle u, v \rangle := u^\top v$ for $u, v \in \mathbb{R}^n$, the linear operator $\mathcal{A}$ can now be viewed as a matrix A in $\mathcal{M}_n(\mathbb{R})$. Its adjoint is nothing other than the transpose $A^\top$, and therefore A is self-adjoint if and only if it is symmetric. In this case, if A is unitary (or orthogonal), then it is invertible with $A^{-1} = A^\top$. Rotation matrices in $\mathbb{R}^2$, which will play an important role later when constructing quantum circuits, are the only unitary maps of $\mathbb{R}^2$ onto itself and are of the form

$$\begin{bmatrix} \cos(\theta) & \delta \sin(\theta) \\ \sin(\theta) & -\delta \cos(\theta) \end{bmatrix}, \tag{1.1.1}$$

for $\theta \in [0, 2\pi)$ and $\delta \in \{-1, +1\}$.

Example (Complex Matrices): If $\mathbf{V} = \mathbb{C}^n$ with inner product $\langle u, v \rangle := v^\dagger u$ for $u, v \in \mathbb{C}^n$, the linear operator $\mathcal{A}$ can now be viewed as a matrix in $\mathcal{M}_n(\mathbb{C})$. The adjoint of such a matrix is then the Hermitian conjugate, $A^\dagger$, and A is called Hermitian if $A = A^\dagger$ and unitary if $A^\dagger A = \mathbf{I}_n$. We shall denote by $\mathcal{U}_n(\mathbb{C})$ the set of unitary matrices in $\mathcal{M}_n(\mathbb{C})$. We will discuss Hermitian matrices over $\mathbb{C}$ in more detail in the Section 1.1.6.

1.1.4   Condition number

In order to manipulate matrices and measure them, we require matrix norms:

Definition 5. A matrix norm $\|\cdot\| : \mathcal{M}_{m,n}(\mathbb{F}) \to \mathbb{R}$ is a function satisfying, for any $\alpha \in \mathbb{F}$ and $A, B \in \mathcal{M}_{m,n}(\mathbb{F})$,

- (positively valued) $\|A\| \geq 0$;
- (definite) $\|A\| = 0$ if and only if $A = \mathbf{0}_{m,n}$;
- (absolutely homogeneous) $\|\alpha A\| = |\alpha| \|A\|$;
- (triangle inequality) $\|A + B\| \leq \|A\| + \|B\|$.

The norm is further called sub-multiplicative if $\|AB\| \leq \|A\| \|B\|$.

The condition number of a matrix is an important tool to understand the stability of linear equations of the form $Ax = b$, for $A \in \mathcal{M}_n(\mathbb{F})$, $b \in \mathbb{F}^n$. Assuming A to be non-singular, the true solution is clearly $x_* := A^{-1}b$. Suppose, however, that the vector b is only known up to some (not necessarily quantum) measurement error, and one observes instead $b + \Delta_b$. The solution is then $A^{-1}(b + \Delta_b) = x_* + \Delta_x$, with $\Delta_x := A^{-1}\Delta_b$. In particular, we can write, for any (sub-multiplicative) matrix norm $\|\cdot\|$:

$$\frac{\|\Delta_x\|}{\|x\|} = \frac{\|A^{-1}\Delta_b\|}{\|A^{-1}b\|} \leq \|A^{-1}\| \frac{\|b\|}{\|A^{-1}b\|} \frac{\|\Delta_b\|}{\|b\|} \leq \|A^{-1}\| \|A\| \frac{\|\Delta_b\|}{\|b\|}.$$

From this inequality, we see that the quantity $\|A^{-1}\| \|A\|$ bounds the relative error in the solution with respect to the relative error in the measurement of the input vector b. This leads to the following terminology:

Definition 6. Given a matrix $A \in \mathcal{M}_n(\mathbb{F})$ and a sub-multiplicative norm $\|\cdot\|$, we call

$$\kappa_{\|\cdot\|}(A) := \|A^{-1}\| \|A\|$$

the *condition number* (with respect to the norm $\|\cdot\|$) of the matrix A (and assign to it infinite value if A is singular).

Remark: The definition of the condition number above holds for any matrix norm $\| \cdot \|$, but admits a more explicit representation in the particular case of the spectral norm $\| \cdot \|_2$, defined as

$$\|A\|_2 := \sup_{x \neq 0} \frac{\|Ax\|_2}{\|x\|_2},$$

where $\|x\|_2 := \left(\sum_{i=1}^n |x_i|^2 \right)^{\frac{1}{2}}$ is the L_2 norm for vectors. If the matrix A is not singular, then

$$\kappa(A) := \frac{|\lambda_{\max}(A)|}{|\lambda_{\min}(A)|},$$

where $\lambda_{\max}(A)$ and $\lambda_{\min}(A)$ denote the largest and smallest eigenvalues of A.

Having defined essential properties of (complex) matrices, we will now introduce several essential tools that allow us to gain a better understanding of their properties.

1.1.5   Matrix decompositions and spectral theorem

The Singular Value Decomposition (SVD) is a key tool to analyse the properties and behaviours of matrices. It is ubiquitous in applied statistics and machine learning and allows us to reduce the explanatory dimension of a large matrix into a small number of meaningful components.

Theorem 1 (Singular Value Decomposition). *Let* $A \in \mathcal{M}_{m,n}(\mathbb{F})$ *and* $p := \min(m, n)$. *There exist* $U \in \mathcal{U}_m(\mathbb{F})$, $V \in \mathcal{U}_n(\mathbb{F})$ *and* $\sigma_1 \geq \cdots \geq \sigma_p \geq 0$ *such that* $A = U\Sigma V^\dagger$, *where* $\Sigma \in \mathcal{M}_{m,n}(\mathbb{F})$ *is diagonal with* $\Sigma_{ii} = \sigma_i$ *for* $i = 1, \ldots, p$ *and* $\Sigma_{ii} = 0$ *for* $i > p$.

The numbers $\{\sigma_1, \ldots, \sigma_p\}$ are called the singular values of A and are uniquely defined. The columns of U and V are the left-singular and right-singular vectors of A, in the sense that, if $\sigma \in \{\sigma_1, \ldots, \sigma_p\}$, then there exist a column u of U and a column v of V such that $Av = \sigma u$ and $A^\dagger u = \sigma v$. Recall that the rank of a matrix is defined as the dimension of the span of its columns. As a corollary of the SVD theorem, the rank of a matrix is therefore equal to the number of non-zero singular values. The SVD is general in the sense that it holds for any matrix. In the particular case of square matrices, the Schur decomposition and the Spectral Theorem provide refinements.

The Spectral Theorem is a cornerstone result in the theory of linear operators, and in particular for (finite-dimensional) matrices. Recall that an operator $\mathcal{A} : \mathbf{V} \to \mathbf{V}$ is called normal if it commutes with its adjoint, namely if $\mathcal{A}\mathcal{A}^\dagger = \mathcal{A}^\dagger\mathcal{A}$. Self-adjoint (or Hermitian) operators are clearly normal, yet the converse is not true in general. Recall further that an eigenvector of $\mathcal{A}$ is a non-zero vector $u \in \mathbf{V}$ such that $\mathcal{A}u = \lambda u$ for some $\lambda \in \mathbb{C}$, and we denote by $\sigma(\mathcal{A})$ the set of eigenvalues of $\mathcal{A}$. The following result, which is more general than the subsequent spectral theorem, allows us to decompose any arbitrary complex square matrix.

Theorem 2 (Schur Decomposition). *For any $A \in \mathcal{M}_n(\mathbb{C})$ there exits a unitary matrix $U \in \mathcal{U}_n(\mathbb{C})$ and an upper triangular matrix T such that $A = UTU^{-1}$.*

Note that since U is unitary, then $U^{-1} = U^\dagger$. We call the matrix T the Schur transform of A and the identity in the theorem means that A and T are similar; so, in particular, they possess the same eigenvalues, all located on the diagonal of T. If A is a normal matrix, then so is T, and therefore T must be diagonal; we write $T = D$ for clarity. In this case, we say that the matrix A is diagonalisable with $A = UDU^\dagger$, where the diagonal entries of D are the eigenvalues of A and the column vectors of U are the orthonormal eigenvectors of A.

Theorem 3 (Spectral Theorem). *The linear operator $\mathcal{A} : \mathbf{V} \to \mathbf{V}$ is normal if and only if there exists an orthonormal basis of $\mathbf{V}$ consisting of eigenvectors of A.*

For each eigenvalue $\lambda \in \sigma(\mathcal{A})$, denote the corresponding eigenspace

$$\mathcal{V}_\lambda := \{u \in \mathbf{V} : \mathcal{A}u = \lambda u\}.$$

Since the vector space $\mathbf{V}$ is the orthogonal direct sum of the eigenspaces (indexed by the eigenvalues of $\mathcal{A}$), we can then write the spectral decomposition

$$\mathcal{A} = \sum_{\lambda \in \sigma(\mathcal{A})} \lambda \mathcal{P}_\lambda,$$

where $\mathcal{P}_\lambda$ is the orthogonal projection operator onto $\mathcal{V}_\lambda$. Note that such an operator is naturally self-adjoint [336, Theorem 2, Section III-1].

1.1.6 Hermitian matrices

We previously introduced Hermitian matrices as the set of matrices A over the complex field $\mathbb{C}$ such that $A = A^\dagger$. As fundamental building blocks of quantum computing, we need to investigate their properties further. Clearly, a real matrix is Hermitian if and only if it is symmetric, in which case $A^\top = A$.

Proposition 1. *The eigenvalues of a Hermitian matrix are real.*

Proof. If $Ax = \lambda x$ for $\lambda \in \mathbb{C}$ and $x \in \mathbb{C}^n$, then

$$\langle Ax, x \rangle = x^\dagger A x = \lambda x^\dagger x = \lambda \|x\|^2,$$
$$\langle x, Ax \rangle = (Ax)^\dagger x = (\lambda x)^\dagger x = \lambda^* x^\dagger x = \lambda^* \|x\|^2.$$

Since both are equal by the Hermitian property, then $\lambda = \lambda^*$, proving the proposition. $\square$

The Singular Value Decomposition (Theorem 1) takes a particular flavour in the case of Hermitian matrices:

Theorem 4. *With the notations of Theorem 1, if $A \in \mathcal{M}_n(\mathbb{C})$ is Hermitian, then the matrices U and V are equal and the matrix Σ is diagonal with real entries.*

Theorem 5. *For a Hermitian matrix $A \in \mathcal{M}_n(\mathbb{C})$, the following are equivalent:*

  (i) *The eigenvalues are non-negative.*
 (ii) *There exists a Hermitian matrix, $B \in \mathcal{M}_n(\mathbb{C})$, such that $A = B^2$.*
(iii) *There exists a matrix, $B \in \mathcal{M}_n(\mathbb{C})$, such that $A = B^\dagger B$.*
 (iv) *For every $x \in \mathbb{C}^n$, $\langle Ax, x \rangle \geq 0$.*

Such a matrix is called positive semi-definite.

Proof. The Spectral Theorem shows that there exist a unitary matrix $U \in \mathcal{U}_n(\mathbb{C})$ and a diagonal matrix $\Sigma \in \mathcal{M}_n(\mathbb{C})$ such that $A = U\Sigma U^\dagger$, where the diagonal elements of Σ are the eigenvalues of A. Assuming (i), we can define $B = U\sqrt{\Sigma}U^\dagger \in \mathcal{M}_n(\mathbb{C})$. Then clearly

$$B^\dagger = B \quad \text{and} \quad B^2 = \left(U\sqrt{\Sigma}U^\dagger\right)\left(U\sqrt{\Sigma}U^\dagger\right) = A,$$

since U is unitary. The equality $A = B^\dagger B$ is also obvious. The latter implies that

$$\langle Ax, x \rangle = \langle B^\dagger Bx, x \rangle = \langle Bx, Bx \rangle = \|Bx\|^2 \geq 0, \quad \text{for any } x \in \mathbb{C}^n.$$

Finally, assume (iv) and let λ be an eigenvalue of A with eigenvector u. Then

$$\langle Au, u \rangle = \langle \lambda u, u \rangle = \lambda \langle u, u \rangle = \lambda \|u\|^2.$$

Since the latter is strictly positive, then clearly $\lambda \geq 0$. $\qquad\square$

The following property lies at the core of the Hamiltonian simulation of quantum systems:

Theorem 6. *If $A \in \mathcal{M}_n(\mathbb{C})$ is Hermitian, then, for any $t \in \mathbb{R}$, e^{itA} is unitary; conversely, every unitary matrix has the form e^{itA} for some Hermitian matrix A.*

Recall that for a matrix $A \in \mathcal{M}_n(\mathbb{C})$, its exponential is given by

$$e^A = \sum_{k \geq 0} \frac{A^k}{k!}.$$

In practice, though, given a Hermitian matrix A, finding the corresponding unitary matrix U is not easy. The Hamiltonian simulation problem is defined as follows.

Hamiltonian Problem: Given a Hermitian matrix $A \in \mathcal{M}_n(\mathbb{C})$, a time $t > 0$, a tolerance level $\varepsilon > 0$, and some matrix norm $\|\cdot\|$, find a unitary matrix U such that $\|U - e^{itA}\| \leq \varepsilon$.

1.1.7 Rotation matrices

Rotation matrices, and later their quantum gate equivalents, will play a key role in building quantum circuits. Let us start with the following lemma:

Lemma 1. *If a matrix* $A \in \mathcal{M}_n(\mathbb{C})$ *is such that* $A^2 = I$, *then for any* $\theta \in \mathbb{R}$,

$$e^{i\theta A} = \cos(\theta)I + i\sin(\theta)A.$$

Proof. This follows directly from the series expansion

$$e^A = \sum_{k \geq 0} \frac{A^k}{k!},$$

which has an infinite radius of convergence.   □

Lemma 1 will prove essential for computational purposes. As simple examples, consider the following:

Exercise: Compute $e^{i\theta A}$ for $A \in \{X, Y, Z\}$ and $\theta \in \mathbb{R}$, where

$$X = \begin{bmatrix} 0 & 1 \\ 1 & 0 \end{bmatrix}, \quad Y = \begin{bmatrix} 0 & -i \\ i & 0 \end{bmatrix}, \quad Z = \begin{bmatrix} 1 & 0 \\ 0 & -1 \end{bmatrix}.$$

For any $\alpha \in [0, 2\pi)$, consider now the map $\mathcal{R}_\alpha : \mathbb{R}^2 \to \mathbb{R}^2$ such that

$$\mathcal{R}_\alpha\Big(r\cos(\theta), r\sin(\theta)\Big) := \Big(r\cos(\theta + \alpha), r\sin(\theta + \alpha)\Big),$$

for any $r \in \mathbb{R}$ and $\theta \in [0, 2\pi)$,

which is basically a rotation of angle α and does not affect the norm of the input vector. To the map $\mathcal{R}_\alpha$, we can associate the (rotation) matrix R_α such that $\mathcal{R}_\alpha(u) = R_\alpha u$ for any $u \in \mathbb{R}^2$. It is easy (exercise) to show the following:

Lemma 2. *The matrix* R_α *has the form*

$$R_\alpha = \begin{bmatrix} \cos(\alpha) & -\sin(\alpha) \\ \sin(\alpha) & \cos(\alpha) \end{bmatrix}.$$

This representation is the general form of a rotation matrix in $\mathbb{R}^2$ (introduced in (1.1.1)).

Exercise: Write the matrices $e^{i\theta A}$ for $A \in \{X, Y, Z\}$ from the previous exercise as rotation matrices.

1.1.8   Polar coordinates

Recall that a point $z = x + iy$, with $x, y \in \mathbb{R}$, lying on the unit circle can be written as $z = e^{i\theta}$ with $\theta \in [0, 2\pi)$. Indeed, simply let $x = r\cos(\theta)$, $y = r\sin(\theta)$ and add the constraint $r = 1$.

Consider now a general vector $u \in \mathbb{C}^2$ of the form

$$u = \alpha e_1 + \beta e_2,$$

with $\alpha, \beta \in \mathbb{C}$ such that $|\alpha|^2 + |\beta|^2 = 1$. Here, (e_1, e_2) forms a basis of $\mathbb{R}^2$:

$$e_1 := \begin{bmatrix} 1 \\ 0 \end{bmatrix}, \quad e_2 := \begin{bmatrix} 0 \\ 1 \end{bmatrix}.$$

In polar coordinates, we can then write

$$u = r_\alpha e^{i\theta_\alpha} e_1 + r_\beta e^{i\theta_\beta} e_2.$$

Note that arbitrary multiplication phases have no influence – a fact of key importance in quantum mechanics – because, for any $\gamma \in \mathbb{R}$,

$$|e^{i\gamma}\alpha|^2 = (e^{i\gamma}\alpha)^* e^{i\gamma}\alpha = \alpha^* e^{-i\gamma} e^{i\gamma}\alpha = \alpha^*\alpha = |\alpha|^2,$$

so that in fact, multiplying u by the global phase $e^{-i\theta_\alpha}$ and letting $\theta := \theta_\beta - \theta_\alpha$, we consider

$$u = r_\alpha e_1 + r_\beta e^{i\theta} e_2.$$

Write temporarily $r_\beta e^{i\theta} = x + iy$. Insisting on u being on the unit sphere further imposes $\|u\|^2 = 1$, namely

$$
\begin{aligned}
1 = \|u\|^2 &= \Big(r_\alpha e_1 + (x + iy)e_2 \Big)^\dagger \Big(r_\alpha e_1 + (x + iy)e_2 \Big) \\
&= \Big(r_\alpha e_1^\top + (x - iy)e_2^\top \Big) \Big(r_\alpha e_1 + (x + iy)e_2 \Big) \\
&= r_\alpha^2 + x^2 + y^2,
\end{aligned}
$$

since (e_1, e_2) is orthonormal. This is nothing more than the equation of the unit sphere. In polar coordinates, we can write

$$
\begin{aligned}
x &= r \sin(\theta) \cos(\phi), \\
y &= r \sin(\theta) \sin(\phi), \\
r_\alpha &= r \cos(\theta),
\end{aligned}
$$

and clearly $r = 1$ since we are on the unit sphere. Therefore

$$
\begin{aligned}
u &= \cos(\theta)e_1 + \Big(\sin(\theta)\cos(\phi) + i\sin(\theta)\sin(\phi) \Big)e_2 \\
&= \cos(\theta)e_1 + \sin(\theta)e^{i\phi}e_2.
\end{aligned}
$$

1.1.9   Dirac notations

Given a vector $v \in \mathbb{C}^n$, Dirac's *ket* and *bra* notations read

$$
|v\rangle := \begin{bmatrix} v_1 \\ v_2 \\ \vdots \\ v_n \end{bmatrix} \quad \text{and} \quad \langle v| := [v_1^*, v_2^*, \ldots, v_n^*].
$$

With these notations, the operation $\langle u, v \rangle := \langle u|v \rangle$ defines an inner product on $\mathbb{C}^n$. The notation for the standard orthonormal basis in $\mathbb{C}^n$ is $(|i\rangle)_{i=0,\ldots,n-1}$, that is,

$$|0\rangle := \begin{bmatrix} 1 \\ 0 \\ \vdots \\ 0 \end{bmatrix}, \quad |1\rangle := \begin{bmatrix} 0 \\ 1 \\ \vdots \\ 0 \end{bmatrix}, \quad \dots \quad |n-1\rangle := \begin{bmatrix} 0 \\ 0 \\ \vdots \\ 1 \end{bmatrix}. \tag{1.1.2}$$

In coordinates, we can write, for any $u, v \in \mathbb{C}^n$,

$$|u\rangle = \sum_i u_i |i\rangle \quad \text{and} \quad |v\rangle = \sum_i v_i |i\rangle,$$

and, therefore,

$$\langle u, v \rangle = \sum_i u_i^* v_i.$$

1.1.10 Quantum operators

In the language of Dirac's notations, we can define the *outer product* $|u\rangle \langle v|$ (for $u \in \mathbf{U}$ and $v \in \mathbf{V}$) as a linear operator from $\mathbf{V}$ to $\mathbf{U}$, two vector spaces, as

$$\left(|u\rangle \langle v| \right) |w\rangle := \langle v|w\rangle |u\rangle, \quad \text{for any } w \in \mathbf{V}.$$

In particular, $|v\rangle \langle v|$ is the projection on the one-dimensional space generated by $|v\rangle$. Any linear operator can be expressed as a linear combination of outer products as

$$\mathcal{A} = \sum_{ij} A_{ij} |i\rangle \langle j|,$$

where $|i\rangle$ and $|j\rangle$ are the standard basis vectors (1.1.2).

Similarly to the linear algebra setting above, we can define an eigenvector of a linear operator $\mathcal{A} : \mathbf{V} \to \mathbf{V}$ as a non-zero vector $|v\rangle$ such that

$$\mathcal{A} |v\rangle = \lambda |v\rangle$$

for some complex eigenvalue λ.

Associated with any linear operator $\mathcal{A}$, the adjoint operator $\mathcal{A}^\dagger$ satisfies

$$\langle u | \mathcal{A} v \rangle = \langle \mathcal{A}^\dagger u | v \rangle .$$

Indeed, in the language of linear operators above, we have

$$\langle u, \mathcal{A} v \rangle = \langle \mathcal{A} v, u \rangle^* = \langle v, \mathcal{A}^\dagger u \rangle^* = \langle \mathcal{A}^\dagger u, v \rangle,$$

by definition of the inner product (Definition 1).

1.1.11   Tensor product

Given two vector spaces $\mathbf{U}$ and $\mathbf{V}$ of dimensions m and n, the tensor product $\mathbf{U} \otimes \mathbf{V}$ is a vector space of dimension mn. For $u \in \mathbf{U}$ and $v \in \mathbf{V}$, we can form the vector $|uv\rangle := |u\rangle \otimes |v\rangle \in \mathbf{U} \otimes \mathbf{V}$ with the following properties:

- $|(u + u')v\rangle = |uv\rangle + |u'v\rangle$, for any $u' \in \mathbf{U}$;
- $|u(v + v')\rangle = |uv\rangle + |uv'\rangle$, for any $v' \in \mathbf{V}$;
- $\alpha |uv\rangle = |(\alpha u)v\rangle = |u(\alpha v)\rangle$, for any $\alpha \in \mathbb{C}$.

Given the linear operators $\mathcal{A} : \mathbf{U} \to \mathbf{U}$ and $\mathcal{B} : \mathbf{V} \to \mathbf{V}$, we can then define their tensor product as an operator $\mathcal{A} \otimes \mathcal{B}$ on $\mathbf{U} \otimes \mathbf{V}$:

$$\left(\mathcal{A} \otimes \mathcal{B} \right) |uv\rangle := |(\mathcal{A}u), (\mathcal{B}v)\rangle ,$$

which can be represented in matrix form as $A \otimes B \in \mathcal{M}_{mn,mn}(\mathbb{C})$.

The Dirac formalism, fully anchored in (classical) linear algebra, opens the gates to a proper dive into the foundations of quantum mechanics.

1.2 Postulates of Quantum Mechanics

Quantum mechanics states several mathematical postulates that a physical theory must satisfy. It turns out that the mathematics of quantum mechanics allows for more general *computation*: more general definition of the *memory state* in comparison with classical digital computing and a wider range of possible *transformations* of such memory states. A natural question arises: what is the reason for this superior mode of computation not being used until very recently? The answer is that although quantum mechanics was formulated almost a century ago (Paul Dirac's seminal work "The Principles of Quantum Mechanics" [86] was published in 1930), the realisation of the rules of quantum mechanics in the computational protocol performed on classical digital computers requires an enormous amount of memory. Exponential gains in computing power are offset by exponential memory requirements.

In order to perform quantum computations efficiently, we need to use actual quantum mechanical systems, with their ability to encode information in their states. To illustrate this point, the state of a quantum system consisting of n quantum bits (qubits) can be described by specifying 2^n probability amplitudes – this is a huge amount of information even for very small systems ($n \sim 100$) and it would be impossible to store this information in classical memory. It took decades of technological progress before quantum processing units (QPUs) – devices that control quantum mechanical systems performing computations – became feasible.

Let us now proceed with the formulation of the mathematical postulates that lie at the foundation of quantum mechanics. These postulates specify a general framework for describing the behaviour of a physical system [197, 272]:

1. How to describe the state of a closed system.
2. How to describe the evolution of a closed system.
3. How to describe the interactions of a system with external systems.
4. How to describe the observables of a system.
5. How to describe the state of a composite system in terms of its component parts.

1.2.1 First postulate – Statics

Postulate 1. *Associated with any physical system is a complex inner product space known as the state space of the system. The system is completely described at any given point in time by its state vector, which is a unit vector in its state space.*

What is the importance of the first postulate from a quantum computing point of view? The answer is that quantum mechanics offers us a straightforward generalisation of the classical *binary digit* (bit). The classical bit is a two-state system with controlled transitions between them. As an example, we can use an electrical switch that can exist in one of the two discrete, stable states ("on" and "off"). Although electrical switches may seem an odd physical realisation of bits in the age of transistors, they illustrate an important point about computation in general: it is substrate independent. Exactly the same computational results can be obtained using electrical relays and CMOS transistors.

The quantum mechanical version of a bit, called a *quantum binary digit* (qubit), is a quantum mechanical two-state system. The first postulate of quantum mechanics tells us that the state of such a system can be represented mathematically by a unit vector in the two-dimensional complex vector space. This also means that such a system can exist in a superposition of basis states. Indeed, any vector $|v\rangle$ in the two-dimensional complex vector space,

$$|v\rangle = \begin{bmatrix} \alpha \\ \beta \end{bmatrix},$$

can be represented as a linear combination of the standard basis vectors:

$$\begin{bmatrix} \alpha \\ \beta \end{bmatrix} = \alpha \begin{bmatrix} 1 \\ 0 \end{bmatrix} + \beta \begin{bmatrix} 0 \\ 1 \end{bmatrix}, \quad |v\rangle = \alpha |0\rangle + \beta |1\rangle.$$

Since the state vector is a unit vector, the coefficients α and β must satisfy

$$|\alpha|^2 + |\beta|^2 = 1.$$

The coefficients α and β are *probability amplitudes*. Even though a qubit can exist in a superposition of basis states, once *measured* (see Postulate 3), its state *collapses* to one of the basis states: $|\alpha|^2$ and $|\beta|^2$ give us the probability of finding the qubit, respectively, in states $|0\rangle$ and $|1\rangle$ after measurement.

One can draw an analogy with how the space of natural numbers, $\mathbb{N}$, can be extended to the space of real numbers, $\mathbb{R}$, and then to the space of complex numbers, $\mathbb{C}$. We have a much wider range of functions that can operate on and take values in $\mathbb{R}$ and $\mathbb{C}$ than in $\mathbb{N}$. Similarly, allowing the two-state system to exist in a superposition of states significantly extends the range of possible operators that can transform such states (i.e., perform computation).

For example, for the classical functions with values from a two-element set ({0, 1} or {True, False}), called Boolean functions, the following statement is true. There is no Boolean function f that, when applied twice to a classical bit, would result in a NOT gate: $f(f(0)) = 1$ and $f(f(1)) = 0$. But there is such an operator in quantum computing. We can easily verify by direct calculations that the matrix

$$M := \frac{1}{2} \begin{bmatrix} 1+i & 1-i \\ 1-i & 1+i \end{bmatrix},$$

applied twice to the basis vector $|0\rangle$ would transform it to the basis vector $|1\rangle$, and applied twice to the basis vector $|1\rangle$ would transform it to the basis vector $|0\rangle$. M is an example of a *quantum logic gate* – an operator that transforms the state of a qubit, thus implementing the computation.

Remark: The state space of a physical system can be infinite-dimensional. The quantum computing paradigm based on infinite-dimensional Hilbert spaces is called *continuous-variable quantum computing*, which is realised in, e.g., some photonic quantum computing systems. However, in the context of digital quantum computing, we will restrict our analysis to finite-dimensional state spaces.

> The state of a qubit (the fundamental memory unit of quantum computing that generalises the concept of a classical bit) can be described mathematically as a unit vector in the two-dimensional complex vector space. Any physical system whose state space can be described by $\mathbb{C}^2$ can serve as an implementation of a qubit.

1.2.2 Second postulate – Dynamics

Postulate 2. *The time evolution of a closed quantum system is described by the Schrödinger equation*

$$i\hbar\frac{d\,|\psi(t)\rangle}{dt} = \mathcal{H}\,|\psi(t)\rangle,\tag{1.2.1}$$

where $\hbar$ is Planck's constant and $\mathcal{H}$ is a time-independent Hermitian operator known as the Hamiltonian of the system.

The Hamiltonian of a quantum system is an operator corresponding to the total energy of that system, and its eigenvalues are the possible energy levels of the system. The knowledge of the Hamiltonian provides all the necessary information about system dynamics.

In the Schrödinger equation (1.2.1), the state $|\psi(t_1)\rangle$ of a closed quantum system at time t_1 is related to the state $|\psi(t_2)\rangle$ at time t_2 by a unitary operator $\mathcal{U}(t_1, t_2)$ that depends only on t_1 and t_2 via

$$|\psi(t_2)\rangle = \mathcal{U}(t_1, t_2)\,|\psi(t_1)\rangle,\tag{1.2.2}$$

where $\mathcal{U}(t_1, t_2)$ is obtained from the Hamiltonian $\mathcal{H}$ as

$$\mathcal{U}(t_1, t_2) = \exp\left(-\frac{i\mathcal{H}(t_2 - t_1)}{\hbar}\right).\tag{1.2.3}$$

Unitary operators preserve the inner product (and therefore norms, lengths, and distances), which means that for two vectors, $|u\rangle$ and $|v\rangle$, if $\mathcal{U}$ is a unitary operator then the inner product between $\mathcal{U}\,|u\rangle$ and $\mathcal{U}\,|v\rangle$ is the same as the inner product between $|u\rangle$ and $|v\rangle$:

$$\langle u|\,\mathcal{U}^\dagger\mathcal{U}\,|v\rangle = \langle u|v\rangle.$$

A unitary operator is a complex generalisation of a rotation: unitary operators take an orthonormal basis to another orthonormal basis, and any operator with this property is unitary. In quantum mechanics, physical transformations such as rotations, translations, and time evolution correspond to maps that take quantum states to other quantum states. These maps should be linear and preserve the inner product. This allows us to look at the unitary operators as the *quantum logic gates* implementing quantum computation protocols. Furthermore, unitary operators are invertible, a key property that ensures that quantum computing is *reversible*.

> Quantum logic gates (quantum counterparts of the Boolean logic gates in classical computing) are unitary operators that transform quantum states, thus implementing the computation.

1.2.3   Third postulate – Measurement

Given a Hermitian operator $\mathcal{A}$, the spectral theorem implies that the state $|\psi\rangle$ of a system can be written as a superposition

$$|\psi\rangle = \sum_{i=1}^{N} \alpha_i |\psi_i\rangle, \tag{1.2.4}$$

where the coefficients $(\alpha_i)_{i=1,...,N}$ are complex *probability amplitudes*, assumed to be normalised with $\sum_{i=1}^{N} |\alpha_i|^2 = 1$, and where $(|\psi_i\rangle)_{i=1,...,N}$ are eigenfunctions of $\mathcal{A}$. The measurement postulate then reads as follows:

Postulate 3. *If we measure the Hermitian operator $\mathcal{A}$ in the state $|\psi\rangle$ given in (1.2.4), the possible outcomes for the measurement are the eigenvalues $(\lambda_i)_{i=1,...,N}$ of $\mathcal{A}$, and the probability p_i to measure λ_i is given by $p_i = |\alpha_i|^2$. After the outcome λ_i, the state of the system becomes*

$$|\psi\rangle = |\psi_i\rangle.$$

An immediate measurement in the same computational basis will deliver the same result without any uncertainty.

The quantum measurements are described by *measurement operators* $(\mathcal{P}_i)_{i=1,\ldots,N}$, acting on the state space of the system with N possible outcomes. If the state of the system is $|\psi\rangle$ before the measurement, then the probability of outcome i is

$$\mathbb{P}(i) = \langle\psi|\,\mathcal{P}_i^\dagger\mathcal{P}_i\,|\psi\rangle.$$

The measurement operators should also satisfy the *completeness condition*

$$\sum_{i=1}^{N}\mathcal{P}_i^\dagger\mathcal{P}_i = \mathcal{I}, \qquad (1.2.5)$$

where $\mathcal{I}$ is the identity operator. This ensures that the sum of the probabilities of all outcomes adds up to 1.

These measurement operators are linear but not unitary. From a quantum computing perspective, we are interested in measurement operators that are *projections* (Definition 2) onto the *computational basis*, such as the standard orthonormal basis given by (1.1.2).

For example, the measurement operators for a single qubit can be defined as

$$\mathcal{P}_0 := |0\rangle\langle0| = \begin{bmatrix} 1 & 0 \\ 0 & 0 \end{bmatrix} \quad \text{and} \quad \mathcal{P}_1 := |1\rangle\langle1| = \begin{bmatrix} 0 & 0 \\ 0 & 1 \end{bmatrix}.$$

We can easily verify that $\mathcal{P}_0^2 = \mathcal{P}_0$ and $\mathcal{P}_1^2 = \mathcal{P}_1$, as should be the case for projection operators, and that the completeness condition (1.2.5) is satisfied. If the qubit is in state $|\psi\rangle = \alpha\,|0\rangle + \beta\,|1\rangle$, then the measurement operator $\mathcal{P}_0$ will give us $|0\rangle$ with probability $|\alpha|^2$, and the measurement operator $\mathcal{P}_1$ will give us $|1\rangle$ with probability $|\beta|^2$. Indeed,

$$\mathcal{P}_0\,|\psi\rangle = |0\rangle\langle0|\left(\alpha\,|0\rangle + \beta\,|1\rangle\right) = \alpha\,|0\rangle\langle0|\,|0\rangle + \beta\,|0\rangle\langle0|\,|1\rangle = \alpha\,|0\rangle,$$
$$\mathcal{P}_1\,|\psi\rangle = |1\rangle\langle1|\left(\alpha\,|0\rangle + \beta\,|1\rangle\right) = \alpha\,|1\rangle\langle1|\,|0\rangle + \beta\,|1\rangle\langle1|\,|1\rangle = \beta\,|1\rangle.$$

The measurement postulate of quantum mechanics states that an immediate measurement in the same computational basis will deliver the same result without any uncertainty. The key words here are "the same computational basis". What would happen if the subsequent measurement is performed in another basis (the basis specified by another set of linearly independent unit vectors from the state space)? For example, assume that the qubit is in state

$$|\psi\rangle = \frac{1}{\sqrt{2}}|0\rangle + \frac{1}{\sqrt{2}}|1\rangle.$$

Measuring $|\psi\rangle$ in the $\{|0\rangle, |1\rangle\}$ computational basis will result in observing states $|0\rangle$ and $|1\rangle$ with equal probability $1/2$. Let us assume that we measured $|0\rangle$. The qubit state is now

$$|\psi'\rangle = 1 \cdot |0\rangle + 0 \cdot |1\rangle.$$

If we repeat the measurement in the same $\{|0\rangle, |1\rangle\}$ computational basis, we obtain state $|0\rangle$ with probability 1 in accordance with the measurement postulate. However, had we measured state $|\psi'\rangle$ in the *Hadamard basis* $\{|+\rangle, |-\rangle\}$, given by

$$|+\rangle := \frac{1}{\sqrt{2}}(|0\rangle + |1\rangle) \quad \text{and} \quad |-\rangle := \frac{1}{\sqrt{2}}(|0\rangle - |1\rangle), \tag{1.2.6}$$

we would have equal probabilities of $|+\rangle$ and $|-\rangle$ outcomes. Let us assume that we measured $|-\rangle$ and the state of the qubit is now

$$|\psi''\rangle = 0 \cdot |+\rangle + 1 \cdot |-\rangle.$$

If we repeat the measurement of state $|\psi''\rangle$ in the Hadamard basis $\{|+\rangle, |-\rangle\}$, we obtain state $|-\rangle$ with probability 1. But the state of the qubit is an equal superposition of states $|0\rangle$ and $|1\rangle$ from the $\{|0\rangle, |1\rangle\}$ computational basis perspective and we have an equal chance of measuring either $|0\rangle$ or $|1\rangle$ in this basis.

Remark: The basis vectors $|0\rangle$ and $|1\rangle$ that form the standard computational basis can be transformed into the basis vectors $|+\rangle$ and $|-\rangle$ that form the Hadamard basis by applying the following unitary operator (rotation), called the *Hadamard gate*:

$$H = \frac{1}{\sqrt{2}} \begin{bmatrix} 1 & 1 \\ 1 & -1 \end{bmatrix}.$$

Chapters 6, 10, and 11 provide examples of applications of the Hadamard gate.

Measurement plays a crucial role in quantum computing. This is the process of *collapsing* a quantum state and reading out the classical information: measuring qubits encoding a quantum state will produce a classical bit string. The measurement process generates probabilistic outcomes. Therefore, we need to perform measurements on the same quantum state multiple times to generate a sufficiently large number of classical bit strings to produce reliable statistics.

> The process of measurement describes the collapse of the quantum state due to contact with the environment. After measurement, the states of the qubits are known without any uncertainty. It is possible to extract at most 1 bit of information from a qubit. In order to extract more information about the probability distribution encoded in a given quantum state, it is necessary to perform measurement of the same state multiple times.

1.2.4   Fourth postulate – Observable

Postulate 4. *For every measurable property of a physical system, there exists a corresponding Hermitian operator. The values of the physical observables correspond to the expectation values of Hermitian operators. The expectation value $\langle A \rangle$ of the Hermitian operator A in the normalised state $|\psi\rangle$ is given by*

$$\langle A \rangle := \langle \psi | A | \psi \rangle. \tag{1.2.7}$$

Let us consider the general case where the expectation value of a Hermitian operator $\mathcal{A}$ is calculated in state $|\psi\rangle$, which is not an eigenfunction of $\mathcal{A}$. According to the Spectral Theorem 3 (see also (1.2.4)), the state $|\psi\rangle$ of a system can be represented as the superposition

$$|\psi\rangle = \sum_{i=1}^{N} \alpha_i |\psi_i\rangle ,$$

where $(|\psi_i\rangle)_{i=1,...,N}$ are the eigenfunctions of $\mathcal{A}$ and $(\alpha_i)_{i=1,...,N}$ are the corresponding probability amplitudes.

Therefore, the *expectation value* of $\mathcal{A}$ in state $|\psi\rangle$, given in (1.2.7), is calculated as

$$\langle \mathcal{A} \rangle = \sum_{i=1}^{N} \sum_{j=1}^{N} \alpha_i^* \alpha_j \langle \psi_i | \mathcal{A} | \psi_j \rangle = \sum_{i=1}^{N} \sum_{j=1}^{N} \alpha_i^* \alpha_j \lambda_j \langle \psi_i | \psi_j \rangle ,$$

where $(\lambda_i)_{i=1,...,N}$ are the eigenvalues of $\mathcal{A}$. The only terms that survive in the expression for $\langle \mathcal{A} \rangle$ are those with $i = j$ due to the orthogonality of the eigenfunctions, so that

$$\langle \mathcal{A} \rangle = \sum_{i=1}^{N} \alpha_i^* \alpha_i \lambda_i = \sum_{i=1}^{N} |\alpha_i|^2 \lambda_i.$$

Therefore, the value of the observable is a weighted average of the eigenvalues of the corresponding Hermitian operator. The weights are the coefficients $(|\alpha_i|^2)_{i=1,...,N}$, which are the probabilities of measuring the corresponding eigenstate of $\mathcal{A}$.

Hermitian operators play an exceptionally important role in quantum mechanics since their expectation values correspond to physical observables.

1.2.5   Fifth postulate – Composite System

Postulate 5. *The state space of a composite physical system is the tensor product of the state spaces of the individual component physical systems.*

If the first component physical system is in state $|\psi_A\rangle$ and the second component physical system is in state $|\psi_B\rangle$, then the state of the combined system, $|\psi\rangle$, is given by the tensor product

$$|\psi\rangle = |\psi_A\rangle \otimes |\psi_B\rangle. \tag{1.2.8}$$

Not all states of a combined system can be separated into the tensor product of states of individual components. If the state of a system cannot be separated into component parts, we say that the component parts are *entangled*.

The entanglement of quantum systems is one of the major sources of computational power of quantum computing. It allows us to store exponentially more information in the correlations between the states of individual subsystems (in the limit – individual qubits) than directly in the states of individual subsystems.

To illustrate this point, we can look at the number of probability amplitudes needed to describe the state of an n-qubit system. An individual qubit can be found in one of the two possible states after measurement – one of the two basis states, $|0\rangle$ or $|1\rangle$. This means that we need to specify two probability amplitudes to fully describe the state of the qubit before measurement. If all our qubits are independent and the state of the system can be represented as a tensor product of individual qubit states

$$|\psi\rangle = |\psi_1\rangle \otimes |\psi_2\rangle \otimes \ldots \otimes |\psi_n\rangle,$$

then we need to specify $2n$ probability amplitudes (two for each individual quantum states) to describe the state $|\psi\rangle$ of the system. If, however, all individual qubits are entangled and the tensor product representation of the system state $|\psi\rangle$ does not exist, we need to specify 2^n probability amplitudes – this is an effective measure of useful information that can be stored in the system.

The power of quantum computing is derived from the principles of super-position and entanglement. Entanglement allows us to store most of the information in correlations between the qubit states.

1.3   Pure and Mixed States

There are situations when we have to deal with either entangled components of a quantum mechanical system or a statistical ensemble of quantum states. In such cases the system cannot be described with the help of a state vector. Here, we look at such situations and provide a mathematical tool for describing them.

1.3.1   Density matrix

Let us start with the state of a combined two-component physical system given by (1.2.8). Let $(|i\rangle)_{i=1,...,N}$ and $(|j\rangle)_{j=1,...,M}$ denote, respectively, the standard orthonormal bases of the Hilbert spaces of systems A and B:

$$|\psi_A\rangle = \sum_{i=1}^{N} \alpha_i |i\rangle, \quad |\psi_B\rangle = \sum_{j=1}^{M} \beta_j |j\rangle, \tag{1.3.1}$$

where $(\alpha_i)_{i=1,...,N}$ and $(\beta_j)_{j=1,...,M}$ are some probability amplitudes. The states that allow the state vector representation (1.3.1) are called *pure* states. In this case, the state of the combined system is

$$|\psi\rangle = |\psi_A\rangle \otimes |\psi_B\rangle = \sum_{i=1}^{N}\sum_{j=1}^{M} \alpha_i \beta_j |i\rangle \otimes |j\rangle.$$

However, in general, the state of the combined system would look like

$$|\psi\rangle = \sum_{i=1}^{N}\sum_{j=1}^{M} \gamma_{ij} |i\rangle \otimes |j\rangle, \tag{1.3.2}$$

where γ_{ij} are probability amplitudes that may not necessarily be factorised as the product of probability amplitudes $(\alpha_i)_{i=1,\ldots,N}$ and $(\beta_j)_{j=1,\ldots,M}$. If γ_{ij} cannot be factorised as $\alpha_i\beta_j$, then the component systems A and B are entangled and their states cannot be represented by the state vectors (1.3.1). Such states of systems A and B are called *mixed* states.

The more general setup is that of an *ensemble* of states of the form $\{p_k, |\psi_k\rangle\}_{k=1,\ldots,K}$, where each $|\psi_i\rangle$ is a quantum state whose wave function is known with certainty (although this does not necessarily provide full knowledge of the measurement statistics), and each p_k is the associated probability (not amplitude) in $[0, 1]$. In order to define properly pure and mixed states, introduce the *density operator* as follows:

Definition 7. A density operator ρ is a positive semidefinite Hermitian operator with unit trace that takes the form

$$\rho := \sum_{k=1}^{K} p_k |\psi_k\rangle \langle\psi_k|, \quad \sum_{k=1}^{K} p_k = 1, \quad p_k \geq 0, \text{ for all } k = 1, \ldots, K.$$

A density operator ρ corresponds to a *density matrix* $(\rho_{kl})_{k,l=1,\ldots,K}$ such that

$$\rho = \rho^\dagger, \quad \text{Tr}(\rho) \equiv \sum_{k=1}^{K} \rho_{kk} = 1, \quad \rho_{kk} \geq 0, \text{ for all } k = 1, \ldots, K.$$

1.3.2 Pure state

A pure state is one that can be represented by a state vector

$$|\psi\rangle = \sum_{i=1}^{N} \alpha_i |i\rangle, \tag{1.3.3}$$

where $(\alpha_i)_{i=1,\ldots,N}$ are probability amplitudes in $\mathbb{C}$ such that $\sum_{i=1}^{N} |\alpha_i|^2 = 1$. In the ensemble setup above, this means that there exists $k^* \in \{1, \ldots, N\}$ such that $p_{k^*} = 1$ and hence $|\psi\rangle = |\psi_{k^*}\rangle$ and, therefore, $\rho = |\psi\rangle \langle\psi|$. The density matrix also allows us to compute expectations of the form (1.2.7).

Lemma 3. *Let ρ be the density matrix associated with the pure state (1.3.3) and let $\mathcal{A}$ be an observable (Hermitian operator), then*

$$\langle \mathcal{A} \rangle := \langle \psi | \, \mathcal{A} \, | \psi \rangle = \mathrm{Tr}(\rho \mathcal{A}).$$

Proof. The lemma follows from the immediate computation:

$$\langle \psi | \, \mathcal{A} \, | \psi \rangle = \langle \psi | \, \mathcal{A} \sum_{i=1}^{N} \alpha_i \, | i \rangle = \sum_{i=1}^{N} \alpha_i \, \langle \psi | \, \mathcal{A} \, | i \rangle$$

$$= \sum_{i=1}^{N} \langle i | \psi \rangle \, \langle \psi | \, \mathcal{A} \, | i \rangle = \sum_{i=1}^{N} \langle i | \, \rho \mathcal{A} \, | i \rangle = \mathrm{Tr}(\rho \mathcal{A}).$$

$\square$

With the state $|\psi\rangle$ given by (1.3.3), we obtain

$$\langle \mathcal{A} \rangle = \sum_{i=1}^{N} \sum_{j=1}^{N} \alpha_i \alpha_j^* \, \langle j | \, \mathcal{A} \, | i \rangle . \tag{1.3.4}$$

At the same time we have

$$\langle \mathcal{A} \rangle = \mathrm{Tr}(\rho \mathcal{A}) = \sum_{i=1}^{N} \sum_{j=1}^{N} \rho_{ij} \, \langle j | \, \mathcal{A} \, | i \rangle . \tag{1.3.5}$$

A comparison of (1.3.4) and (1.3.5) yields the following expression for the density matrix of a pure state:

$$\rho_{ij} = \alpha_i \alpha_j^*, \quad \rho = \sum_{i=1}^{N} \sum_{j=1}^{N} \alpha_i \alpha_j^* \, | i \rangle \, \langle j | = | \psi \rangle \, \langle \psi | .$$

Example: An example of a pure state is the Hadamard state

$$| + \rangle = \frac{1}{\sqrt{2}}(|0\rangle + |1\rangle) = \frac{1}{\sqrt{2}} \begin{bmatrix} 1 \\ 1 \end{bmatrix} ,$$

with the corresponding density matrix

$$\rho = |+\rangle \langle +| = \frac{1}{2} \begin{bmatrix} 1 & 1 \\ 1 & 1 \end{bmatrix}.$$

1.3.3   Mixed state

A mixed state is one that cannot be represented by a single pure state vector, and is therefore represented as a statistical distribution of pure states in the form of an ensemble of quantum states $\{p_k, |\psi_k\rangle\}_{k=1,\dots,N}$, where $\sum_{k=1}^{N} p_k = 1$ and $p_k \in [0, 1]$ for each k. The density of a mixed state therefore reads

$$\rho = \sum_{k=1}^{N} p_k |\psi_k\rangle \langle \psi_k|. \tag{1.3.6}$$

Similarly to Lemma 3, we can write expectations of observables with respect to mixed states using the density matrix:

Lemma 4. *Let ρ be the density matrix associated with the mixed state (1.3.6) and let $\mathcal{A}$ be an observable (Hermitian operator), then*

$$\mathrm{Tr}(\rho \mathcal{A}) = \sum_{k=1}^{N} p_k \langle \psi_k| \mathcal{A} |\psi_k\rangle.$$

Proof. The lemma follows from the immediate computation

$$\mathrm{Tr}(\rho \mathcal{A}) = \sum_{i=1}^{N} \langle i| \rho \mathcal{A} |i\rangle = \sum_{i=1}^{N} \langle i| \left(\sum_{k=1}^{N} p_k |\psi_k\rangle \langle \psi_k| \right) \mathcal{A} |i\rangle$$

$$= \sum_{k=1}^{N} p_k \left(\sum_{i=1}^{N} \langle i|\psi_k\rangle \langle \psi_k| \mathcal{A} |i\rangle \right) = \sum_{k=1}^{N} p_k \langle \psi_k| \mathcal{A} |\psi_k\rangle.$$

$\square$

Let us see now how the density matrix formalism can help us describe the state of a combined system. Consider an entangled state of two systems, A and B, given by (1.3.2), and a Hermitian operator $\mathcal{A}$ that only acts within the Hilbert space of system A. What would be the expectation value of $\mathcal{A}$ in this state? Starting with (1.2.7), we obtain

$$\langle \mathcal{A} \rangle = \sum_{i=1}^{N} \sum_{j=1}^{M} \sum_{k=1}^{N} \sum_{l=1}^{M} \gamma_{ij} \gamma_{kl}^{*} \langle k| \mathcal{A} |i \rangle \langle l|j \rangle. \tag{1.3.7}$$

Since only terms with $l = j$ survive in (1.3.7) due to the orthogonality of the basis states, we have

$$\langle \mathcal{A} \rangle = \sum_{i=1}^{N} \sum_{k=1}^{N} \left(\sum_{j=1}^{M} \gamma_{ij} \gamma_{kj}^{*} \right) \langle k| \mathcal{A} |i \rangle.$$

Thus, the density matrix that describes the mixed state of system A is

$$\rho_{ik} = \sum_{j=1}^{M} \gamma_{ij} \gamma_{kj}^{*}.$$

Note that in the case where the probability amplitudes γ_{ij} can be factorised as the product of probability amplitudes $(\alpha_i)_{i=1,\dots,N}$ and $(\beta_j)_{j=1,\dots,M}$, we obtain

$$\rho_{ik} = \sum_{j=1}^{M} \alpha_i \beta_j \alpha_k^{*} \beta_j^{*} = \alpha_i \alpha_k^{*} \sum_{j=1}^{M} |\beta_j|^2 = \alpha_i \alpha_k^{*},$$

which describes a pure state.

Lemma 5. *Let ρ be a density matrix. The inequality $\mathrm{Tr}(\rho^2) \leq 1$ always holds and $\mathrm{Tr}(\rho^2) = 1$ if and only if ρ corresponds to a pure state.*

Proof. Consider an ensemble of pure states $\{p_i, |\psi_i\rangle\}_{i=1,\dots,N}$. From (1.3.6) we get

$$\mathrm{Tr}(\rho^2) = \mathrm{Tr} \left(\left(\sum_{i=1}^{N} p_i |\psi_i\rangle \langle \psi_i| \right) \left(\sum_{j=1}^{N} p_j |\psi_j\rangle \langle \psi_j| \right) \right)$$

$$= \mathrm{Tr} \left(\sum_{i=1}^{N} \sum_{j=1}^{N} p_i p_j |\psi_i\rangle \langle \psi_i| |\psi_j\rangle \langle \psi_j| \right),$$

and, due to orthogonality, we obtain

$$\text{Tr}(\rho^2) = \text{Tr}\left(\sum_{i=1}^{N} p_i^2 |\psi_i\rangle \langle\psi_i|\right)$$

$$= \sum_{i=1}^{N} p_i^2 \text{Tr}\left(|\psi_i\rangle \langle\psi_i|\right)$$

$$= \sum_{i=1}^{N} p_i^2 \langle\psi_i|\psi_i\rangle = \sum_{i=1}^{N} p_i^2,$$

which is smaller than 1 since the p_i are probabilities in $[0, 1]$ summing up to 1. Assume now that $\text{Tr}(\rho^2)$ equals 1, then so does $\sum_{i=1}^{N} p_i^2$. If $p_i \in (0, 1)$ for all $i = 1, \ldots, N$, then

$$1 = \sum_{i=1}^{N} p_i^2 < \sum_{i=1}^{N} p_i = 1,$$

which is a contradiction, and therefore there exists $i^* \in \{1, \ldots, N\}$ such that $p_{i^*} = 1$, so that $\rho = |\psi_{i^*}\rangle \langle\psi_{i^*}|$ is a pure state. Conversely, if $\rho = |\psi_i\rangle \langle\psi_i|$ for some $i \in \{1, \ldots, N\}$ represents a pure state, then

$$\text{Tr}(\rho^2) = \text{Tr}(|\psi_i\rangle \langle\psi_i| |\psi_i\rangle \langle\psi_i|) = \text{Tr}(|\psi_i\rangle \langle\psi_i|) = \langle\psi_i|\psi_i\rangle = 1.$$

$\square$

Example: An example of a mixed state is a *statistical ensemble* of states $|0\rangle$ and $|1\rangle$. If a physical system is prepared to be in either state $|0\rangle$ or state $|1\rangle$ with equal probability, it can be described by the following mixed state:

$$\rho = \frac{1}{2} |0\rangle \langle 0| + \frac{1}{2} |1\rangle \langle 1| = \frac{1}{2} \begin{bmatrix} 1 & 0 \\ 0 & 1 \end{bmatrix}. \tag{1.3.8}$$

Note that this is different from the density matrix of the pure state

$$|\psi\rangle = \frac{1}{\sqrt{2}}(|0\rangle + |1\rangle),$$

which reads

$$\rho_\psi = |\psi\rangle\langle\psi| = \frac{1}{2}(|0\rangle + |1\rangle)(\langle 0| + \langle 1|)$$

$$= \frac{1}{2}(|0\rangle\langle 0| + |1\rangle\langle 0| + |0\rangle\langle 1| + |1\rangle\langle 1|) = \frac{1}{2}\begin{bmatrix} 1 & 1 \\ 1 & 1 \end{bmatrix}.$$

Unlike pure quantum states, mixed quantum states cannot be described by a single state vector. However, pure states and mixed states can be described using the density matrix formalism.

Summary

In this chapter, we learned the key principles of quantum mechanics, starting with a review of the basic elements of linear algebra, followed by an introduction to Dirac notations.

We then covered the main postulates of quantum mechanics and their relevance to quantum computing. We learned how to describe the state (statics) and the evolution (dynamics) of a closed system, the interactions of a system with external systems (measurement), observables, as well as the state of a composite system in terms of its component parts.

We finally introduced the density operator, which allows us to describe both pure and mixed quantum states, contrasting with the state vector, which can only represent pure quantum states.

The following table summarises some of the key rules and definitions:

---

1. Any *linear operator* $\mathcal{A}$ can be expressed as a linear combination of outer products of *basis vectors* $|i\rangle$:

$$\mathcal{A} = \sum_{ij} A_{ij} |i\rangle \langle j| \,.$$

2. A linear operator is *diagonalisable* if

$$\mathcal{A} = \sum_{i} \lambda_i |\psi_i\rangle \langle \psi_i| \,,$$

    where λ_i and $|\psi_i\rangle$ are, respectively, the *eigenvalues* and *eigenvectors* of $\mathcal{A}$.

3. *Adjoint* operator, $\mathcal{A}^\dagger$, is defined as

$$\mathcal{A}^\dagger = (\mathcal{A}^*)^\top \,.$$

    We have:

$$\langle v | \mathcal{A}u \rangle = \langle \mathcal{A}^\dagger v | u \rangle \,.$$

4. A linear operator $\mathcal{A}$ is *normal* if

$$\mathcal{A}\mathcal{A}^\dagger = \mathcal{A}^\dagger \mathcal{A} \,.$$

    A linear operator is diagonalisable if, and only if, it is normal.

5. A linear operator $\mathcal{A}$ is *Hermitian* if

$$\mathcal{A} = \mathcal{A}^\dagger \,.$$

    A normal operator is Hermitian if, and only if, it has real eigenvalues.

6. A linear operator $\mathcal{A}$ is *unitary* if

$$\mathcal{A}\mathcal{A}^\dagger = \mathcal{A}^\dagger \mathcal{A} = \mathcal{I}.$$

Unitary operators are normal and, therefore, diagonalisable.

7. Unitary operators are *invertible*: and *norm preserving*:

$$\langle \mathcal{A}u | \mathcal{A}v \rangle = \langle u | v \rangle .$$

8. All eigenvalues of a unitary operator have modulus 1. If a linear operator is both unitary and Hermitian, its igenvalues are $+1$ and -1.

9. Hermitian operators can represent physical observables.

10. If $\mathcal{A}$ is Hermitian, the state $|\psi\rangle$ of a system can be written as a superposition

$$|\psi\rangle = \sum_i \alpha_i |\psi_i\rangle ,$$

where the coefficients α_i are complex *probability amplitudes* and $|\psi_i\rangle$ are eigenvectors of $\mathcal{A}$.

11. A value of a physical observable represented by Hermitian operator $\mathcal{A}$ in state $|\psi\rangle$ is given by the *expectation value* of $\mathcal{A}$ in state $|\psi\rangle$:

$$\langle \mathcal{A} \rangle = \langle \psi | \mathcal{A} | \psi \rangle = \sum_i \lambda_i \alpha_i^* \alpha_i = \sum_i \lambda_i |\alpha_i|^2.$$

12. A *density operator* ρ is a positive semidefinite Hermitian operator with unit trace that takes the form

$$\rho = \sum_i p_i |\psi_i\rangle \langle \psi_i| , \quad \sum_i p_i = 1, \quad p_i \geq 0, \ \forall i.$$

A density operator corresponds to a *density matrix* such that

$$\rho = \rho^\dagger, \quad \mathrm{Tr}(\rho) \equiv \sum_i \rho_{ii} = 1, \quad \rho_{ii} \geq 0, \; \forall i.$$

13. A *pure state* is one that can be represented by a state vector

$$|\psi\rangle = \sum_i \alpha_i \, |i\rangle ,$$

where $|i\rangle$ are the basis vectors and α_i are probability amplitudes. The density operator of a pure state is given by

$$\rho = |\psi\rangle \langle\psi| .$$

14. A state ρ is pure if, and only if, any one of the following conditions holds:

$$\text{a) } \rho^2 = \rho, \qquad \text{b) } \mathrm{Tr}(\rho^2) = 1.$$

15. A *mixed state* is one that cannot be represented by a single pure state vector, and is therefore represented as a statistical ensemble of quantum states $|\psi_i\rangle$ with positive weights p_i that add up to one. The density operator of a mixed state reads

$$\rho = \sum_i p_i \, |\psi_i\rangle \langle\psi_i| .$$

16. The state space of a composite physical system is the tensor product of the state spaces of the individual component physical systems. If the state of a system cannot be separated into component parts, we say that the component parts are *entangled*.

17. For the density operator ρ of a mixed or entangled state we have

$$\text{a) } \rho^2 \neq \rho, \qquad \text{b) } \mathrm{Tr}(\rho^2) < 1.$$

18. A value of a physical observable represented by Hermitian operator $\mathcal{A}$ in a state described by the density matrix ρ is given by

$$\langle \mathcal{A} \rangle = \mathrm{Tr}(\rho \mathcal{A}).$$

19. If a state $|\psi(t_1)\rangle$ of a *closed quantum system* described by the density matrix $\rho(t_1)$ at time t_1 evolves into a state $|\psi(t_2)\rangle$ described by the density matrix $\rho(t_2)$ at time t_2 under the action of a unitary operator $\mathcal{U}$, then

$$|\psi(t_2)\rangle = \mathcal{U}\,|\psi(t_1)\rangle, \qquad \rho(t_2) = \mathcal{U}\rho(t_1)\mathcal{U}^\dagger.$$

20. If $\mathcal{A}$ is a *commutator* of two operators $\mathcal{B}$ and $\mathcal{C}$,

$$\mathcal{A} = [\mathcal{B}, \mathcal{C}] = \mathcal{B}\mathcal{C} - \mathcal{C}\mathcal{B},$$

then $\mathrm{Tr}(\mathcal{A}) = 0$.

PART I

ANALOG QUANTUM COMPUTING – QUANTUM ANNEALING

2

Adiabatic Quantum Computing

Search algorithms are among the most important and fundamental algorithms in computer science, the most basic example being that of finding one special item among a list of N items. Classical algorithms are known to solve this problem in time proportional to the problem size, N, which becomes highly untractable when the latter grows large. In 1996, Grover [124] devised a quantum algorithm to solve such search problems with a quadratic speedup, with the obvious caveat that quantum computers did not exist at the time. Soon after, Farhi, Goldstone, Gutmann, and Sipser [98] recast the Grover problem as a satisfiability problem in the context of quantum computation by adiabatic evolution.

Another class of problems that are hard to solve classically is that of combinatorial optimisation problems. The truck dispatching problem, originally proposed by Dantzig and Ramser [78], searches the optimal routing of delivery trucks, and is a generalisation of the famous travelling salesman problem. Other well-known optimisation problems of the same class include partitioning problems and binary integer linear programming, to name just a few. The exact algorithms are only efficient for small-scale instances. Heuristics and metaheuristics (e.g., evolutionary search heuristics such as Genetic Algorithm and Particle Swarm Optimisation) are often more suitable for practical applications because real-world problems are considerably larger in scale. Since search time scales exponentially with problem size, there has always been a strong motivation for finding alternative approaches to solve such problems with a wide range of practical applications, not least in finance. Adiabatic Quantum Computing (AQC) was ready to enter the stage.

2.1 Complexity of Computational Problems

In this and the following chapters, we will often speak about computational problems that are hard for classical computers but can be solved efficiently using quantum algorithms and hardware. How can we quantify the hardness of a computational problem? One way to answer this is to analyse problems from the computational resource perspective: how much time and memory are needed to solve them? This leads to the concept of *complexity classes*. Important examples are as follows:

- The class P (*polynomial*) is the set of decision problems solvable by a deterministic Turing machine in polynomial time.

- The class NP (*non-deterministic polynomial*) is the set of decision problems solvable by a non-deterministic Turing machine in polynomial time.

These definitions, in turn, require us to specify the following objects:

- *A decision problem* is a computational problem that can be posed as a Yes-No question of the input values.

- *Polynomial time* means that the running time of the algorithm is bounded above by a polynomial expression in the size of the input for the algorithm.

- *A Turing machine* is an abstract model of computation that is general enough to embody any computer problem.

- *A deterministic Turing machine* is the most basic type of Turing machine that uses a fixed set of rules to determine its future actions.

- A *non-deterministic Turing machine* is a Turing machine that is able to explore multiple alternative future actions from a given state.

In terms of computational hardness, we will pay special attention to the decision problems that are *NP-complete* and *NP-hard.*

A decision problem is NP-complete when

  i) the correctness of each solution can be verified in polynomial time and a brute-force search algorithm can find a solution by trying all possible solutions;

  ii) it can be used to simulate every other problem for which we can verify in polynomial time that a solution is correct.

NP-complete problems are the hardest, for which solutions can be verified quickly (in polynomial time). If we could find solutions to some NP-complete problems quickly, we could quickly find the solutions to every other problem to which a given solution can be easily verified.

A problem is NP-hard when every problem in NP can be reduced to it in polynomial time. Alternatively, a problem is NP-hard when every NP-complete problem can be reduced to it in polynomial time. Since every problem in NP reduces to an NP-complete problem in polynomial time, the second definition implies the first one.

The NP-hard class is not restricted to decision problems and also includes search problems and optimisation problems. This means that NP-hard problems do not have to be elements of the complexity class NP.

Arguably, the most important open problem in computer science is whether $P = NP$. It is widely believed, although it has not been proven yet, that $P \neq NP$. Figure 2.1 shows the relationship between complexity classes for both scenarios.

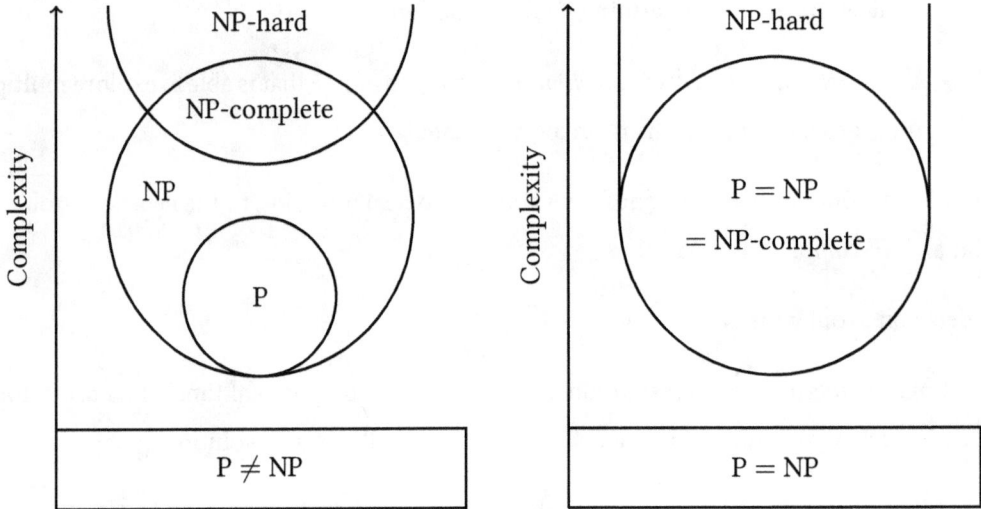

Figure 2.1: Schematic illustration of the relationship between P, NP, NP-complete, and NP-hard sets of problems.

2.2   Principles of Adiabatic Quantum Computing

Adiabatic quantum optimisation is a promising approach to solving NP-complete and NP-hard problems [97]. Assume that a solution to the optimisation problem is encoded in the ground state (i.e., the quantum state corresponding to the lowest eigenvalue) of a quantum Hamiltonian $\mathcal{H}_F$. By the second postulate of quantum mechanics (Section 1.2.2), the dynamics of a quantum system is fully specified by its Hamiltonian. If we know how to encode the objective function that we want to minimise in the Hamiltonian of a quantum system, then finding the ground state of the Hamiltonian is equivalent to finding the set of decision variables that minimises the objective function.

As a simple example of equivalence between the minimum of a function and the ground state of a Hamiltonian, consider a function $f : \{0,1\}^n \to \mathbb{R}$ that needs to be minimised and take the Hamiltonian

$$\mathcal{H}_F := \sum_{z \in \{0,1\}^n} f(z) \, |z\rangle \, \langle z| \, .$$

Clearly, for any $z_0 \in \{0, 1\}^n$,

$$\mathcal{H}_F |z_0\rangle = \left(\sum_{z \in \{0,1\}^n} f(z) |z\rangle \langle z| \right) |z_0\rangle = f(z_0) |z_0\rangle \langle z_0|z_0\rangle = f(z_0) |z_0\rangle,$$

since the computational basis $(|z\rangle)_{z \in \{0,1\}^n}$ is orthonormal. Therefore, any $z_0 \in \{0, 1\}^n$ is an eigenstate of $\mathcal{H}_F$ with eigenvalue $f(z_0)$. Minimising f is thus clearly equivalent to finding the lowest eigenvalue of the Hamiltonian $\mathcal{H}_F$.

Let us further assume that we have another quantum Hamiltonian, $\mathcal{H}_0$, whose ground state is easy to find and easy to prepare in an experimental setup. Then, if we prepare a quantum system to be in the ground state of $\mathcal{H}_0$, and then adiabatically (slowly) change the system Hamiltonian, $\mathcal{H}(t)$, from $\mathcal{H}_0$ at $t = 0$ to $\mathcal{H}_F$ at $t = \tau$ according to the following time evolution:

$$\mathcal{H}(t) = \left(1 - \frac{t}{\tau} \right) \mathcal{H}_0 + \frac{t}{\tau} \mathcal{H}_F, \tag{2.2.1}$$

then if τ is large enough, and $\mathcal{H}_0$ and $\mathcal{H}_F$ do not commute, the quantum system will remain in the ground state at all times according to the quantum adiabatic theorem. Measuring the quantum state at $t = \tau$ will produce a solution of our problem (a bitstring that encodes an optimal configuration of binary decision variables).

We provide a detailed explanation of the adiabatic quantum optimisation algorithm as well as the quantum adiabatic theorem in Section 2.2.1. In terms of potentially achievable quantum speedup in comparison with the best classical algorithms, for a problem of size N, quantum optimisers solve NP-hard combinatorial optimisation problems in time proportional to

$$\exp(\beta N^\gamma), \tag{2.2.2}$$

as N tends to infinity, for positive coefficients β and γ, which may be smaller than known classical algorithms [212]. In fact, early experiments on quantum annealers that realised the principles of AQC demonstrated several orders of magnitude quantum speedup (ignoring various computational overheads that are likely to be reduced as the technology matures) [189, 216, 320]. The coefficient γ is by far the most important. This can be illustrated by the following table, which provides the estimates of a hypothetical computation time as a function of problem size. Assuming that a single operation takes 1 microsecond, and the number of operations scales as either 2^N ("classical benchmark") or $e^{\sqrt{N}}$ ("quantum optimisation" with $\gamma = 0.5$), we obtain the following results:

N	2^N	$e^{\sqrt{N}}$
10	1 millisecond	0.024 milliseconds
50	35.7 years	1.2 milliseconds
100	4×10^{16} years	22 milliseconds
500	10^{137} years	1.4 hours

Table 2.1: Computational time as a function of the problem size.

The asymptotic estimate of the duration of adiabatic system evolution, T, which is exponential in problem size, is a consequence of the requirement that the system should always stay in the ground state of the local Hamiltonian. As the gap between the ground state and the first excited state becomes small at some point, the system evolution process should slow down accordingly.

However, if we only are interested in an approximate solution (and are willing to accept some deterioration in the quality of the obtained solution), we can expect the NP-hard combinatorial problems to be solved in polynomial time proportional to N^γ for $\gamma > 0$, as N becomes large [22, 276].

> The power of adiabatic quantum computing lies in its ability to solve hard computational problems through the natural evolution of a physical system.

2.2.1   The Quantum Adiabatic Theorem

In the Schrödinger equation (1.2.1) (normalised with $\hbar = 1$) with constant Hamiltonian $\mathcal{H}$, if the system starts in $|\psi(0)\rangle$ then the solution moves to

$$|\psi(t)\rangle = e^{-i\mathcal{H}t} |\psi(0)\rangle$$

at time $t \geq 0$. This, in particular, implies that any eigenstate $|\psi_0\rangle$ of $\mathcal{H}$, satisfying $\mathcal{H} |\psi_0\rangle = \lambda_0 |\psi_0\rangle$ for some eigenvalue λ_0, will evolve through the Schrödinger equation from $|\psi_0\rangle$ to

$$|\psi(t)\rangle = e^{-i\mathcal{H}t} |\psi_0\rangle = e^{-i\lambda_0 t} |\psi_0\rangle ,$$

namely, the eigenstate only gains a phase $e^{-i\lambda_0 t}$ and there is no transition over time between different eigenstates. The more interesting case, which we consider now, is that of a time-dependent Hamiltonian.

Consider again the Schrödinger equation (1.2.1) (normalised with $\hbar = 1$) over the time interval $[0, \tau]$, where now the Hamiltonian $\mathcal{H}$ is a function of time. The time change $t(\cdot)$, such that $t(0) = 1$ and $t(1) = \tau$, yields

$$i\frac{\mathrm{d} |\psi(s)\rangle}{\mathrm{d}s} = t'(s)\mathcal{H}(s) |\psi(s)\rangle \tag{2.2.3}$$

over the unit time interval $[0, 1]$. It is important to note here that the Hamiltonian $\mathcal{H}$ has no dependence on the time horizon τ itself. This excludes, in particular, Hamiltonians with multiple timescales as in [221]. Here, we are chiefly interested in Hamiltonians of the (slightly generalised) form (2.2.1),

$$\mathcal{H}(s) = r(s)\mathcal{H}_0 + (1 - r(s))\mathcal{H}_F, \tag{2.2.4}$$

for two given Hamiltonians $\mathcal{H}_0$ and $\mathcal{H}_F$, where $r(\cdot)$ is a continuous adiabatic evolution path decreasing from $r(0) = 1$ to $r(1) = 0$. The standard adiabatic schedule is given by $r(s) = 1 - s$.

The gist of the quantum adiabatic theorem is the following. Assume that the system starts from the ground state of $\mathcal{H}_0$. If the time evolution of the Hamiltonian is sufficiently slow, then the system remains in the ground state of the evolving Hamiltonian up to time 1. It was originally proposed by Born and Fock [41] and generalised by Kato [168] using the theory of perturbation of linear operators.

In order to state it properly, denote by $|\psi(\cdot)\rangle$ the solution to the Schrödinger equation (2.2.3), so that for any $s \in [0, 1]$, there exists a unitary operator $\mathcal{U}$ for which

$$|\psi(s)\rangle = \mathcal{U}(s) |\psi(0)\rangle .$$

Consider a Hamiltonian of the form (2.2.4) and a time-change $t(s) = s\tau$ (hence, $t'(s) = \tau$), so that the Schrödinger evolution dynamics reads

$$\mathrm{i}\frac{\mathrm{d}\,|\psi(t)\rangle}{\mathrm{d}t} = \tau\mathcal{H}(t)\,|\psi(t)\rangle , \tag{2.2.5}$$

over the interval $[0, 1]$. For each $t \in [0, 1]$, we denote $|\phi(t)\rangle$ the ground state of $\mathcal{H}_t$.

We finally present the following version of the quantum adiabatic theorem, due to Jansen, Seiler, and Ruskai [161]. We recall that an eigenvalue is called non-degenerate if there exists only one eigenstate associated with this eigenvalue. For any $t \in [0, 1]$, given the Hamiltonian $\mathcal{H}(t)$, we denote by Δ_t the (strictly positive) energy gap between the lowest eigenvalue of $\mathcal{H}_t$ and the next one.

Theorem 7 (Quantum Adiabatic Theorem). *Assume that, for any $t \in [0, 1]$ the Hamiltonian $\mathcal{H}(t)$ admits a non-degenerate ground state and that there exists $\varepsilon > 0$ such that*

$$\frac{2}{\varepsilon}\left\{ c_0\frac{\|\mathcal{H}'(0)\|}{\Delta_0^2} + c_1\frac{\|\mathcal{H}'(1)\|}{\Delta_1^2} + \int_0^1 \left[(3c_1^2 + c_1 + c_3)\frac{\|\mathcal{H}'(s)\|^2}{\Delta_s^3} + c_2\frac{\|\mathcal{H}''(s)\|}{\Delta_s^2} \right] \mathrm{d}s \right\} \leq \tau.$$

Then, starting the system (2.2.5) in the state $|\psi(0)\rangle = |\phi(0)\rangle$, the Schrödinger evolution yields at time 1 a state $|\psi(1)\rangle$ satisfying

$$\||\phi(1)\rangle - |\psi(1)\rangle\| \leq \varepsilon.$$

This quantitative version of the adiabatic theorem provides an estimate on how large the time horizon τ needs to be be in order to achieve sufficient accuracy ε. Consider, for example, the interpolation scheme (2.2.4) with $r(s) = 1 - s$, so that

$$\mathcal{H}(s) = (1 - s)\mathcal{H}_0 + s\mathcal{H}_F,$$

and, therefore, $\mathcal{H}'(s) = -\mathcal{H}_0 + \mathcal{H}_F$ and $\mathcal{H}''(s) = 0$. In that case, the quantitative estimate in Theorem 7 simplifies to the following:

Corollary 1. *With the same assumptions as in Theorem 7 and with the preceding interpolation scheme, the quantitative estimate*

$$\tau \geq \frac{2}{\varepsilon}\left\{c_0\frac{\|\mathcal{H}_F - \mathcal{H}_0\|}{\overline{\Delta}^2} + \left(3c_1^2 + c_1 + c_3\right)\frac{\|\mathcal{H}_F - \mathcal{H}_0\|^2}{\overline{\Delta}^3}\right\},$$

with $\overline{\Delta} := \min_{s\in[0,1]} \Delta_s$, ensures again that $\||\phi(1)\rangle - |\psi(1)\rangle\| \leq \varepsilon$.

This corollary in particular highlights the importance of the spectral gap $\overline{\Delta}$. The smaller it is, the longer one has to wait to see the adiabatic property become efficient. There exist different versions of the quantum adiabatic theorem, each with slightly different assumptions. A weak form was proved by Avron and Elgart [19] and by Bornemann [40] without the gap condition.

The proof of the quantum adiabatic theorem is rather technical and many versions exist, with slightly different proofs. They all rely, however, on analysing the evolution operator $\mathcal{U}(\cdot)$ corresponding to the Hamiltonian, which clearly solves

$$\frac{d\mathcal{U}(t)}{dt} = -i\tau\mathcal{H}(t)\mathcal{U}(t)$$

starting from the identity. In particular, one needs to construct an adiabatic operator $\mathcal{U}_A(\cdot)$ solving the same Schrödinger equation, replacing $\mathcal{H}(\cdot)$ with its adiabatic version

$$\mathcal{H}_A(t) = \mathcal{H}(t) + \frac{i}{\tau}[\mathcal{P}'(t), \mathcal{P}(t)],$$

where $\mathcal{P}(t)$ denotes the projection operator onto the desired eigenstate of $\mathcal{H}(t)$. The proof then follows from showing that $\mathcal{U}_A$ and $\mathcal{U}$ are close enough as τ becomes large.

By connecting the geometric properties of the adiabatic limit of the quantum system to parallel transport in a vector bundle, Berry's [33] and Simon's [291] works gave rise to Geometric Quantum Computing. Van Dam, Mosca, and Vazirani [316] showed that it is possible to construct a discrete-time approximation of the evolution operator $\mathcal{U}$ with only polynomial time overheads.

> The quantum adiabatic theorem provides the theoretical background for adiabatic quantum computing.

2.2.2   Optimisation and metaheuristics

Metaheuristics are used to find "good" approximate solutions to general optimisation problems. In plain terms, a metaheuristic is a search policy that explores the optimisation function $f(\cdot)$ by evaluating it at certain points. There are myriad metaheuristic algorithms that decide where next (at which value of x) to evaluate $f(x)$ given the history of function evaluations, but all are based on the same essential principle that good solutions are likely to be near other good solutions, or, in other words, that the optimisation surface has some smoothness. This in turn reveals the exploration versus exploitation trade-off that all metaheuristics must make.

A metaheuristic can exploit its "current" position, by descending incrementally. The risk is that this returns a (possibly not very good) local minimum. Alternatively, a metaheuristic can explore the optimisation surface by making "large movements" to discover whether another part of the optimisation surface returns smaller values of $f(x)$. In this case, the global minimum may be found, but the value of x returned may only be a fairly poor approximation of the actual global minimum.

Simulated annealing

Simulated annealing is a metaheuristic inspired by thermal annealing. Consider the minimisation of a given function $f : D \to \mathbb{R}$ over some domain $D \subset \mathbb{R}^n$. The algorithm works as follows:

1: Start with an initial value $x \in D$ and compute $f(x)$.

2: Randomly choose a neighbour y of x and evaluate $f(y)$.

3: If $f(y) < f(x)$, then set $x = y$.

4: Else, either keep x as is or set $x = y$.

5: Repeat until an end criterion is attained.

The crucial step is the random choice in Step 4, designed to avoid being stuck in a local minimum and to favour, at least at the beginning of the algorithm, exploration rather than exploitation. In the case where $f(y) \geq f(x)$, we shall hence make the switch $y \mapsto x$ with the probability

$$\mathbb{P}(\text{switch}) = \exp\left(-\frac{f(y) - f(x)}{\tau}\right),$$

where τ plays the role of the thermal annealing temperature: when the system is hot, particles move (exploration), and it cools down when refinement (exploitation) is required.

Quantum annealing and quantum tunnelling

Quantum annealing combines the idea of simulated annealing with the quantum adiabatic theorem by considering the time-dependent Hamiltonian

$$\mathcal{H}(t) = \mathcal{H}_F + \Gamma(t)\mathcal{H}_0,$$

where $\mathcal{H}_F$ is the final, *longitudinal field* Hamiltonian whose ground state encodes the optimal solution of the minimisation problem and $\mathcal{H}_0$ is the initial, *transverse field* Hamiltonian, assumed not to commute with $\mathcal{H}_F$. The function Γ is the *transverse field coefficient*, playing the role of the thermal temperature, namely, a continuous decreasing function of t converging to zero as t approaches the final time horizon. By the quantum adiabatic theorem (Theorem 7), if the system is in the ground state of $\mathcal{H}(0)$, and $\mathcal{H}(\cdot)$ evolves slowly with time, then it will remain in the ground state of $\mathcal{H}(t)$ for each t, and therefore, will converge to the ground state of $\mathcal{H}_F$ (hence, to the optimal solution of the problem). We note in passing that the quantum equivalent of the jump over the local hills in the simulated annealing framework is now the quantum tunnelling through the hill.

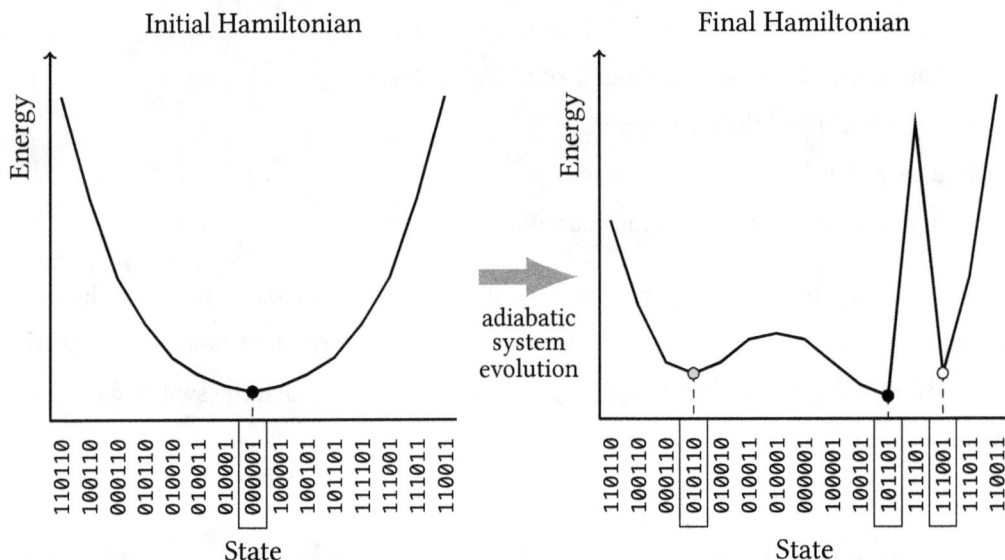

Figure 2.2: Schematic illustration of solving optimisation problems with quantum annealing. The objective is to keep the system in the ground state of $\mathcal{H}(t)$ through adiabatic evolution.

Figure 2.2 illustrates the practical application of quantum annealing to solving optimisation problems. We start by putting the system in the ground state of some easy-to-prepare transverse field Hamiltonian $\mathcal{H}_0$. This initial Hamiltonian is then slowly (adiabatically) transformed into the final longitudinal field Hamiltonian $\mathcal{H}_F$. If the system stays in the ground state of the local Hamiltonian throughout the quantum annealing process, then the readout will give us the optimal configuration of the binary decision variables (qubit values) that correspond to the global minimum of the objective function encoded in the final Hamiltonian.

Recall the following definition of the Hamming distance between two sets of (bit) strings:

Definition 8. Let $a := (a_1, \ldots, a_n)$ and $b := (b_1, \ldots, b_n)$ denote two bit strings in $\{0, 1\}^n$.

The Hamming distance between a and b is defined as

$$\sum_{i=1}^{n} |a_i - b_i|.$$

There are two important mechanisms in finding the global minimum: thermal annealing and quantum tunnelling. In Figure 2.2, the final Hamiltonian has a global minimum at state 101101 and two local minima at states 010110 and 111001. Although the energy of both local minimum states is the same, the Hamming distances from the global minimum are different. The Hamming distance between state 010110 and the global minimum state 101101 is 5: it is necessary to flip 5 bits to get from this local minimum state to the global minimum. The Hamming distance between state 111001 and the global minimum is 2. However, it is more difficult to get from state 111001 to the global minimum state 101101 because they are separated by the tall energy barrier, while getting from state 010110 to the global minimum is straightforward: a series of sequential bit flips over a relatively low and flat energy barrier can be achieved with the help of thermal annealing. Fortunately, we also have a quantum tunnelling effect that allows the system to go through the tall but narrow energy barriers. Flipping 5 bits (wide barrier) in one go could be a challenging task for the quantum tunnelling but flipping 2 bits (narrow barrier) is something that can happen with sufficiently high probability.

Quantum annealing is a practical implementation of the principles of adiabatic quantum computing and can be benchmarked against its classical counterpart – simulated annealing. Quantum annealing derives its power from two sources: thermal annealing and quantum tunnelling. It is the combination of these classical and quantum effects that allows quantum annealing to achieve superior performance.

2.3   Implementations of AQC

How do we build quantum annealers? What is their physical realisation? Can we find a suitable metric that would objectively quantify their performance? Do we observe steady progress in their development? In this section, we provide answers to these questions that are the results of two decades of intensive research.

2.3.1   The short history of quantum annealing

The first major patent was devised by Amin and Steininger [11] for D-Wave Systems, leading to the seminal paper [163]. Since then, numerous works investigated the value of D-Wave computers, in particular, McGeoch and Wang [226], who proved significant speedup of specific problems on one of the earlier D-Wave processors (D-Wave Two with 512 qubits). More experiments followed this early research with two results that, arguably, stand out from many other interesting findings.

First, the contribution of quantum tunnelling to the performance of D-Wave quantum annealers has been established through experiments on D-Wave Two [38]. Multi-qubit tunnelling has been observed and has been experimentally shown to play a computational role in a programmable quantum annealer.

Then an 8-order of magnitude speedup has been established on a 1,152-qubit D-Wave 2X processor relative to the classical benchmark (simulated annealing) for a crafted problem designed to have tall and narrow energy barriers separating local minima by Denchev *et al.* [83]. For instances with 945 variables, the D-Wave 2X quantum annealer achieved a time-to-99%-success-probability that is 100 million times faster than simulated annealing running on a single processor core.

Additionally, a fully connected graph problem has been addressed with the help of forward and reverse quantum annealing performed on the more recent 2,048-qubit D-Wave 2000Q processor by Venturelli and Kondratyev [189,320]. Chapter 3 provides a detailed description of this use case.

At the time of writing, the most advanced D-Wave machine is the 5,760-qubit Advantage system, which also boasts lower noise and much better qubit connectivity – a physical qubit can be directly connected with up to 15 other physical qubits in comparison with D-Wave 2000Q, where a physical qubit could only be connected with, at most, 6 other physical qubits [92]. Therefore, logical qubits that represent binary variables consist of shorter and more stable chains of physical qubits. For example, for the problem size $N = 64$ (which corresponds to the largest fully connected graph that can be embedded on D-Wave 2000Q), the chain length is 17 for the D-Wave 2000Q *Chimera* graph but only 7 for the Advantage *Pegasus* graph.

Interestingly, quantum computing algorithms prompted intensive improvement of their classical counterparts and the progress achieved in optimised classical algorithms proved [283] them to be able to at least match D-Wave machines.

2.3.2 Inter-generational comparison of D-Wave quantum annealers

The technological progress in the development of D-Wave quantum annealers has been investigated by Pokharel *et al* [257]. The performance of four generations of D-Wave quantum annealers has been studied on the task of solving an identical ensemble of a parameterised family of scheduling problems. These problems are NP-hard and find numerous practical applications [321].

The performance of quantum annealers was measured by a metric known as Time-to-Solution (TTS). In benchmarking studies, the data collected from multiple runs of the quantum annealer was used to calculate the probability of finding the ground state solution for the given configuration of adjustable parameters. This probability is given by the following expression:

$$p = \frac{\text{Number of ground state solutions}}{\text{Total number of runs}}.$$

The equivalent TTS is defined as the expected time to obtain the ground state solution at least once with α success probability and is computed as

$$\text{TTS} = t_{\text{run}} \frac{\log(1-\alpha)}{\log(1-p)},$$

where t_{run} is the annealing time for a single run of the quantum annealer, and α is by default taken equal to 99%.

Since scheduling problems are NP-hard, TTS should scale exponentially with the problem size N in the asymptotic limit, as shown by (2.2.2) with $\gamma = 1$. It is a question of what value of parameter β (the scaling exponent) would best fit the experimental results

$$\text{TTS} = T_0 \exp(\beta N), \tag{2.3.1}$$

for some constant $T_0 > 0$. As established in [257], the scaling exponent ranges from 1.01 for D-Wave Two to 0.17 for D-Wave Advantage.

To illustrate the magnitude of this improvement, let us set $N = 10$ and TTS = 100 microseconds for both cases ($\beta = 1.01$ and $\beta = 0.17$). This produces the following impressive results for TTS as a function of N, as shown in Table 2.2:

N	TTS ($\beta = 1.01$)	TTS ($\beta = 0.17$)
10	100 microseconds	100 microseconds
15	16 milliseconds	234 microseconds
20	2.4 seconds	547 microseconds
25	6.3 minutes	1.3 milliseconds
30	16.5 hours	3.0 milliseconds

Table 2.2: TTS as a function of problem size, N, for different values of scaling exponent.

2.3.3 Physical realisations of quantum annealers

Being adiabatic, the quantum annealer is a device that works by first specifying an initial Hamiltonian with an easily computable ground state, and then evolving it slowly to the final problem Hamiltonian. On the hardware side, to ensure quantum coherence, the system lives in a closed environment, away from external magnetic sources, and is kept at a very low temperature. The states of the system are viewed as superpositions of qubits, each represented as a superconducting loop in such a way that the state of the qubit is defined as the electric current in the loop. During computation, the direction of the current is unknown, but when the system decoheres (due to observation or noise), it becomes known. The direction of the spin of eachqubit is physically represented by a magnetic field applied to the loop, acting as a noise that may stir the qubit from its current spin. The qubits are also connected to each other via ferromagnetic (antiferromagnetic) couplings forcing their spins in the same or opposite directions.

The D-Wave quantum annealers rely on the Ising formulation of the Hamiltonian. This is justified by the fact that Barahona [23] showed that finding the ground state of an Ising spin glass is NP-hard. It means that any NP-complete problem can be reduced to an Ising spin glass problem with polynomial overhead (see Section 2.1). Mathematically, the optimisation problem is formulated as the minimisation of the cost function

$$L_{\text{Ising}}(\mathbf{s}) = \sum_{i=1}^{N} g_i s_i + \sum_{i=1}^{N} \sum_{j=i+1}^{N} J_{ij} s_i s_j, \tag{2.3.2}$$

where $\mathbf{s} := (s_1, \ldots, s_N)$ is a vector of binary decision variables (classical spin variables taking values $\{-1, +1\}$), and $(g_i)_{i=1,\ldots,N}$ and $(J_{ij})_{1 \leq i < j \leq N}$ are coefficients encoding the parameters of the optimisation problem.

According to the quantum mechanical description of a spin, we need to replace the classical spin variables with the corresponding operators – the Pauli operators σ_x, σ_y, and σ_z (see Section 6.3.3 for their matrix representations).

The *problem* or *final* Hamiltonian then takes the following Ising form [159]:

$$\mathcal{H}_F = \sum_{i=1}^{N} g_i \sigma_z^i + \sum_{i=1}^{N} \sum_{j=i+1}^{N} J_{ij} \sigma_z^i \sigma_z^j, \tag{2.3.3}$$

where g_i is the bias applied to qubit i and J_{ij} is the coupling between qubits i and j.

The *initial* Hamiltonian has the form

$$\mathcal{H}_0 = \sum_{i=1}^{N} \sigma_x^i, \tag{2.3.4}$$

where the operator σ_x (see Section 6.3.3) is the quantum NOT gate that flips the qubit state:

$$\text{NOT}\,|0\rangle = |1\rangle \quad \text{and} \quad \text{NOT}\,|1\rangle = |0\rangle.$$

Recalling the quantum states $|+\rangle$ and $|-\rangle$ defined in (1.2.6), we have

$$\text{NOT}\,|+\rangle = \frac{1}{\sqrt{2}}\text{NOT}\Big(|0\rangle + |1\rangle\Big) = \frac{1}{\sqrt{2}}\Big(|1\rangle + |0\rangle\Big) = |+\rangle$$

$$\text{NOT}\,|-\rangle = \frac{1}{\sqrt{2}}\text{NOT}\Big(|0\rangle - |1\rangle\Big) = \frac{1}{\sqrt{2}}\Big(|1\rangle - |0\rangle\Big) = -|-\rangle,$$

so that $|+\rangle$ and $|-\rangle$ are the two eigenstates.

The eigenstate of the initial Hamiltonian (2.3.4) is the equal superposition of the states $|0\rangle$ and $|1\rangle$ of all individual qubits:

$$\left[\frac{1}{\sqrt{2}}(|0\rangle + |1\rangle)\right]^{\otimes N}.$$

The Hamiltonian $\mathcal{H}_0$ is the *transverse field* Hamiltonian. Its role is to create disorder and to prevent spins from aligning with each other (along the z axis). The Hamiltonian $\mathcal{H}_F$ is the *longitudinal field* Hamiltonian. At the end of the annealing process, when the transverse field and the σ_x terms go to zero, spins should be aligned either parallel or anti-parallel with each other along the z direction (depending on the values of the corresponding coupling factors and their individual biases).

The architecture of D-Wave quantum annealers is based on a particular graph (the *Chimera* graph in the case of D-Wave 2000Q or the *Pegasus* graph in the case of the Advantage system) that realises a particular connectivity pattern between the physical qubits. These graphs are not fully connected as mentioned earlier, which means that a concrete optimisation problem to be solved on the quantum annealer must first be transformed (*embedded*) into a graph. We refer the reader to [63, 64] for a theoretical description of these embeddings. Here, we only present a schematic rendering of the *Chimera* graph and show how logical qubits can be constructed from the chains of multiple physical qubits.

Apart from D-Wave, Steffen [297] built an adiabatic quantum computation using nuclear magnetic resonance, available at room temperature, based on the discrete-time approximation of the quantum adiabatic theorem. Another nuclear magnetic resonance implementation was developed in [331] for integer factoring.

Finally, it is necessary to mention a perspective approach to building analog quantum computers based on the neutral atoms technology [139, 177]. Neutral atom quantum annealers have the potential to outperform quantum annealers built using other technologies and to bridge the gap between NISQ and fault-tolerant advantage.

2.3.4 Chimera graph and embedding of the logical qubits

The *Chimera* graph topology [76, 91] has a recurring structure of a bipartite graph, called a *unit cell*, as shown in Figure 2.3. The unit cell consists of two groups of four qubits each with pairwise connections between qubits from different groups and no connections between qubits from the same group. Thus, each unit cell graph consists of 8 vertices (physical qubits) and 16 edges (connections between physical qubits). These physical qubits (superconducting loops) are connected via *internal couplers*. Connections between unit cells are achieved via *external couplers*.

Each physical qubit in a *Chimera* graph is connected to six other physical qubits (via four internal couplers and two external couplers). This puts severe limitations on the type of problems that can be embedded in the graph if we would like to establish a one-to-one mapping between the binary decision variable (logical qubit) and the corresponding physical qubit. However, the effective connectivity between logical qubits can be improved if the logical qubit can be represented by a chain of physical qubits. This would allow us to reach distant corners of the graph and solve the fully connected graph problems. Figure 2.3 displays an example of a qubit chain formed by qubits A, B, C, and D (dark shaded qubits connected by bold lines). While individual qubits have only six connections, the qubit chain A-B-C-D has 18 external connections.

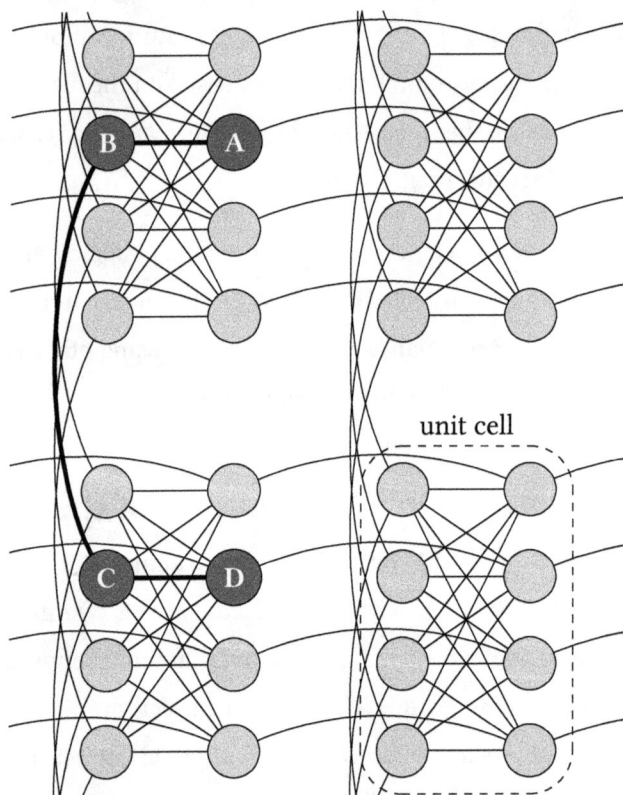

Figure 2.3: Chimera *graph. The figure displays a cropped view of four unit cells. Individual physical qubits are shown as circles, internal couplers are shown as straight lines, and external couplers are shown as curved lines. Qubits A, B, C, and D form a qubit chain (logical qubit).*

To ensure that qubits A, B, C, and D operate as a single logical qubit, we need to make connections between them strong enough such that their states are aligned at all times and they would flip simultaneously if the logical qubit is flipped. Since now they represent a single binary decision variable (logical qubit), the corresponding bias g (in (2.3.2) and (2.3.3)) will be shared equally across all four physical qubits in the chain. We expect all physical qubits in the chain to have the same value after measurement but this is not always the case. When the qubit chain becomes too long with too many internal and external couplers between the physical qubits, it is likely that the chain would be broken at some point. This will be seen as a disagreement between the physical qubits forming the chain – some of them will be measured as -1 and some as $+1$. The solution is to apply the *majority voting* rule: the value of a logical qubit is given by the mode of all constituent physical qubit values. For example, if physical qubits A, B, and C are measured as -1 and qubit D is measured as $+1$, the value of the logical qubit is assumed to be -1. The longer the chain, the easier it is to break it. This is why the improved connectivity of the new *Pegasus* graph in comparison with the old *Chimera* graph that results in shorter chains for the fully connected graph problems leads to a significant reduction in TTS.

> Quantum annealing has been successfully implemented using superconducting qubits controlled by pulses of microwave radiation. With significant progress in improving connectivity and scaling up the systems observed over the last several years, quantum annealers have demonstrated their potential as powerful optimisers.

2.4 Universality of AQC

Quantum computing has the theoretical advantage of being able to represent an exponential number of states at once, thereby proving exponential speedup compared to classical algorithms. The historically important examples are Shor's integer factorisation [289] and Grover's database search [124]. On a practical level, though, it remains unclear whether this speedup is actually within reach, in particular for NP-hard problems.

Adiabatic quantum computing is a completely different paradigm to gate model quantum computing – although they bear some resemblance – and rests upon the quantum adiabatic theorem. Contrary to gate model quantum computing, AQC has inherent fault tolerance, as proved in [61]. Since its performance depends on the spectral gap from the adiabatic theorem, environmental decoherence can be minimised by running an AQC device at temperatures much lower than this gap.

Key insights connecting gate model quantum computing and its adiabatic version were provided by Aharonov *et al.* [7], and by Farhi *et al.* [97] The former proved that AQC can simulate any algorithm with only polynomial overheads compared to gate model quantum computing while the latter showed that gate model quantum computers can reproduce any AQC computation. These two results therefore imply that AQC and gate model quantum computing are in fact polynomially equivalent. The proof in [7] assumes that the initial and final Hamiltonians in the adiabatic formulation are so-called k-local, meaning that they can be represented as sums of independent Hamiltonians, each acting on k qubits only. Existing AQC algorithms rely on a certain number of assumptions regarding the final Hamiltonian, the main one being that the latter has null off-diagonal elements with a 2-local connectivity structure, as in the case of the D-Wave quantum annealer. Unfortunately, this currently falls outside the scope of [7], leaving a question mark about the universality of this setup. Other AQC algorithms exist, in particular, involving stoquastic Hamiltonians [45] with real and non-positive off-diagonal elements or even more restrictive considerations [35].

The standard background for comparing algorithms – classical, quantum, or adiabatic – is that of complexity classes. We shall not delve into too many technicalities here though, but refer to [212] for details. There, the mathematical formulations of Ising problems (the standard problems solved by AQC) for a large number of NP-complete and NP-hard problems are presented, including precise formulations of Karp's 21 fundamental NP-complete problems [167], a perfect playground to compare quantum algorithms.

Summary

At the beginning of this chapter, we introduced several basic complexity classes and discussed their relationships. The time needed to solve NP-hard problems grows exponentially with the problem size, which is a strong motivation for exploring alternative approaches, such as analog adiabatic quantum computing. Even though quantum optimisers also solve NP-hard combinatorial optimisation problems in time that grows exponentially with problem size, the prefactor in the exponent may be smaller than for known classical algorithms. Additionally, we can expect to find an approximate solution in polynomial time, which provides a strong motivation for many practical applications of adiabatic quantum computing.

We then introduced the principles of AQC based on the adiabatic quantum theorem. The physical realisation of AQC – quantum annealing – has been contrasted with its classical counterpart – simulated annealing. We highlighted two main sources of computational power of quantum annealing: thermal annealing, which helps us find the minima of the objective function separated by wide but moderate energy barriers; and quantum tunnelling, which allows us to penetrate through the narrow and steep energy barriers. Their combination ensures efficient convergence to the global minimum of the objective function for many practical problems.

Having established the principles of AQC and the corresponding theoretical framework, we looked at the practical implementation of AQC in the form of quantum annealers based on superconducting qubits. We also introduced the Ising model, which provides a mathematical description of the problems solvable on quantum annealers.

We learned about the importance of the quantum chip layout (graph) and how several physical qubits can be coupled together to form a chain representing a single logical qubit. Finally, we touched on the universality of AQC.

In the next chapter, we will learn how quantum annealing can be used to solve practical NP-hard optimisation problems such as discrete portfolio optimisation.

3

Quadratic Unconstrained Binary Optimisation

Undoubtedly, Quadratic Unconstrained Binary Optimisation (QUBO) is a flagship use case of quantum annealing. We only need to have a closer look at the name of this class of optimisation problems to see why:

- Quantum annealers operate on binary spin variables. It is straightforward to perform mapping between binary decision variables (represented by the logical qubits) and spin variables.

- The objective functions of quadratic optimisation problems have only linear and quadratic terms. This significantly simplifies the models and allows their embedding on existing quantum annealing hardware.

- Unconstrained optimisation means that although QUBO allows us to specify conditions that must be satisfied, they are not hard constraints. The violation of constraints is penalised through the additional terms in the QUBO objective function, but it is still possible to find solutions that violate specified constraints.

All these features make QUBO problems solvable on quantum annealers. At the same time, the QUBO formulation exists for many important NP-hard combinatorial optimisation problems, such as graph partitioning, job-shop scheduling, binary integer linear programming, and many others. This class also includes the discrete portfolio optimisation problem, which we consider in this chapter. We should also mention here some of the recent attempts to address the problem of discrete portfolio optimisation using classical methods, such as the knapsack problem formulation by Vaezi *et al.* [314] and the application of evolutionary search methods such as genetic algorithms by Anagnostopoulos and Mamanis [12]. Both the knapsack problem with integer weights and genetic algorithms are discussed in this chapter.

3.1   Principles of Quadratic Unconstrained Binary Optimisation

QUBO represents optimisation problems in which a quadratic function of N binary variables, $q_1, \ldots, q_N$, has to be minimised over all possible 2^N assignments of its variables. The function to be minimised is referred to as the *cost function*, and it can be written as

$$L_{\text{QUBO}}(\mathrm{q}) = \sum_{i=1}^{N} a_i q_i + \sum_{i=1}^{N} \sum_{j=i+1}^{N} b_{ij} q_i q_j, \qquad (3.1.1)$$

where $\mathrm{q} := (q_1, \ldots, q_N) \in \{0, 1\}^N$ represents the assignment of N binary decision variables.

A broad class of optimisation problems with many practical applications admits a QUBO formulation [212]. To solve hard QUBO instances in an exact way, known classical algorithms require exponential time (in the problem size defined as the number of binary decision variables, N) [127]. Several approximate classical methods have been devised to reduce the computational cost; however, it is the fast maturing quantum annealing that aspires to demonstrate material speedup on the hardest QUBO instances, such as NP-hard discrete portfolio optimisation problems [189, 320].

3.1.1   QUBO to Ising transformation

A QUBO problem can be easily translated into a corresponding Ising problem solvable on a quantum annealer. The Ising cost function of N spin variables $s := (s_1, \ldots, s_N) \in \{-1, +1\}^N$ is given by

$$L_{\text{Ising}}(s) = \sum_{i=1}^{N} g_i s_i + \sum_{i=1}^{N} \sum_{j=i+1}^{N} J_{ij} s_i s_j.$$

The Ising and QUBO models are related through the transformation $s_i = 2q_i - 1$, hence the relationship with (3.1.1) is

$$J_{ij} = \frac{1}{4} b_{ij} \quad \text{and} \quad g_i = \frac{1}{2} a_i + \frac{1}{4} \left(\sum_{k=1}^{i-1} b_{ki} + \sum_{l=i+1}^{N} b_{il} \right)$$

disregarding an unimportant constant offset for the optimisation.

3.1.2   QUBO problem examples

There are many examples of important QUBO problems that can be directly applied to practical use cases arising in finance. Here, following Lucas [212], we mention a few of them in their traditional formulation.

Number Partitioning

Given a set of N positive numbers $\{n_1, \ldots, n_N\}$, is there a partition of this set of numbers into two disjoint subsets such that the sum of the elements in both sets is the same? For example, we can think about the set $\{n_1, \ldots, n_N\}$ as a collection of assets that must be divided equally between two parties. The Ising formulation of this problem is

$$L(s) = \left(\sum_{i=1}^{N} s_i n_i \right)^2,$$

where spin variables $(s_i)_{i=1,\ldots,N} \in \{-1, +1\}$ are the decision variables. If there is a solution to the Ising model with $L = 0$, then there is a configuration of spins where the sum of the n_i for the $+1$ spins is the same for the sum of the n_i for the -1 spins. The number partitioning problem finds numerous applications in economics and finance, from routing and scheduling problems [71] to signal detection and time series analysis [160].

Graph Partitioning

Consider an undirected graph

$$G = (V, E),$$

where V stands for the set of vertices and E stands for the set of edges, with an even number N of vertices. The task is to partition the set V into two subsets of equal size $N/2$, such that the number of edges connecting the two subsets is minimised. This problem has many applications: finding these partitions can allow us to run some graph algorithms in parallel on the two partitions, and then make some modifications due to the few connecting edges at the end [36]. The spin variables represent the graph vertices, with values $+1$ and -1 denoting the vertex being in either the $\{+\}$ set or the $\{-\}$ set. The problem is solved with the cost function consisting of two components:

$$L(\text{s}) = L_A(\text{s}) + L_B(\text{s}),$$

where

$$L_A(\text{s}) = A \sum_{i=1}^{N} s_i$$

provides a penalty if the number of elements in the $\{+\}$ set is not equal to the number in the $\{-\}$ set, and

$$L_B(\text{s}) = B \sum_{(u,v) \in E} \frac{1 - s_u s_v}{2}$$

is a term that provides a penalty for each time that an edge connects vertices from different subsets. If $B > 0$, then we wish to minimise the number of edges between the two subsets; if $B < 0$, we will choose to maximise this number. Should we choose $B < 0$, we must ensure that B is small enough so that it is never favourable to violate the L_A constraint. The graph partitioning problem can be applied to studying clustering in financial markets [295].

Both problems, number and graph partitioning, are NP-hard problems [167].

Binary Integer Linear Programming

Let $q := (q_1, \ldots, q_N)$ be a vector of N binary variables. The task is to maximise $c \cdot q$, for some vector c, given the constraint

$$Sq = b$$

with $S \in \mathcal{M}_{m,N}(\mathbb{R})$ and $b \in \mathbb{R}^m$. Many problems can be posed as binary integer linear programming, for example, a maximisation of profit subject to regulatory constraints [278].

The cost function $L(q)$ corresponding to this problem can be constructed as a sum of two terms, $L(q) = L_A(q) + L_B(q)$, where the first term is

$$L_A(q) = A \sum_{j=1}^{m} \left(b_j - \sum_{i=1}^{N} S_{ij} q_i \right)^2 ,$$

for some constant $A > 0$. Note that $L_A = 0$ enforces the constraint $Sq = b$. The second term is

$$L_B(q) = -B \sum_{i=1}^{N} c_i q_i,$$

with another positive constant $B < A$.

Knapsack with Integer Weights

We have a list of N objects, labelled by indices $i = 1, \ldots, N$, with the weight of each object given by $w_i \in \mathbb{N}$, and its value given by c_i, and we have a knapsack that can only carry some weight up to $W_{\max} \in \mathbb{N}$. If q_i is a binary variable denoting whether ($q_i = 1$) or not ($q_i = 0$) object i is contained in the knapsack, the total weight in the knapsack is

$$W = \sum_{i=1}^{N} w_i q_i,$$

and the total value is

$$C = \sum_{i=1}^{N} c_i q_i.$$

The task is to maximise C subject to the constraint that $W \leq W_{\max}$. The knapsack problem finds multiple applications in economics and finance [172, 314].

We introduce a binary variable, y_n, for each $n = 1, \ldots, W_{\max}$, which is 1 if the final weight of the knapsack is n, and 0 otherwise. As before, the cost function consists of two terms, $L(q) = L_A(q) + L_B(q)$, with

$$L_A(q) = A \left(1 - \sum_{n=1}^{W_{\max}} y_n \right)^2 + A \left(\sum_{n=1}^{W_{\max}} n y_n - \sum_{i=1}^{N} w_i q_i \right)^2,$$

which enforces that the weight can only take on one value and that the weight of the objects in the knapsack equals the value we claimed it did, and

$$L_B(q) = -B \sum_{i=1}^{N} c_i q_i.$$

As we require that it should not be possible to find a solution where L_A is weakly violated at the expense of L_B becoming more negative, we require $0 < B \max_{i=1,\ldots,N} c_i < A$ (adding an item to the knapsack, which makes it too heavy, is not allowed).

Many other famous NP-hard optimisation problems can be solved on quantum annealers. Here, we can mention the Map Colouring Problem [76] and the Job-Shop Scheduling Problem [321], which were successfully solved on the D-Wave quantum annealers. In this chapter, we provide detailed description and analysis of an important finance-related QUBO problem – a discrete portfolio optimisation problem investigated by Venturelli and Kondratyev [189, 320]. This is a hard, fully connected graph problem that can be best addressed using a newly developed technique of reverse quantum annealing.

> Many famous NP-hard problems can be solved with practical efficiency on quantum annealers in their QUBO formulation, up to a non-trivial number of variables. This makes quantum annealing a useful tool in dealing with classically hard optimisation problems.

3.2   Forward and Reverse Quantum Annealing

Having defined the QUBO problem, we now review how quantum annealing can act as an efficient solver.

3.2.1   Forward quantum annealing

The quantum annealing protocol, inspired by the adiabatic principle of quantum mechanics detailed in Chapter 2, dictates driving the system from an easy-to-prepare ground state of an initial Hamiltonian $\mathcal{H}_0$ to the unknown low-energy subspace of states of the problem Hamiltonian $\mathcal{H}_F$, ideally to the lowest energy state corresponding to the global minimum of the objective function. This *forward* quantum annealing procedure can be ideally described as attempting to drive the evolution of the time-dependent Hamiltonian

$$\mathcal{H}(t) = A(t)\mathcal{H}_0 + B(t)\mathcal{H}_F, \tag{3.2.1}$$

starting from

$$\mathcal{H}_0 = \sum_{i=1}^{N} \left(\sum_{c=1}^{N_c} \sigma_x^{ic} \right), \tag{3.2.2}$$

where $\mathcal{H}_0$ is a Hamiltonian describing an independent collection of local transverse fields for each spin of the system (σ_x is the Pauli X spin operator, or the quantum NOT gate, detailed in Section 6.3.3).

In expression (3.2.2), the first sum runs over all *logical qubits* representing binary decision variables in the QUBO formulation of the optimisation problem, and the second sum runs over all *physical qubits* in the chain that represents a logical qubit. The construction of a logical qubit from a chain of physical qubits is explained in Chapter 2 and a sample embedding of the qubit chain on the quantum chip is shown in Figure 2.3 for the *Chimera* graph. No physical qubit can be an element of more than one qubit chain. Assuming the most dense embedding scheme, we have $N \times N_c = K$, where K is the total number of physical qubits in the quantum chip. However, in practice, we have $N \times N_c < K$ for most problems of realistic size that require full connectivity between the logical qubits. For example, in the case of a D-Wave 2000Q *Chimera* graph consisting of $K = 2{,}048$ physical qubits, the maximum number of fully connected logical qubits is 64 and the maximum number of physical qubits in a chain is 17, meaning that $N \times N_c = 1{,}088$.

Figure 3.1 shows how $A(t)$ and $B(t)$ vary over the scale of total *annealing time* τ.

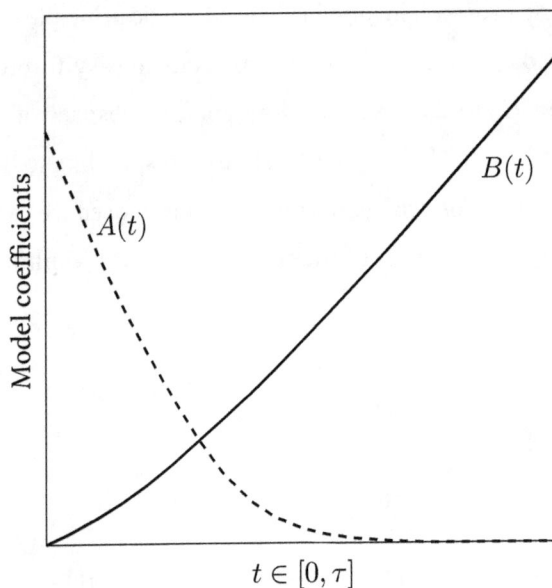

Figure 3.1: Schematic forward annealing schedule. $A(t)$ indicates the strength of the transverse magnetic field and $B(t)$ indicates the strength of the longitudinal magnetic field.

At the end of the annealing run, $A(\tau) = 0$ and the system is projected on the computational basis by a measurement of each qubit magnetisation. The duration of the anneal, τ, is a free parameter, hence it is often useful to define the fractional completion of the annealing schedule $s = t/\tau$.

3.2.2 Reverse quantum annealing

Figure 3.2 illustrates the quantum annealing protocol when the quantum annealer is set to operate as a reverse annealer.

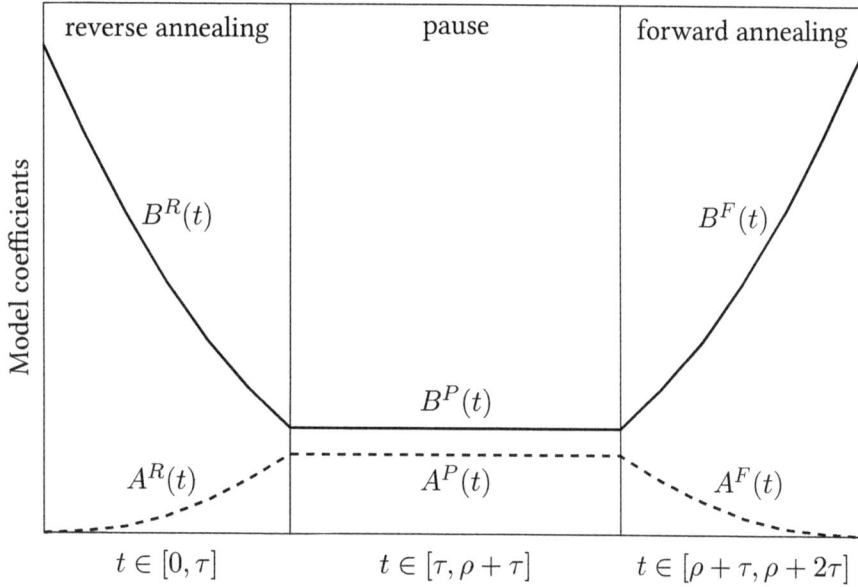

Figure 3.2: Schematic reverse annealing schedule. $A(t)$ indicates the strength of the transverse magnetic field and $B(t)$ indicates the strength of the longitudinal magnetic field.

The system is initialised with $B(0) = \max\{B\}$ and $A(0) = 0$, with spins set to a classical bitstring. The evolution then undergoes an inverse schedule up to a point where the Hamiltonian time-dependence is temporarily paused. With reference to the Hamiltonian evolution in (3.2.1), the transverse field evolution that we program for this protocol is the following three-phase function (analogous equations for $B(t)$):

$$A^R(t) := A(\tau + (s_p - 1)t) \qquad \text{Reverse Annealing: } t \in [0, \tau],$$

$$A^P(t) := A(s_p\tau) \qquad \text{Annealing Pause: } t \in [\tau, \tau + \rho],$$

$$A^F(t) := A((1 - s_p)(t - \rho) - (1 - 2s_p)\tau) \qquad \text{Forward Annealing: } t \in [\tau + \rho, 2\tau + \rho],$$

where ρ is the duration of the pause and $s_p \in [0, 1]$ indicates the location of the forward schedule where the pause is implemented. The total duration of the selected reverse anneal protocol is $2\tau + \rho$ as opposed to τ for the forward anneal. While the theory of reverse annealing is just starting to be investigated, the physics rationale of reverse annealing is to be found in the oversimplified idea that if the system is initialised in a state $\mathcal{S}$ corresponding to a local minimum of the objective function, the interplay of quantum and thermal fluctuations might help the state *tunnel* out of the energy trap during the reverse annealing, while the annealing pause (and, to some extent, the final forward annealing) allows the system to thermalise and relax in the neighbourhood of the newfound minimum. The quality of the initial state $\mathcal{S}$ is likely to influence dramatically the reverse annealing process. For the portfolio optimisation use case presented in this chapter, a classical greedy algorithm to set $\mathcal{S}$ can be used as described in [320].

> The combination of reverse quantum annealing with a classical greedy search algorithm has the potential to massively speed up a QUBO solver, thus realising a promising hybrid quantum-classical algorithm.

3.3   Discrete Portfolio Optimisation

The optimal portfolio construction problem is one of the most extensively studied problems in quantitative finance. The Modern Portfolio Theory (MPT) [217] has laid the foundation for highly influential mean-variance portfolio optimisation approach. According to the MPT, a typical portfolio optimisation problem can be formulated as follows. Let N be the number of assets, μ_i be the expected return of asset i, Σ_{ij} be the covariance between the returns for assets i and j, and R be the target portfolio return. Then the decision variables are the weights w_i, that is, the investment associated with the asset i ($w_i \in \mathbb{R}$). The standard Markowitz mean-variance approach consists in the constrained, quadratic optimisation problem

$$\min \sum_{i=1}^{N} \sum_{j=1}^{N} w_i w_j \Sigma_{ij}, \quad \text{subject to} \quad \sum_{i=1}^{N} w_i = 1 \quad \text{and} \quad \sum_{i=1}^{N} w_i \mu_i = R. \quad (3.3.1)$$

Quadratic problems of this form are efficiently solvable by standard computational means (e.g., quadratic programming with linear constraints) if the covariance matrix is positive definite. However, related discrete portfolio optimisation problems (with discrete weights w_i) are much harder to solve. In fact, they are known to be NP-complete [171].

Interestingly, the problem can also be cast into an unconstrained quadratic optimisation problem, which is a suitable formulation for quantum annealers [121, 220, 255, 274]. The problem we are trying to solve here is a construction of the optimal portfolio from the universe of assets with known characteristics, such as asset returns, volatilities, and pairwise correlations. A stylised portfolio optimisation problem consists of selecting M assets from the universe of N investable assets. These M assets should ideally be the best possible choice according to some criteria.

The scenario we target is a Fund of Funds portfolio manager who is facing the task of selecting the best funds that follow particular trading strategies in order to maximise the risk-adjusted returns according to some model [189] with a constraint that the assets are selected with equal preference weights [82]. Should we want to generalise the portfolio with larger allocation to a given asset, we could allow for multiples of the reference weight by cloning an asset and treating it as a new one.

3.3.1 QUBO encoding

The task of encoding the relationship among the choices of M funds (*without replacement*) from the universe of N funds can then be formulated as a quadratic form:

$$L(q) = \sum_{i=1}^{N} a_i q_i + \sum_{i=1}^{N} \sum_{j=i+1}^{N} b_{ij} q_i q_j, \qquad (3.3.2)$$

where $q_i = 1$ means that asset i is selected and $q_i = 0$ means that asset i is not selected. The task is then to find a configuration of $q := (q_1, \ldots, q_N)$ that minimises $L(q)$ subject to satisfying the cardinality constraint (i.e., selection of exactly M assets). A common way to deal with the cardinality constraint would be to add a term $L_{\text{penalty}}(q)$ to the cost function given by (3.3.2) such that the unsatisfying selections would be penalised by a large value

$P \gg 1$, which would force the global minimum to be such that $\sum_{i=1}^{N} q_i = M$:

$$L_{\text{penalty}}(\mathsf{q}) = P \left(M - \sum_{i=1}^{N} q_i \right)^2 . \tag{3.3.3}$$

The coefficients $(a_i)_{1 \leq i \leq N}$ reflect asset attractiveness on a standalone basis and can be derived from the individual assets' expected risk-adjusted returns. Assets with large expected risk-adjusted returns should be rewarded with negative values of a_i, while assets with small expected risk-adjusted returns should be penalised with positive values of a_i. The coefficients $(b_{ij})_{1 \leq i < j \leq N}$ reflect pairwise diversification penalties (positive values) and rewards (negative values). These coefficients can be derived from the pairwise correlations.

The minimisation of the QUBO cost function given by (3.3.2) and (3.3.3) should optimise the risk-adjusted returns by the use of the metrics of the Sharpe ratio. The Sharpe ratio (excess asset return measured in the asset volatility units) is calculated as $(r - r_0)/\sigma$, where r is the expected annualised asset return, r_0 is the applicable risk-free interest rate, and σ is asset volatility (annualised standard deviation of the asset returns). The higher the fund's Sharpe ratio, the better the fund's returns have been relative to the risk it has taken on. Volatility can be estimated as the historical annualised standard deviation of the net asset value returns (per share). Expected return can be either estimated as the historical return on fund investment or derived independently by the analyst/portfolio manager taking into account different considerations about the future fund performance.

3.3.2   The coarse-grained encoding scheme

Instead of using the raw real numbers obtained from the financial data for the QUBO coefficients, we opt to coarse-grain the individual fund's Sharpe ratios and their mutual correlations down to integer values by grouping intervals in buckets (the sample mapping schemes are shown in Table 3.1). By using bucketed values, we define a scorecard, which is loosely based on the past fund performances but can be easily adjusted by portfolio managers according to their personal views and any new information not yet reflected in the funds reports.

Sharpe ratio bucket	Coefficient a_i		
Equally spaced buckets,	Mapping scheme:		
from worst to best	A	B	C
1st	15	25	30
2nd	12	16	24
3rd	9	9	18
4th	6	4	12
5th	3	1	6
6th	0	0	0
7th	−3	−1	−6
8th	−6	−4	−12
9th	−9	−9	−18
10th	−12	−16	−24
11th	−15	−25	−30

Correlation bucket	Coefficient b_{ij}		
	Mapping scheme:		
	D	E	F
$-1.00 \leq \rho_{ij} < -0.25$	−5	−9	−10
$-0.25 \leq \rho_{ij} < -0.15$	−3	−4	−6
$-0.15 \leq \rho_{ij} < -0.05$	−1	−1	−2
$-0.05 \leq \rho_{ij} < 0.05$	0	0	0
$0.05 \leq \rho_{ij} < 0.15$	1	1	2
$0.15 \leq \rho_{ij} < 0.25$	3	4	6
$0.25 \leq \rho_{ij} \leq 1.00$	5	9	10

Table 3.1: Specification of the sample QUBO coefficients.

The choice of QUBO coefficients as small integer numbers is dictated by the technical realisation of the existing quantum annealer architecture (precision of the superconducting chip circuitry). Within this restriction, the portfolio manager may choose any linear or non-linear scale for QUBO coefficients. For example, the quadratic mapping scheme B strongly penalises low Sharpe ratio funds and strongly rewards high Sharpe ratio funds. The linear mapping schemes A and C distinguish better between funds with average performances. Similarly, the mapping scheme E penalises large positive correlations and rewards large negative correlations stronger than the mapping scheme D.

3.3.3   Construction of the instance set for numerical experiments

The instance set used for our case study is obtained by simulating asset values with the help of correlated geometric Brownian motion processes with a constant correlation ρ, drift μ, and log-normal volatility σ. The specific values of these parameters were derived from a wide range of fund industry researches (see [79] for the Sharpe ratio distributions) and, therefore, can be viewed as representative of the industry. The simulation time horizon was chosen to be one year and the time step was set at one month.

Every simulated (or "realised") portfolio scenario consists of 12 monthly returns for each asset. From these returns, we calculated the total realised return and realised volatility for each asset (which, obviously, differ from their expected values μ and σ) and for the portfolio as a whole. We also calculated realised pairwise correlations between all assets according to the input uniform correlation ρ. Finally, we calculated individual assets and portfolio Sharpe ratios. For reference, with $\rho = 0.1$, $\mu = 0.075$, $\sigma = 0.15$, and the constant risk-free interest rate set at $r_0 = 0.015$, the expected Sharpe ratio for each asset in the portfolio is 0.4. The expected Sharpe ratio for the portfolio of N assets is significantly larger due to the diversification and low correlation between the assets, for example, or a 48-asset portfolio, we would expect Sharpe ratio values from 0.5 (25th percentile) to 2.1 (75th percentile) with a mean around 1.4.

3.3.4   Classical benchmark – Genetic Algorithm

We selected Genetic Algorithm (GA) as a classical benchmark heuristic, which is a popular choice for solving hard combinatorial optimisation problems. GAs are adaptive methods of searching a solution space by applying operators modelled after natural genetic inheritance and simulating the Darwinian struggle for survival. There is a rich history of GA applications to portfolio optimisation problems, including [186, 194] recently.

In the case of portfolio optimisation, the solution (chromosome) is a vector $q := (q_1, \ldots, q_N)$ consisting of N elements (genes) that can take binary values in $\{0, 1\}$. The task is to find a combination of genes that minimises the cost (fitness) function $L(q)$. Due to the solution being represented by a relatively short bit string, we do not use the crossover recombination mechanism as it provides very little value in improving the algorithm convergence. Algorithm 1 provides a detailed description of the GA procedure.

Algorithm 1: GA − portfolio optimisation with cardinality constraint

---

Result: Optimal portfolio.

1: Generate L initial solutions by populating solutions with random draws from a pool of possible element values {0, 1}, with the restriction that "1" is assigned to the values of exactly M elements and "0" is assigned to the values of remaining $N - M$ elements.

2: Evaluate the cost (fitness) function for each solution.

3: Rank the solutions from "best" to "worst" according to the cost function evaluation results.

for i **from** *0* **to** *number of iterations* − *1* **do**

    a) Select K best solutions from the previous generation and produce L new solutions by randomly swapping the values of two elements with opposite values. With $L = mK$, every one of the "best" solutions will be used to produce m new solutions.

    b) Evaluate the cost (fitness) function for each solution.

    c) Rank the solutions from "best" to "worst" according to the cost function evaluation results.

end

The optimal values of the parameters L and K depend on the problem size and the specific QUBO coefficient mapping scheme, and can be found through trial and error. The objective here is to achieve the target convergence with the smallest possible number of objective function calls.

Our first task is to verify that the proposed mapping schemes are sensible in the sense that the minimisation of the objective functions (3.3.2) and (3.3.3) indeed leads to the construction of optimal portfolios. One possible approach to the problem of selecting M best assets from the universe of N investable assets is to select M individually best assets according to their individual Sharpe ratios. This approach would ignore the potential negative impact on diversification due to a probable large positive correlation between some of individually best assets, and there is no reason to believe that such a portfolio would be optimal. Therefore, we should demand that the optimal portfolio constructed by minimising $L(q) + L_{penalty}(q)$ should outperform the portfolio of M individually best assets.

For example, we can compare the results for the optimal 24-asset portfolio selected from the universe of 48 assets (for 10,000 portfolio instances simulated with $\rho = 0.1$, $\mu = 0.075$, $\sigma = 0.15$; as before, we assumed a constant risk-free interest rate $r_0 = 0.015$). Table 3.2 displays the Sharpe ratio distribution statistics obtained for the sample QUBO coefficient mapping schemes provided by Table 3.1 [320]. The results are presented in the format: **mean** (25th percentile; 75th percentile).

		Mapping schemes for b_{ij}		
		D	E	F
Mapping	A	**4.7** (2.5; 6.4)	**4.5** (2.1; 6.1)	**4.0** (1.7; 5.5)
schemes	B	**4.8** (2.7; 6.5)	**4.3** (2.0; 5.9)	**4.2** (2.0; 5.8)
for a_i	C	**5.0** (3.0; 6.7)	**4.8** (2.6; 6.3)	**4.6** (2.5; 6.1)

Table 3.2: Sharpe ratio distribution statistics for different mapping schemes.

For the portfolio of individually best assets, the Sharpe ratio distribution statistics looks as follows:

mean: 3.8   25th percentile: 2.6   75th percentile: 4.7

On average, the Sharpe ratio of the optimal portfolio is larger than that of the portfolio of individually best assets by 0.8, although some QUBO coefficient mapping schemes produce better results than others. Figure 3.3 illustrates a better performance of the optimal portfolio found through the minimisation of the cost function $L(q) + L_{penalty}(q)$ in comparison with the portfolio consisting of 24 individually best assets for the mapping schemes A and D.

Our second task is to understand how the time to solution scales with the problem size and whether quantum annealing can demonstrate a material speedup compared with the classical algorithm. It would be interesting to see what happens when we remove the constraint on the number of assets in the optimal portfolio. The portfolio optimisation results shown in Figure 3.3 were obtained for $M = N/2$, which is arguably the hardest combinatorial problem with a constraint on the number of assets, based on the size of the search space [189]. From a brute force approach perspective, the problem becomes even harder computationally if we remove this constraint due to the solution space growing with N as 2^N instead of $\frac{N!}{M!(N-M)!}$.

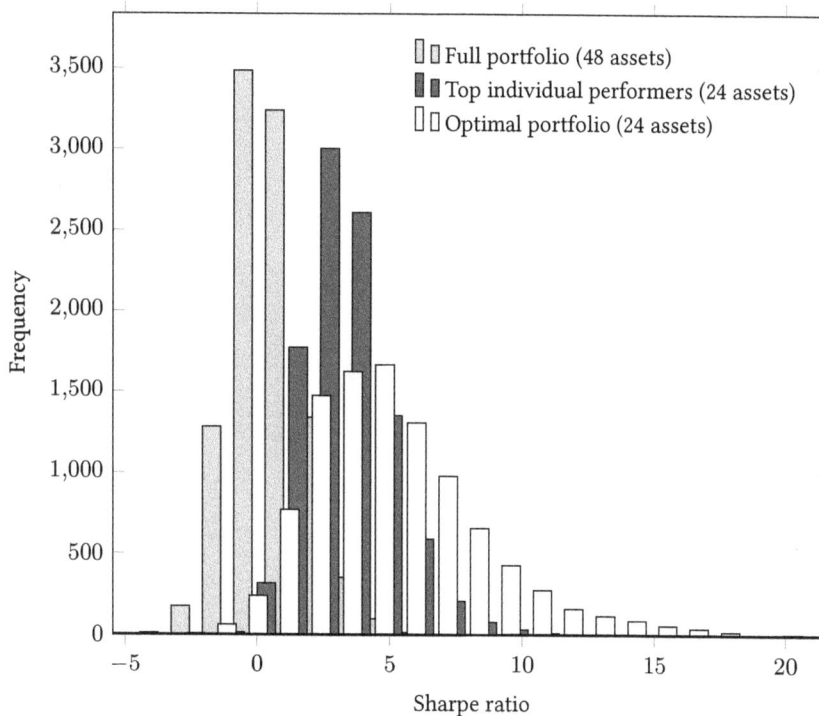

Figure 3.3: Sharpe ratio histogram (QUBO coefficients mapping schemes A and D). $M = N/2$.

The removal of the cardinality constraint is also warranted by the fact that a large energy scale P in expression (3.3.3) is typically associated with precision issues connected to the analog nature of the quantum annealing machine and the fact that there is a physical maximum to the energy that can be controllably programmed on local elements of the quantum chip. However, several hybrid quantum-classical strategies can be put in place to overcome this limitation.

For instance, we observe that artificially shifting the Sharpe ratio values by a constant amount $\pm\Delta$ (and adding buckets according to the prescription chosen, e.g., Table 3.1) will essentially amount to forcing the ground state solution of the unconstrained problem to more or less have a desired number of assets selected. Hence, while not solving the same problem, we could imagine a solver of a similarly constrained problem that will enclose the quantum annealing runs in a classical loop that checks for the number of selected assets $m(\Delta)$ in the best found solution with $\Delta = 0$, then increases or decreases the individual desirability of the assets according to whether m is larger or smaller than M and runs again until $m(\Delta) = M$ for $\Delta = \Delta^\star$. While, in this case, it is an approximation of the original problem, this sort of hybridisation scheme is not uncommon for quantum-assisted solvers [310] and the number of expected rounds of runs should scale proportionally to $\log_2(\Delta^\star)$ as per a binary search, introducing a prefactor over the time-to-solution complexity that should stay manageable. Other hybrid approaches could also be put forward to deal with the constraint, such as fixing some asset selections in preprocessing via sample persistence [166].

As per the preceding arguments, in our benchmark case study, we focus on running the unconstrained problem, setting $\Delta = 0$. Table 3.3 provides the characteristics of the benchmark instance set [320]. For a problem of the given size, the table reports the median number of assets in the optimal portfolio (and the minimum and the maximum in parenthesis), over 30 instances, for the unconstrained portfolio optimisation problem.

Problem size N	Number of assets in the optimal portfolio (unconstrained problem)
42	16 $(-7, +6)$
48	17 $(-6, +5)$
54	19 $(-7, +12)$
60	23 $(-13, +15)$

Table 3.3: Benchmark instance set characterisation.

3.3.5   Establishing quantum speedup

The aim is to solve representative portfolio instances at the limit of programmability for the D-Wave 2000Q quantum annealer. D-Wave 2000Q features 2,048 physical qubits; however, due to the limited connectivity of the D-Wave 2000Q *Chimera* graph, we can embed a maximum of 64 logical binary variables on a fully connected graph. Practically, we limit ourselves by working with up to 60 logical qubits, which means that the largest search space for our benchmarks is around $60!/(30!)^2 \simeq 10^{17}$ if $M = N/2$. This constraint dictates the configuration of the instance set, which consists of 30 randomly generated instances for $N = \{42, 48, 54, 60\}$ assets.

As mentioned in Chapter 2, a common metric to benchmark the performance of non-deterministic iterative heuristics against quantum annealing is the Time-to-Solution (TTS) [273]. The latter is defined as the expected number of independent runs of the annealer in order to find the ground state with probability (confidence level) $\alpha \in (0, 1)$:

$$\text{TTS} = t_{\text{run}} \frac{\log(1 - \alpha)}{\log(1 - p)},$$

where t_{run} is the running time elapsed for a single run – either τ for forward annealing (see Section 3.2.1), or $2\tau + \rho$ for reverse annealing (see Section 3.2.2) – and p is the probability of finding the optimum of the objective function in that single shot.

Figure 3.4 displays the TTS results for the GA, the forward QA solver, and the reverse QA solver for the unconstrained portfolio optimisation problem encoded using the mapping schemes A and D [320]. In the figure, the markers are the median values and the error bars indicate the 30th and the 70th percentiles over the 30-instance set. All TTSs are measured not counting the time required to run the greedy descent that initialises the initial ansatz S nor the overhead times for operating the quantum annealer.

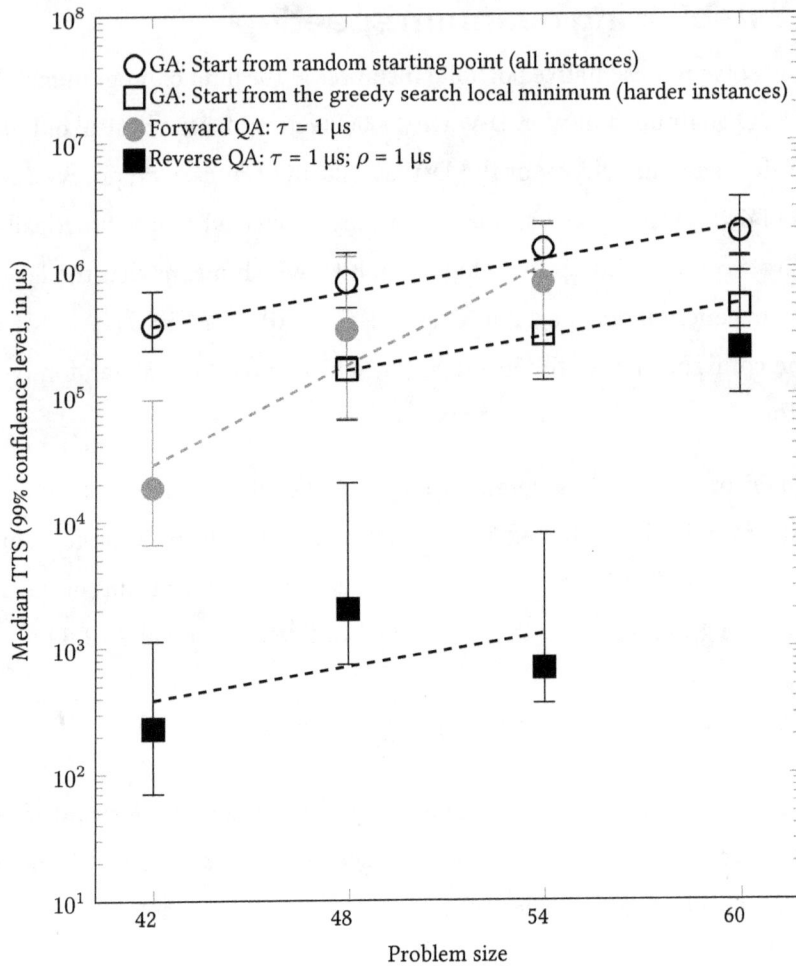

Figure 3.4: Time-to-solution (99% confidence level): GA, forward, and reverse quantum annealing. Unconstrained portfolio optimisation problem encoded using the mapping schemes A and D.

The GA can also be initialised by the greedy search heuristics, and this also decreases the TTS required for the GA to find the global minimum. As was experimentally established in [320], the best results are obtained for the smallest possible annealing time and pause time (1 μs). In the median case, we observe one to three orders of magnitude in speedup when applying reverse quantum annealing with respect to forward quantum annealing or classical benchmark.

It is likely that the non-monotonic behaviour of the reverse quantum annealing TTS for $N = 54$ is not of fundamental significance but is due to the noise associated with the finite, small size of our instance set. Although the small size of the instance set makes it difficult to draw a definite conclusion about the scaling of QA with the problem size, it appears that reverse quantum annealing displays similar scaling to GAs for portfolios of up to $N = 54$ assets – as illustrated by the dashed regression lines. A worse scaling for the limit case of $N = 60$ assets is probably due to the fact that, in this case, the physical qubit chains become too long and more likely to be broken. It is worth noting that for the same reason, $N = 60$ is also a very hard case for forward annealing. Reverse annealing displays significantly better scaling in comparison with forward annealing. While parameter β in (2.3.1) for forward annealing is equal to 0.3, it becomes 0.1 for reverse annealing.

Discrete portfolio optimisation is an NP-hard problem that can be solved on a quantum annealer using a hybrid quantum-classical reverse annealing technique with several orders of magnitude in quantum speedup (ignoring the measurement and system reset computational overheads). Although it is too early to say whether quantum annealing can become a widely adopted portfolio optimisation tool, there are indications that as technology and theory progress, it could represent a viable choice.

Summary

In this chapter, we applied quantum annealing to solving a discrete portfolio optimisation problem. We started with the principles of quadratic unconstrained binary optimisation and looked at several examples of NP-hard optimisation problems and their QUBO formulations.

Then, we introduced the concept of a quantum annealing protocol and specified two such protocols: forward annealing and reverse annealing. We also specified the classical benchmark: the genetic algorithm, an evolutionary search heuristic ideally suited for operations on binary variables.

Once we had all the necessary building blocks, we translated a sample discrete portfolio optimisation problem into QUBO and experimented with solving its instances on a D-Wave 2000Q quantum annealer. We collected sufficient statistics on various problem sizes to compare the performance of forward and reverse quantum annealing with the classical benchmark. We obtained encouraging results in terms of TTS, especially for the reverse quantum annealing protocol.

In the next chapter, we will learn how to apply quantum annealing to the problem of construction of a robust classifier. The proposed solution – quantum boosting – is a hybrid quantum-classical classifier (trained quantumly and run classically) that performs on a par with the standard classical models and, sometimes, outperforms them.

4

Quantum Boosting

In this chapter, we consider a quantum version of the classical boosting meta-algorithm –
a family of machine learning algorithms that convert weak classifiers into strong ones.
Classically, boosting consists of two main operations: i) adaptive (iterative) training of the
weak classifiers, thus improving their individual performance, and ii) finding an optimal
configuration of weights applied to the individual weak learners when combining them
into a single strong one.

Adaptive learning consists of iterative re-weighting of the samples from the training dataset,
forcing the model to improve its performance on the difficult-to-classify samples by giving
them heavier weights. These weights are adjusted at each algorithm iteration. Arguably,
the best-known and most successful example of such algorithms is the popular *adaptive
boosting* (AdaBoost) model. It was first formulated in 1997 by Freund and Schapire [107],
whose work has been recognised by the awarding of the prestigious Gödel Prize in 2003.

The main principle of AdaBoost is that the base classifiers (weak learners) are trained
in sequence and each base classifier is trained using a weighted form of the dataset: the

weighting coefficient associated with each sample depends on the performance of the previous classifiers. Samples that are misclassified by one of the base classifiers are given larger weights when used to train the next base classifier in the sequence. Once all base classifiers have been trained, their predictions are combined through some kind of weighted majority voting scheme [37]. Therefore, AdaBoost can be seen as a general framework that allows many possible realisations with various degrees of sophistication rather than a narrowly defined algorithm.

In contrast to AdaBoost, a boosting approach that consists of finding an optimal set of weights for the individual weak learners (with the weak learners being trained in the usual way) is straightforward to implement and relies on standard optimisation routines. However, this task becomes a hard combinatorial problem when an additional set of constraints is introduced. When the weights are only allowed to take binary values, the problem naturally lends itself to being formulated as a QUBO problem.

This is where quantum annealing has a role to play, as we have seen in Chapter 3. For a large enough number of weak classifiers, the search space becomes enormous, and classical algorithms (such as various evolutionary search heuristics) may take a non-trivial amount of time to find an optimal configuration of weights (or, at least, a good approximation). This is an ideal scenario for quantum annealing to demonstrate its strong points, including the possibility of achieving a material quantum speedup.

> Quantum boosting is a QUBO-based technique combining individual weak learners into a single strong classifier by constructing an optimal linear combination of binary classifiers. It is transparent, easy to interpret, and resistant to overfitting.

4.1   Quantum Annealing for Machine Learning

Quantum boosting is the first QML algorithm we will consider in this book. This is also the algorithm that plays to the natural strengths of quantum annealing.

4.1.1 General principles of the QBoost algorithm

We start with the general principles of the Quantum Boosting (QBoost) algorithm before exploring a specific finance-related application. In the formulation of QBoost, we will be using the following definitions and notations:

Object	Definition
$\mathbf{x}_\tau = (x_1(\tau), x_2(\tau), \ldots, x_N(\tau))$	Vector of N variables (features)
$y_\tau = \pm 1$	Binary label indicating whether $\mathbf{x}_\tau$ corresponds to Class 0 $(y = -1)$ or Class 1 $(y = +1)$
$\{\mathbf{x}_\tau, y_\tau\}_{\tau=1,\ldots,M}$	Set of training events
$c_i(\mathbf{x}_\tau) = \pm \frac{1}{N}$	Value of the weak classifier i on the event τ
$\mathbf{q} := (q_1, q_2, \ldots, q_N)$	Vector of binary (0 or 1) weights associated with each weak classifier

Table 4.1: QBoost algorithm notations.

Note that the $1/N$ weight associated with each individual classifier (signal, strategy) is a natural choice in the absence of a priori knowledge of the ensemble of classifiers performance and works well in practice [82].

We first specify the classification error for sample τ, which is given by the squared error

$$\left(\sum_{i=1}^{N} c_i(\mathbf{x}_\tau) q_i - y_\tau \right)^2 .$$

The total cost function to minimise is then the sum of squared errors across all samples:

$$L(\mathbf{q}) = \sum_{\tau=1}^{M} \left(\sum_{i=1}^{N} c_i(\mathbf{x}_\tau) q_i - y_\tau \right)^2$$

$$= \sum_{\tau=1}^{M} \left(\sum_{i=1}^{N} c_i(\mathbf{x}_\tau) q_i \sum_{j=1}^{N} c_j(\mathbf{x}_\tau) q_j - 2 y_\tau \sum_{i=1}^{N} c_i(\mathbf{x}_\tau) q_i + y_\tau^2 \right) .$$

Note that y_τ^2 does not depend on q and therefore has no influence on the minimisation of L. Adding a penalty $\lambda > 0$ to prevent overfitting, the cost function to minimise is thus

$$
\begin{aligned}
\tilde{L}(\mathbf{q}) &= \sum_{\tau=1}^{M}\left(\sum_{i=1}^{N} c_i(\mathbf{x}_\tau)q_i \sum_{j=1}^{N} c_j(\mathbf{x}_\tau)q_j - 2y_\tau \sum_{i=1}^{N} c_i(\mathbf{x}_\tau)q_i\right) + \lambda \sum_{i=1}^{N} q_i \\
&= \sum_{\tau=1}^{M}\left(\sum_{i=1}^{N}\sum_{j=1}^{N} q_i q_j c_i(\mathbf{x}_\tau)c_j(\mathbf{x}_\tau) - 2\sum_{i=1}^{N} q_i c_i(\mathbf{x}_\tau)y_\tau\right) + \lambda \sum_{i=1}^{N} q_i \\
&= \sum_{i=1}^{N}\sum_{j=1}^{N}\left(\sum_{\tau=1}^{M} c_i(\mathbf{x}_\tau)c_j(\mathbf{x}_\tau)\right) q_i q_j - 2\sum_{i=1}^{N}\left(\sum_{\tau=1}^{M} c_i(\mathbf{x}_\tau)y_\tau\right) q_i + \lambda \sum_{i=1}^{N} q_i \\
&= \sum_{i=1}^{N}\sum_{j=1}^{N} C_{ij} q_i q_j + \sum_{i=1}^{N} (\lambda - 2C_i)\, q_i,
\end{aligned}
$$

with

$$
C_{ij} := \sum_{\tau=1}^{M} c_i(\mathbf{x}_\tau)c_j(\mathbf{x}_\tau) \quad \text{and} \quad C_i := \sum_{\tau=1}^{M} c_i(\mathbf{x}_\tau)y_\tau.
$$

Remark: Adding a penalty term controlled by the coefficient λ is analogous to the LASSO regression method [6] with L_1 penalty, which is ubiquitous in machine learning. Ridge regression [264] with L_2 penalty could also be used and would also lead to another QUBO problem.

4.1.2 QUBO to Ising

As developed in Chapter 3.1.1, we now perform a transformation from QUBO to Ising from the binary decision variables $\mathbf{q} := (q_1, \ldots, q_N) \in \{0,1\}^N$ to spin variables $\mathbf{s} := (s_1, \ldots, s_N) \in \{-1,+1\}^N$ using

$$
\mathbf{s} = 2\mathbf{q} - 1 \quad \text{or} \quad \mathbf{q} = \frac{1}{2}(\mathbf{s}+1).
$$

Therefore, the Ising problem to be solved on the quantum annealer reads

$$\mathcal{H} = \tilde{L}(s) = \sum_{i,j=1}^{N} \left(\frac{1}{2}s_i + \frac{1}{2}\right) \left(\frac{1}{2}s_j + \frac{1}{2}\right) C_{ij} + \sum_{i=1}^{N} \left(\frac{1}{2}s_i + \frac{1}{2}\right) (\lambda - 2C_i)$$

$$= \frac{1}{4} \sum_{i,j=1}^{N} s_i s_j C_{ij} + \frac{1}{2} \sum_{i,j=1}^{N} s_i C_{ij} + \frac{1}{4} \sum_{i,j=1}^{N} C_{ij}$$

$$+ \frac{1}{2} \sum_{i=1}^{N} s_i \lambda + \frac{\lambda N}{2} - \sum_{i=1}^{N} s_i C_i - \sum_{i=1}^{N} C_i.$$

Since the three terms

$$\frac{1}{4} \sum_{i,j=1}^{N} C_{ij}, \quad \frac{\lambda N}{2}, \quad \text{and} \quad \sum_{i=1}^{N} C_i$$

do not depend on s, they can be removed from the cost function. The substitution $\bar{\lambda} = \frac{1}{2}\lambda$ then yields the final Ising problem

$$\mathcal{H} = \frac{1}{4} \sum_{i,j=1}^{N} s_i s_j C_{ij} + \frac{1}{2} \sum_{i,j=1}^{N} s_i C_{ij} + \sum_{i=1}^{N} s_i (\bar{\lambda} - C_i).$$

The problem that quantum annealing attempts to solve is to minimise $\mathcal{H}$ and to return the minimising, ground-state spin configuration $(s_i^g)_{i=1,...,N}$. The strong classifier is then built as

$$R(x) = \sum_{i=1}^{N} s_i^g c_i(x) \in [-1, 1], \tag{4.1.1}$$

for each new event x that we wish to classify [233].

4.2 QBoost Applications in Finance

Quantum Annealing for Machine Learning (QAML) has been applied productively to a wide range of financial and non-financial use cases. It demonstrated a performance advantage in comparison with standard classical machine learning models such as the binary decision tree-based Extreme Gradient Boosting (XGBoost) and Deep Neural Network (DNN) classifiers, especially on relatively small datasets. The QAML use cases come from such diverse fields as high-energy physics (the Higgs boson detection [233]) and computational biology (the classification and ranking of transcription factor binding [200]). In finance, the most obvious application of QAML is to credit scoring and fraud detection as well as to the construction of strong trading signals from large numbers of weak binary (buy/sell) trading signals.

In this section, we analyse QBoost performance on the more conventional binary classification problem – forecasting credit card client defaults. We also provide classical benchmarks (gradient boosting and feedforward neural network classifiers) and analyse QBoost performance from different angles. The chosen dataset is relatively large, with tens of thousands of samples, which should help standard classical classifiers avoid overfitting and demonstrate their best qualities.

It has been established in [233] that the QBoost algorithm is resistant to overfitting because it involves an explicit linearisation of correlations (hence its better performance on the smaller dataset in comparison with classical benchmarks). Another useful aspect of the model is that it is interpretable directly, with each weak classifier corresponding to a specific feature or a combination of features (or their functions), and the strong classifier being a simple linear combination thereof. This compares favourably with the "black box" machine learning discriminants, such as when using gradient boosting or DNNs. This is especially important for financial products aimed at retail customers.

4.2.1   Credit card defaults

The default of credit card clients (DCCC) dataset is available from the UCI Machine Learning Repository [334, 335]. The dataset consists of 30,000 samples with binary classification: a client defaults on the credit card payment (Class 1) and a client does not default (Class 0). There are 23 features (F1-F23) that have at least some predictive power and can be used for the classification decision:

- F1: Amount of the given credit (NT dollar): it includes both the individual consumer credit and his/her family (supplementary) credit.

- F2: Gender (1 = male; 2 = female).

- F3: Education (1 = graduate school; 2 = university; 3 = high school; 4 = others).

- F4: Marital status (1 = married; 2 = single; 3 = others).

- F5: Age (years).

- F6-F11: History of past payments. F6 – the repayment status in the previous month, F7 – the repayment status two months ago, and so on. The measurement scale for the repayment status is: -1 = pay duly; 1 = payment delay for one month; 2 = payment delay for two months; ...; 8 = payment delay for eight months; 9 = payment delay for nine months and above.

- F12-F17: Amount of bill statement (NT dollar). F12 – amount of bill statement previous month, F13 – amount of bill statement two months ago, and so on.

- F18-F23: Amount of previous payment (NT dollar). F18 – amount paid last month, F19 – amount paid two months ago, and so on.

The weak classifiers were constructed in the following way: each feature was used separately as an input into the logistic regression classifier using one-hot-encoding with the aim of making a binary prediction: $-1/N$ for Class 0 (no default) and $+1/N$ for Class 1 (default), where $N = 23$ is the total number of weak classifiers (number of features in the dataset). Note that this is not the only possible approach. It is perfectly feasible to build the weak classifiers through some (possibly non-linear) combination of original features, which is the usual approach in machine learning. This should be done every time we have a clear understanding of which combination of features would produce a more meaningful and insightful result. However, in this particular example, our objective is to illustrate the general principles of the QBoost algorithm and we do not assume any subject matter expertise that would allow us to construct better derived features.

We have used `sklearn.linear_model.LogisticRegression` from the `scikit-learn` package [249] for the weak classifiers. The dataset was split into training and testing datasets at a 70:30 ratio with the help of the `sklearn.model_selection.train_test_split` module.

The class labels were encoded as -1 for Class 0 and $+1$ for Class 1, as per the QBoost algorithm requirements.

The following configuration of the `LogisticRegression` model was used in constructing the weak classifiers dataset (all other parameters were set at their default values):

- penalty = 'l2'
- C = 1.0
- solver = 'lbfgs'
- max_iter = 1000

Therefore, we have a training dataset (21,000 samples) and a testing dataset (9,000 samples), each consisting of the predictions of 23 weak classifiers (taking values $\{-1/23, +1/23\}$) and class labels (taking values in $\{-1, +1\}$). If the prediction of the strong classifier is given by the sum of predictions of the weak classifiers (a simple majority voting approach), its value will be in the $[-1, 1]$ range, with the values of -1 and $+1$ achieved if all the weak classifiers are in perfect agreement with each other.

QBoost provides an improvement on this approach by finding an optimal configuration of the weak classifiers such that the majority voting is performed on a subset of available weak classifiers. In other words, a majority voting performed on all weak classifiers is just a special case of the QBoost (one of the possible configurations to explore). Therefore, it is necessary to compare QBoost performance with more advanced classical machine learning models such as gradient boosting and neural networks. We provide this comparison in Section 4.3.

4.2.2   QUBO classification results

Each feature in the DCCC dataset is uniquely mapped to the corresponding (weak) logistic regression classifier and associated binary decision variable $(q_i)_{i=1,\dots,23}$. These decision variables are represented by the logical qubits/spin variables in the QUBO/Ising formulation of the optimisation problem. The number of non-zero decision variables (weights) depends on the degree of regularisation we would like to impose. Table 4.2 shows the optimal configurations of the weights as a function of the penalty λ obtained for the training dataset. Given the relatively small number of weak classifiers in our example, the optimal configuration can be found by an exhaustive search. As one would expect, the larger the value of the penalty λ, the smaller the number of non-zero weights.

λ	non-zero weights
500	$\{q_1, q_6, q_7, q_8, q_9, q_{10}, q_{11}\}$
600	$\{q_6, q_7, q_8, q_9, q_{10}, q_{11}\}$
700	$\{q_6, q_7, q_{10}, q_{11}\}$
800	$\{q_6, q_{10}, q_{11}\}$
900	$\{q_6, q_{11}\}$
1000	$\{q_6\}$

Table 4.2: Optimal configurations of QUBO weights q for various values of the penalty λ. The optimal configurations list all non-zero weights.

Given a configuration of weights, we can build the strong classifier as per (4.1.1). Then, we can compare the performance of the obtained strong classifier on both training (in-sample) and testing (out-of-sample) datasets. The performance metrics of choice are *accuracy*, *precision*, and *recall*. The classifier performance can also be visualised with the help of a *confusion matrix*. Here are their definitions:

- **Accuracy** is the ratio of correctly predicted observations to the total observations. Accuracy is a good metric for classes of roughly the same size and equivalent importance. However, it is a poor metric for the dataset in our example: the Class 0 samples (no default) are far more numerous but the relative importance of Class 1 samples (default) is much higher.

- **Precision** is the ratio of correctly predicted positive observations to the total predicted positive observations. High precision corresponds with a low false positive rate. This is a metric we would like to maximise in the context of credit card defaults if there is a high cost associated with the incorrect default predictions.

- **Recall** is the ratio of correctly predicted positive observations to all observations in the positive class. In the context of credit card defaults, this metric shows how many of the actual defaults were predicted by the classifier. We would like to maximise this metric from the risk management perspective.

- **Confusion matrix** for a binary classifier is a 2×2 matrix whose elements are the counts of the true positive (TP), true negative (TN), false positive (FP), and false negative (FN) predictions of a classifier, as shown in Figure 4.1.

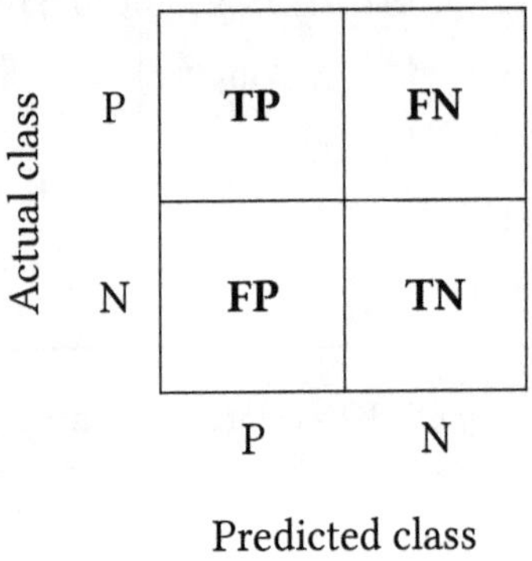

Figure 4.1: Confusion matrix for a binary classifier.

Accuracy, precision, and recall are then defined as follows:

$$\text{Accuracy} := \frac{\text{TP} + \text{TN}}{\text{TP} + \text{TN} + \text{FP} + \text{FN}},$$
$$\text{Precision} := \frac{\text{TP}}{\text{TP} + \text{FP}},$$
$$\text{Recall} := \frac{\text{TP}}{\text{FN} + \text{TP}}.$$

Figure 4.2 displays in-sample and out-of-sample confusion matrices for the strong QBoost classifier assuming that Class 1 (default) is the positive class and Class 0 (no default) is the negative class. The penalty was set at $\lambda = 10^3$, thus enforcing strong regularisation.

The in-sample and out-of-sample results are quite close, as one would expect from a strongly regularised classifier. Table 4.3 summarises the results.

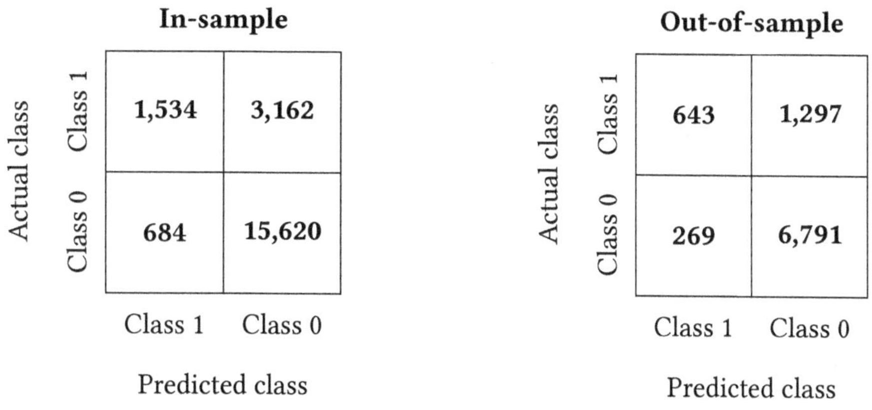

In-sample

Out-of-sample

		Class 1	Class 0
Actual class	Class 1	1,534	3,162
	Class 0	684	15,620

Predicted class

		Class 1	Class 0
Actual class	Class 1	643	1,297
	Class 0	269	6,791

Predicted class

Figure 4.2: Confusion matrices for the QBoost classifier (DCCC dataset).

	Accuracy	Precision	Recall
In-sample	0.82	0.69	0.33
Out-of-sample	0.83	0.71	0.33

Table 4.3: Accuracy, precision, and recall for the QBoost classifier trained and tested on the DCCC dataset.

4.3   Classical Benchmarks

Classical benchmarking is an important element of the testing of quantum algorithms. Small-scale (or even stylised) problems are ideally suited for this task. Let us see how the QBoost model performs in comparison with the standard classical ML classifiers: neural networks and gradient boosting.

4.3.1   Artificial neural network

An Artificial Neural Network (ANN) is a network of interconnected *activation units* (or *artificial neurons*), where each activation unit performs three main functions (Figure 4.3):

- Summation of the input signals $(x_i)_{i=1,...,N}$, from all the upstream units to which it is connected with multiplication by the corresponding weights $(w_i)_{i=1,...,N}$;

- Non-linear transformation of the aggregated input;
- Sending the result to the downstream units to which it is connected.

Sometimes the activation unit also performs binarisation (or, more generally, digitisation) of the output – typically, this is a task of the activation units in the output layer of an ANN trained as a classifier.

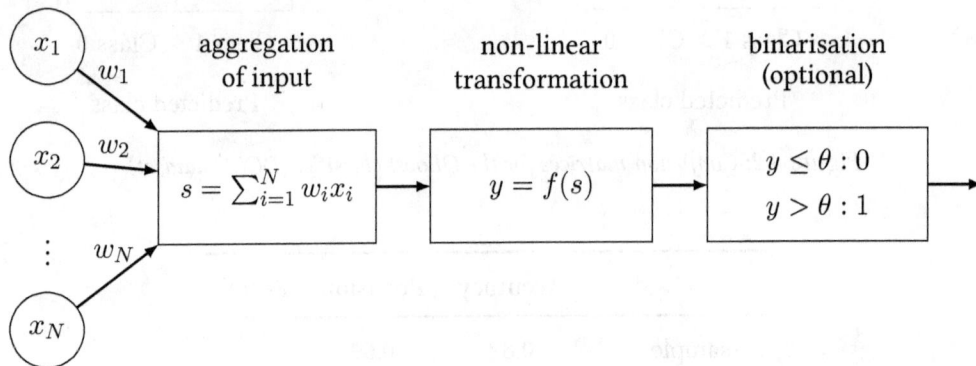

Figure 4.3: Schematic representation of an artificial neuron (perceptron).

In its simplest form, an ANN is organised as layers of activation units: an input layer, an output layer, and one or several hidden layers, as schematically pictured in Figure 4.4.

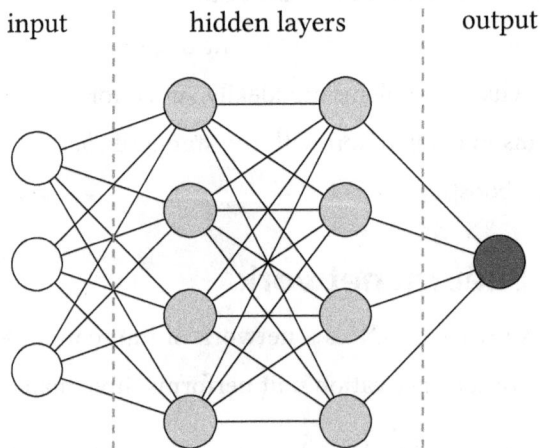

Figure 4.4: Schematic representation of a feedforward ANN.

The activation unit in Figure 4.3 is known as a *perceptron*, and the ANNs consisting of layers of perceptrons are known as Multi-Layer Perceptrons (MLPs). MLPs are *feedforward* neural networks: the signal travels in one direction from the input layer to the output layer. ANNs can be organised differently with a signal travelling back and forth between the layers, and we will explore one such model in the next chapter. However, when it comes to building a classifier, the simple feedforward architecture works well in practice.

The practical approach to the ANN architecture is based on the fundamental result obtained by Cybenko [75]. It states that arbitrary decision regions can be arbitrarily well approximated by continuous feedforward neural networks with only a single hidden layer and any continuous sigmoidal non-linearity. This result was further generalised to the wider range of activation functions by Hornik, Stinchcombe, and White [150]. It was established that multilayer feedforward networks with only a single hidden layer and an appropriately smooth hidden layer activation function are capable of arbitrarily accurate approximating of any arbitrary function and its derivatives. In fact, these networks can even approximate functions that are not differentiable in the classical sense, but possess only generalised derivatives [239].

4.3.2 Training artificial neural networks

The process of training an ANN consists of finding an optimal configuration of network parameters (weights and biases) such that the new unseen input is transformed in the desired way. The network is trained on what is known as a *training dataset*. The samples from the training dataset can be *labelled* (each sample is assigned a class label, either numerical or categorical). In this case, we can perform *supervised learning*, where the network is tasked with learning the mapping between the features and the class labels – an ANN trained in the supervised learning mode becomes a classifier. When the samples are not labelled, we can train the network as a regressor. Although ANNs trained as classifiers may seem to be the most obvious practical decision-making tools, regressors too find numerous applications in various fields of quantitative finance – for example, in learning the natural dynamics and transformations of interest rate curves [183].

However, we would like to focus here on the labelled datasets since our objective is to consider a classical counterpart of the QBoost classifier. The standard approach to training a feedforward ANN is the *backpropagation* of error with gradient descent [118]. We will briefly explain the main idea of this method.

The starting point is the specification of some suitable cost function that indicates how far we are from the correct classification. Without loss of generality, assume that we work with a training dataset consisting of M samples, where each sample is a pair of an N-dimensional vector of features and a binary class label:

$$\{x^j, y^j\}_{j=1,\ldots,M}, \quad \text{with} \quad x^j := (x_1^j, \ldots, x_N^j) \quad \text{and} \quad (y^j)_{j=1,\ldots,M} \in \{0, 1\}. \quad (4.3.1)$$

Let $(\hat{y}^j)_{j=1,\ldots,M}$ be the class labels assigned to the corresponding training samples by the ANN for some configuration of the network weights $\mathrm{w} = (w_1, \ldots, w_K)$. Then, we can define the cost function as

$$L(\mathrm{w}) := \sum_{j=1}^{M} g\left(y^j, \hat{y}^j(\mathrm{w})\right), \quad (4.3.2)$$

where $g(y^j, \hat{y}^j(\mathrm{w}))$ is the estimation error for sample j. There are many possible ways of specifying the error function, the most popular being the squared error

$$g(y^j, \hat{y}^j) := \left(y^j - \hat{y}^j\right)^2. \quad (4.3.3)$$

Given the cost function $L(\cdot)$, we can calculate its sensitivities (derivatives) $\partial L(\mathrm{w})/\partial w_k$, for each $k = 1, \ldots, K$, with respect to the network weights. We can then *update* the weights by changing them in the direction that would reduce the estimation error, i.e., by moving in the opposite direction of the corresponding gradients:

$$w_k \longleftarrow w_k - \eta \frac{\partial L(\mathrm{w})}{\partial w_k}, \quad (4.3.4)$$

where the coefficient η is called the *learning rate*, which can be either constant or dynamic.

We then iterate the procedure given by (4.3.2), (4.3.3), and (4.3.4) until either the estimation error drops below a predefined threshold or a maximum number of iterations is reached. Often, the learning rate is set initially at some relatively large value and then decays exponentially with the number of iterations.

The gradients can be calculated numerically (e.g., using the finite difference method) or analytically, the latter being obviously preferable. The most widely used non-linear activation functions and their gradients are listed in Table 4.4 and their plots are shown in Figure 4.5:

Activation function	Notation	Function	Derivative
Logistic sigmoid	$\sigma(x)$	$\left(1 + e^{-x}\right)^{-1}$	$\sigma(x)(1 - \sigma(x))$
Hyperbolic tangent	$\tanh(x)$	$\dfrac{e^x - e^{-x}}{e^x + e^{-x}}$	$1 - \tanh^2(x)$
Rectified Linear Unit	$\mathrm{ReLU}(x)$	$\max(0, x)$	0 if $x < 0$; 1 if $x > 0$

Table 4.4: Activation functions.

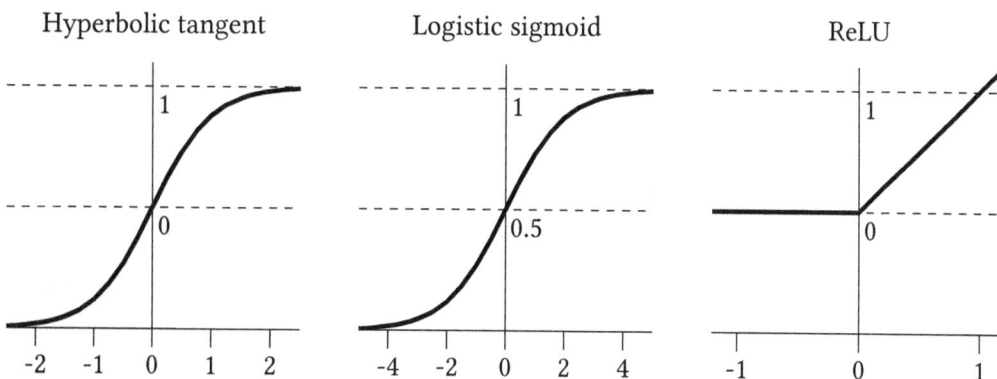

Figure 4.5: Activation functions.

Remark: The sigmoid activation functions, such as logistic sigmoid and hyperbolic tangent, are the activation functions of choice for shallow neural networks with only a couple of hidden layers. In this case, it is possible to exploit the smoothness of the sigmoid functions in order to achieve the best possible approximation of the function we are trying to learn. However, in the case of deep neural networks with a large number of hidden layers, we face the problem of vanishing gradients – gradients of $\sigma(x)$ and $\tanh(x)$ become null as $x \to \pm\infty$. At the same time, ReLU always has a non-zero gradient for all $x > 0$, which makes it the activation function of choice for deep neural networks whenever it makes sense to sacrifice the smoothness of the activation function for non-zero gradients.

Finally, the problem of overfitting can be addressed by adding a regularisation penalty term to (4.3.2) – for example, the following L_2 penalty, which discourages large network weights associated with strong non-linearity:

$$L(\mathrm{w}) := \sum_{j=1}^{M} g\left(y^j, \hat{y}^j(\mathrm{w})\right) + \lambda ||\mathrm{w}||^2,$$

where the parameter λ controls the degree of regularisation.

4.3.3 Decision trees and gradient boosting

The decision tree approach to classification is based on the concept of splitting a dataset on the available features in order to maximise the *information gain*, defined as

$$G(D, f) = I(D) - \sum_{j=1}^{M} \frac{N_j}{N} I(d_j),$$

where D is the dataset of the parent node, $(d_j)_{j=1,\ldots,M}$ are the datasets of the child nodes into which the parent node is split, N is the number of samples in the parent node, $(N_j)_{j=1,\ldots,M}$ are the number of samples in the child nodes, and I is the chosen *impurity measure*. The latter indicates the presence of the samples from the different classes in the same node: it is zero if the node holds samples from a single class and is maximal if the node holds an equal number of samples from the available classes. Therefore, maximisation of the information gain is achieved through minimisation of the child node impurities. Figure 4.6 provides a schematic representation of a decision tree based on the binary ("rainy/not rainy") and continuous ("wind speed") features. The decision tree algorithm starts at the *root*, which is shown in the figure as a shaded box. Splitting the dataset on the root feature results in the largest information gain. The splitting leads to the creation of *branches* (shown in the figure as arrows going from the parent node to the child nodes) and *leaves* (shown in the figure as white boxes). The terminal leaves (classes) are represented as dashed boxes. The splitting continues until either no more branches can be created or the maximum allowed depth is reached. It is good practice to avoid the construction of a too deep tree by imposing *pruning* – a strict limit on the maximum depth of the tree, in order to avoid overfitting.

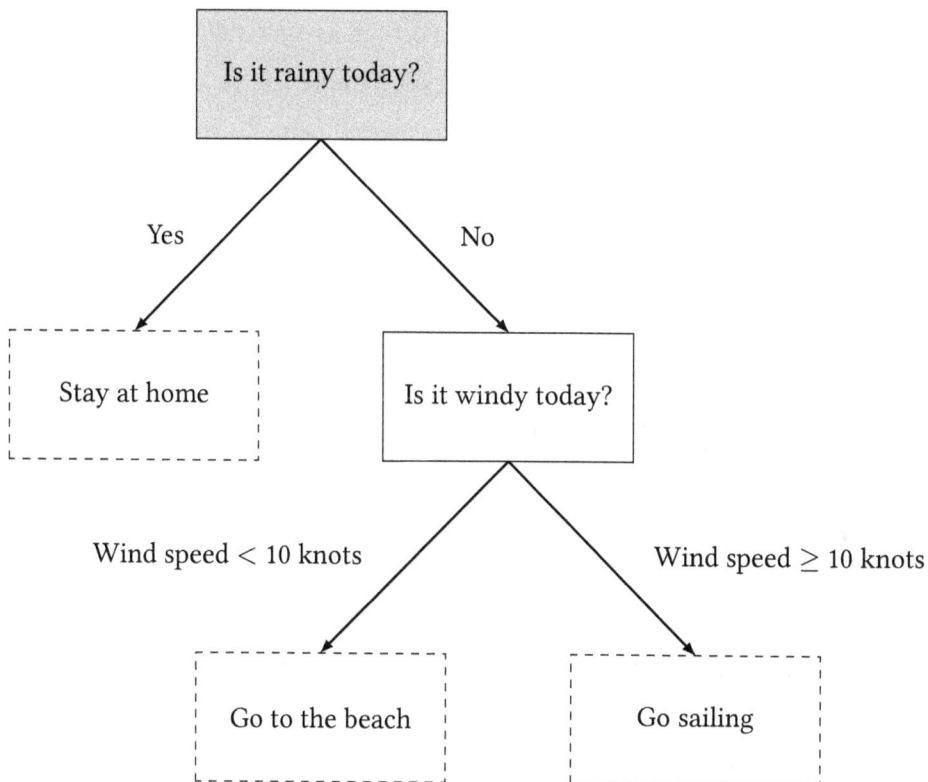

Figure 4.6: Schematic representation of a decision tree.

The most widely used impurity measures are *Gini impurity* and *entropy*. Let $(p_i^l)_{i=1,\ldots,C}$ be the proportion of the samples that belong to class i for node l. Then the impurity measures are defined as

$$I_{\text{Gini}} := \sum_{i=1}^{C} p_i^l(1 - p_i^l) \quad \text{and} \quad I_{\text{Entropy}} := -\sum_{i=1}^{C} p_i^l \log_2(p_i^l).$$

Decision trees can be seen as weak learners that can be *boosted* to be strong learners. One of the most popular methods of combining weak classifiers into a single strong classifier is *gradient boosting*. The main principle of gradient boosting is as follows [135, Section 10.10].

The objective is to improve the weak classifier through an iterative process with the improvement measured as a minimisation of the estimation error (for example, the squared error given by (4.3.3)). As before, without loss of generality, we assume that we are dealing with the binary classification problem (4.3.1). Further, assume that at the k-th iteration, the weak learner returns the estimate $\hat{y}_k(x^j)$ for sample x^j. In order to improve the classification results, the algorithm should add some estimator h_k, such that for the given sample x^j we have

$$\hat{y}_{k+1}(x^j) := \hat{y}_k(x^j) + h_k(x^j) = y^j,$$

where y^j is the correct class label for sample x^j. In other words, the task is to fit the new estimator h_k to the residuals $y^j - \hat{y}_k(x^j)$, $j = 1, \ldots, M$. We also notice that the estimator h_k is proportional to the negative gradient of the squared error (4.3.3) with respect to $\hat{y}_k$:

$$h_k(x^j) := y^j - \hat{y}_k(x^j) = -\frac{1}{2}\frac{\partial g(y^j, \hat{y}_k(x^j))}{\partial \hat{y}_k}.$$

Therefore, gradient boosting combines boosting with the gradient descent algorithm.

4.3.4 Benchmarking against standard classical classifiers

The classical benchmarks of choice are the MLP classifier (`sklearn.neural_network.MLPClassifier`) and the gradient boosting classifier (`sklearn.ensemble.GradientBoostingClassifier`). Table 4.5 holds weakly optimised model parameters: we did not search for the absolute best set of model parameters but tried just a few configurations. We can think of it as a very rough grid search method that produces a viable configuration of model parameters but is not necessarily optimal. All other model parameters were set at their default values.

Gradient Boosting Classifier	MLP Classifier
loss = 'deviance'	hidden_layer_sizes = (20)
learning_rate = 0.1	activation = 'tanh'
n_estimators = 1000	solver = 'adam'
criterion = 'friedman_mse'	alpha = 0.1
max_depth = 3	max_iter = 5000
	alpha = 0.01

Table 4.5: Model parameters for classical benchmarks.

Figure 4.7 displays out-of-sample confusion matrices for the classical benchmarks and Table 4.6 provides a direct comparison of the out-of-sample results for the QBoost and classical classifiers.

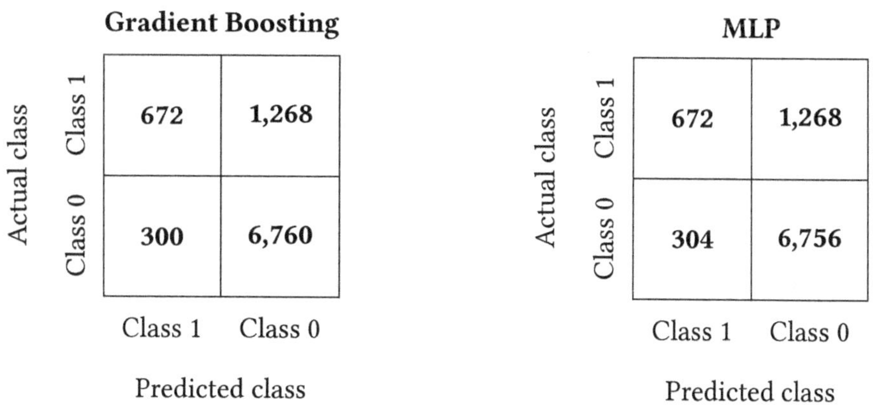

Figure 4.7: Confusion matrices for the gradient boosting and MLP classifiers (DCCC dataset, out-of-sample results).

	Accuracy	Precision	Recall
Gradient Boosting	0.83	0.69	0.35
MLP	0.83	0.69	0.35
QBoost	0.83	0.71	0.33

Table 4.6: Out-of-sample accuracy, precision, and recall for the QBoost, gradient boosting, and MLP classifiers (DCCC dataset).

QBoost achieves similar out-of-sample results to those of gradient boosting and MLP classifiers. A comparison of in-sample and out-of-sample QBoost performance confirms QBoost's ability to impose strong regularisation and avoid overfitting. At the same time, QBoost provides full transparency in terms of which features contribute to the strong classifier. We also obtain an explicit optimal configuration of features for any given degree of regularisation. This is not the case when we deal with the conventional machine learning models, which may require an extensive analysis of sensitivities and feature importances to unpack their "black boxes".

Quantum boosting can be applied to financial optimisation problems where the emphasis is on transparency, interpretability, and robustness.

Summary

In this chapter, we learned how to apply quantum annealing to build a strong classifier from several weak ones. We started with the general principles of quantum boosting and its corresponding QUBO formulation.

We then illustrated the application of the QBoost algorithm to solving a practical real-world financial problem, namely predicting credit card clients defaulting on their payments. The chosen dataset is reasonably large and complex enough to provide a meaningful challenge while remaining easy to understand and interpret the obtained results.

It is important to have an objective comparison with the corresponding classical counter-parts. With this in mind, we introduced several classical classifiers based on the concepts of a feedforward neural network and a decision tree. We benchmarked QBoost against the MLP and gradient boosting models using such metrics as accuracy, precision, and recall.

In the next chapter, we will learn how quantum annealing can assist in training powerful generative machine learning models.

5

Quantum Boltzmann Machine

As we saw in Chapters 3 and 4, quantum annealing can be used to solve hard optimisation problems. However, the range of possible applications of quantum annealing is much wider than that. In this chapter, we will consider two distinct but related use cases that go beyond solving optimisation problems: sampling and training deep neural networks. Specifically, we will focus on the Quantum Boltzmann Machine (QBM) – a generative model that is a direct quantum annealing counterpart of the classical Restricted Boltzmann Machine (RBM), and the Deep Boltzmann Machine (DBM) – a class of deep neural networks composed of multiple layers of latent variables with connections between the layers but not between units within each layer.

We start by providing detailed descriptions of the classical RBM, including the corresponding training algorithm. Due to the fact that an RBM operates on stochastic binary activation units, one can establish the correspondence between the RBM graph and the QUBO graph embedded onto the quantum chip.

This provides the main motivation for performing Boltzmann sampling (the key stage in training RBMs and DBMs) using quantum annealing. DBMs can be trained as both generative and discriminative models. In both cases, since a DBM can be constructed by stacking together layers of RBMs, efficient Boltzmann sampling is the key element of the training process. Quantum annealing, which can be integrated into the hybrid quantum-classical training routine, has the potential to improve speed and accuracy. Quantum speedup is an especially appealing element of the envisaged quantum advantage since it can be achieved not only during the RBM training stage but also during the process of generating new samples.

5.1   From Graph Theory to Boltzmann Machines

We provide here a short self-contained review of graph theory in order to introduce Boltzmann machines (or energy-based models), which one can view as particular types of connected graphs or networks.

A *graph* is a set of vertices (points or nodes) and edges that connect the vertices. A *directed graph* is a type of graph that contains ordered pairs of vertices, while an *undirected graph* is a type of graph that contains unordered pairs of vertices.

We consider a graph $\mathcal{G} = (\mathcal{V}, \mathcal{E})$ characterised by a finite number of vertices $\mathcal{V}$ and undirected edges $\mathcal{E}$. For a given vertex $v \in \mathcal{V}$, its neighbourhood is defined as the set of all vertices connected to it by some edge, or

$$\mathcal{N}(v) := \{w \in \mathcal{V} : \{v, w\} \in \mathcal{E}\}.$$

Finally, a *clique* $\mathcal{C}$ is a subset of $\mathcal{V}$ such that all vertices in $\mathcal{C}$ are pairwise connected by some edge in $\mathcal{E}$.

To each vertex $v \in \mathcal{V}$, we associate a random variable X_v taking values in some space $\mathcal{X}$. The vector $X \in \mathcal{X}^{|\mathcal{V}|}$ is called a Markov random field if

$$\text{Law}\left(X_v | (X_w)_{\{w \in \mathcal{V} \setminus \{v\}\}}\right) = \text{Law}\left(X_v | (X_w)_{\{w \in \mathcal{N}(v)\}}\right),$$

where Law means all the properties of the probability distribution. The following theorem, originally proved by Hammersley and Clifford [132] (see also [181, Theorem 4.2]), provides a way to express the law of Markov random fields over graphs in a convenient form. The Markovian property is fundamental here, as dynamics (for example, the passing of a signal from a hidden layer to a visible layer of an RBM network) should only depend on the current state and not on the whole path followed by the system.

Theorem 8 (Hammersley-Clifford Theorem). *A strictly positive distribution satisfies the Markov property with respect to an undirected graph if and only if it factorises over it.*

Phrased differently, the theorem says that X is Markovian over $\mathcal{G}$ if its distribution can be written as

$$\mathbb{P}_X(\mathrm{x}) := \mathbb{P}(X = \mathrm{x}) = \frac{1}{Z} \prod_{C \in \mathcal{C}} \psi_C(\mathrm{x}_C), \quad \text{for all } \mathrm{x} \in \mathcal{X}^{|\mathcal{V}|}, \tag{5.1.1}$$

for a set $\{\psi_C\}_{C \in \mathcal{C}}$ of functions called the potential over all the cliques $C \in \mathcal{C}$ and where Z is a normalisation constant such that the probabilities integrate to unity. Here, x_C naturally corresponds to the elements of the vector x over the clique C. The factorisation is often taken over the so-called *maximal cliques*, namely the cliques that are no longer cliques if any node is added. If the distribution of X is strictly positive, then so are the functions $\{\psi_C\}_{C \in \mathcal{C}}$ and therefore (5.1.1) can be written as

$$\mathbb{P}_X(\mathrm{x}) = \frac{1}{Z} \exp\left(\sum_{C \in \mathcal{C}} \log(\psi_C(\mathrm{x}_C)) \right) =: \frac{1}{Z} e^{-E(\mathrm{x})}, \tag{5.1.2}$$

for all $\mathrm{x} \in \mathcal{X}^{|\mathcal{V}|}$. The function

$$E(\mathrm{x}) := - \sum_{C \in \mathcal{C}} \log(\psi_C(\mathrm{x}_C))$$

is called the *energy* function. Because of their uses in statistical physics, strictly positive distributions of Markov random fields, taking the form (5.1.2), are also called Boltzmann or Gibbs distributions.

Energy-based models are generative models that discover data dependencies by applying a measure of compatibility (scalar energy) to each configuration of the observed and latent variables. The inference consists of finding the values of latent variables that minimise the energy given the values of the observed variables. Energy-based models possess many useful properties (simplicity, stability, flexibility, compositionality) – this makes them models of choice for learning complex multivariate probability distributions, though the training of energy-based models is often more involved in comparison with feedforward neural networks, as we shall see below.

5.2   Restricted Boltzmann Machine

5.2.1   The RBM as an energy-based model

The RBM corresponds to a special structure of such a graph, called bipartite, where the set V of vertices can be split into two groups of visible vertices $\mathcal{V}_V$ and hidden vertices $\mathcal{V}_H$ such that the set $\mathcal{E}$ of edges only consists of elements of the form $\{v, h\} \in \mathcal{V}_V \times \mathcal{V}_H$. Figure 5.1 provides a schematic representation of the RBM that implements the bipartite graph structure. This in particular implies that cliques can only be of size one (all the singleton nodes) or two (all the pairs (v, h) in $\mathcal{V}_V \times \mathcal{V}_H$). For simplicity, we shall denote v an element of $\mathcal{X}^{|\mathcal{V}_V|}$ and h an element of $\mathcal{X}^{|\mathcal{V}_H|}$, and identify the random variable X with the vertices. The following lemma gives us the general form of the energy function (5.1.2) for RBMs:

Lemma 6 (RBM Energy Lemma). *In a Restricted Boltzmann Machine, the energy function takes the form*

$$E(\text{v}, \text{h}) = \sum_{i=1}^{N} E_v(v_i) + \sum_{j=1}^{M} E_h(h_j) + \sum_{i=1}^{N} \sum_{j=1}^{M} E_{v,h}(v_i, h_j),$$

for any $\text{v} := (v_1, \ldots, v_N) \in \mathcal{X}^{|\mathcal{V}_V|}$, $\text{h} := (h_1, \ldots, h_M) \in \mathcal{X}^{|\mathcal{V}_H|}$. *Here, N is the number of visible vertices and M is the number of hidden vertices.*

Proof. By the Hammersley-Clifford theorem, for any $v \in \mathcal{X}^{|\mathcal{V}_V|}$, $h \in \mathcal{X}^{|\mathcal{V}_H|}$, we have the factorisation

$$\mathbb{P}(v, h) = \frac{1}{Z} \prod_{C \in \mathcal{C}} \psi_C((v_C, h_C) \in C)$$

$$= \frac{1}{Z} \prod_{\{\{v\}:v \in \mathcal{V}_V\}} \psi_{\{v\}}(v) \prod_{\{\{h\}:h \in \mathcal{V}_H\}} \psi_{\{h\}}(h) \prod_{\{\{v,h\} \in \mathcal{V}_V \times \mathcal{V}_H\}} \psi_{\{v,h\}}(v, h)$$

$$= \frac{1}{Z} \exp\{-E(v, h)\},$$

over all singletons (cliques of size one) and couples (cliques of size two), where the term $-E(v, h)$ reads

$$-E(v, h) = \log \left(\prod_{\{\{v\}:v \in \mathcal{V}_V\}} \psi_{\{v\}}(v) \prod_{\{\{h\}:h \in \mathcal{V}_H\}} \psi_{\{h\}}(h) \prod_{\{\{v,h\} \in \mathcal{V}_V \times \mathcal{V}_H\}} \psi_{\{v,h\}}(v, h) \right)$$

$$= \log \left(\prod_{\{\{v\}:v \in \mathcal{V}_V\}} \psi_{\{v\}}(v) \right)$$

$$+ \log \left(\prod_{\{\{h\}:h \in \mathcal{V}_H\}} \psi_{\{h\}}(h) \right)$$

$$+ \log \left(\prod_{\{\{v,h\} \in \mathcal{V}_V \times \mathcal{V}_H\}} \psi_{\{v,h\}}(v, h) \right)$$

$$= \sum_{\{\{v\}:v \in \mathcal{V}_V\}} \log\left(\psi_{\{v\}}(v)\right)$$

$$+ \sum_{\{\{h\}:h \in \mathcal{V}_H\}} \log\left(\psi_{\{h\}}(h)\right)$$

$$+ \sum_{\{\{v,h\} \in \mathcal{V}_V \times \mathcal{V}_H\}} \log\left(\psi_{\{v,h\}}(v, h)\right)$$

$$= -\sum_{i=1}^{N} E_v(v_i) - \sum_{j=1}^{M} E_h(h_j) - \sum_{i=1}^{N}\sum_{j=1}^{M} E_{v,h}(v_i, h_j),$$

which concludes the proof of the lemma. $\square$

The standard example of an RBM is when the random variables follow Bernoulli distribution, i.e., with $\mathcal{X} = \{0,1\}^{|\mathcal{V}|}$. In this case, their energies read

$$E_v(v_i) = -a_i v_i, \quad E_h(h_j) = -b_j h_j, \quad E_{v,h}(v_i, h_j) = -w_{ij} v_i h_j, \tag{5.2.1}$$

for some parameters $a_i, b_j, w_{ij}, i = 1, \ldots, N, j = 1, \ldots, M$. In particular, for a given v_i, we can write, using Bayes' formula,

$$
\begin{aligned}
\mathbb{P}(v_i = 1 | \mathbf{v}_{v_i}, \mathbf{h}) &= \frac{\mathbb{P}(v_i = 1, \mathbf{v}_{v_i}, \mathbf{h})}{\mathbb{P}(v_i = 1, \mathbf{v}_{v_i}, \mathbf{h}) + \mathbb{P}(v_i = 0, \mathbf{v}_{v_i}, \mathbf{h})} \\
&= \frac{\exp\left(-E(v_i = 1, \mathbf{v}_{v_i}, \mathbf{h})\right)}{\exp\left(-E(v_i = 1, \mathbf{v}_{v_i}, \mathbf{h})\right) + \exp\left(-E(v_i = 0, \mathbf{v}_{v_i}, \mathbf{h})\right)}.
\end{aligned}
\tag{5.2.2}
$$

where we denote $\mathbf{v}_{v_i}$ the states of all the nodes in $\mathcal{V} \setminus \{v_i\}$. Now, using the RBM energy lemma, we can single out the energy arising from the particular node v using (5.2.1) as

$$E(v_i, \mathbf{v}_{v_i}, \mathbf{h}) = -\Phi_{\mathrm{v}}(v_i) - \Psi_{\mathrm{v}}(\mathbf{v}_{v_i}, \mathbf{h}),$$

where

$$\Phi_{\mathrm{v}}(v_i) := a_i v_i + \sum_{j=1}^{M} w_{ij} v_i h_j = \left[a_i + \sum_{j=1}^{M} w_{ij} h_j \right] v_i,$$

$$\Psi_{\mathrm{v}}(\mathbf{v}_{v_i}, \mathbf{h}) := \sum_{k=1 (k \neq i)}^{N} a_k v_k + \sum_{j=1}^{M} b_j h_j + \sum_{k=1 (k \neq i)}^{N} \sum_{j=1}^{M} w_{kj} v_k h_j.$$

Plugging this into (5.2.2) then yields

$$
\begin{aligned}
\mathbb{P}(v_i = 1 | \mathbf{v}_{v_i}, \mathbf{h}) &= \frac{\exp\left(\Phi_{\mathrm{v}}(v_i = 1) + \Psi_{\mathrm{v}}(\mathbf{v}_{v_i}, \mathbf{h})\right)}{\exp\left(\Phi_{\mathrm{v}}(v_i = 1) + \Psi_{\mathrm{v}}(\mathbf{v}_{v_i}, \mathbf{h})\right) + \exp\left(\Phi_{\mathrm{v}}(v_i = 0) + \Psi_{\mathrm{v}}(\mathbf{v}_{v_i}, \mathbf{h})\right)} \\
&= \frac{\exp\left(\Phi_{\mathrm{v}}(v_i = 1)\right)}{\exp\left(\Phi_{\mathrm{v}}(v_i = 1)\right) + 1} \\
&= \sigma\left(\Phi_{\mathrm{v}}(v_i = 1)\right),
\end{aligned}
$$

since $\Phi_v(v_i = 0) = 0$, where

$$\sigma(x) := \frac{1}{1 + e^{-x}} \tag{5.2.3}$$

is the sigmoid function.

Similarly, we can single out the contribution of the energy on a given hidden node h_j using the RBM energy lemma:

$$E(v, h_j, \mathrm{h}_{h_j}) = -\Phi_\mathrm{h}(h_j) - \Psi_\mathrm{h}(v, \mathrm{h}_{h_j}),$$

where

$$\Phi_\mathrm{h}(h_j) := b_j h_j + \sum_{i=1}^{N} w_{ij} v_i h_j = \left[b_j + \sum_{i=1}^{N} w_{ij} v_i \right] h_j,$$

$$\Psi_\mathrm{h}(v, \mathrm{h}_{h_j}) := \sum_{i=1}^{N} a_i v_i + \sum_{k=1(k \neq j)}^{M} b_k h_k + \sum_{i=1}^{N} \sum_{k=1(k \neq j)}^{M} w_{ik} v_i h_k.$$

Plugging this into (5.2.2) then yields

$$
\begin{aligned}
\mathbb{P}(h_j = 1 | v, \mathrm{h}_{h_j}) &= \frac{\exp\left(\Phi_\mathrm{h}(h_j = 1) + \Psi_\mathrm{h}\left(v, \mathrm{h}_{h_j}\right)\right)}{\exp\left(\Phi_\mathrm{h}(h_j = 1) + \Psi_\mathrm{h}\left(v, \mathrm{h}_{h_j}\right)\right) + \exp\left(\Phi_\mathrm{h}(h_j = 0) + \Psi_\mathrm{h}\left(v, \mathrm{h}_{h_j}\right)\right)} \\
&= \frac{\exp\left(\Phi_\mathrm{h}(h_j = 1)\right)}{\exp\left(\Phi_\mathrm{h}(h_j = 1)\right) + 1} \\
&= \sigma\left(\Phi_\mathrm{h}(h_j = 1)\right),
\end{aligned}
$$

since again $\Phi_\mathrm{h}(h_j = 0) = 0$.

5.2.2 RBM network architecture

As shown above, an RBM is thus a shallow two-layer neural network that operates on stochastic binary activation units. The network forms a bipartite graph connecting stochastic binary inputs (visible units) to stochastic binary feature detectors (hidden units) with no connections between the units within the same layer, as shown in Figure 5.1 [102].

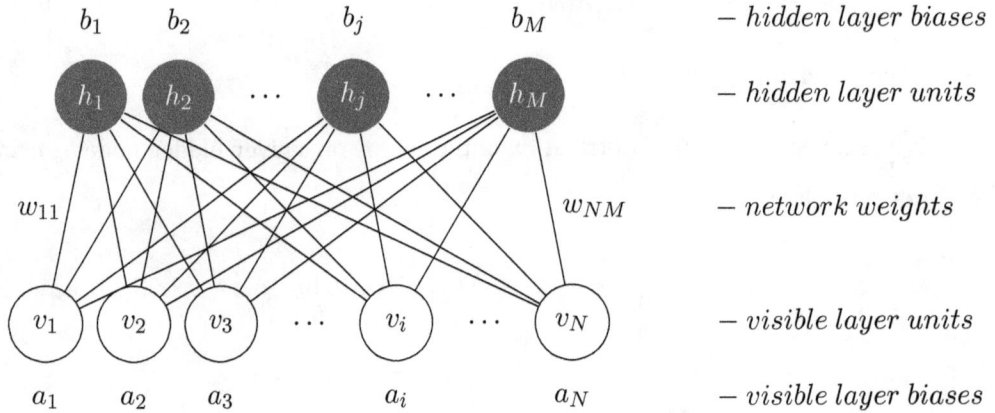

Figure 5.1: Schematic representation of an RBM with the visible layer units (white) and hidden layer units (dark) forming a bipartite graph.

Only the visible layer of the network is exposed to the training dataset and its inputs $v := (v_1, \ldots, v_N)$ flow through the network (forward pass) to the hidden layer, where they are aggregated and added to the hidden layer biases $b := (b_1, \ldots, b_M)$. The hidden layer sigmoid activation function (5.2.3) converts aggregated inputs into probabilities. Each hidden unit then "fires" randomly and outputs a $\{0, 1\}$ Bernoulli random variable with the associated probabilities:

$$\mathbb{P}(h_j = 1 | v) = \sigma \left(b_j + \sum_{i=1}^{N} w_{ij} v_i \right) \quad \text{and} \quad \mathbb{P}(h_j = 0 | v) = 1 - \sigma \left(b_j + \sum_{i=1}^{N} w_{ij} v_i \right).$$

The outputs from the hidden layer $h := (h_1, \ldots, h_M)$ then flow back (backward pass) to the visible layer, where they are aggregated and added to the visible layer biases $a := (a_1, \ldots, a_N)$. Similar to the hidden layer, the visible layer sigmoid activation function first translates aggregated inputs into probabilities and then into Bernoulli random variables:

$$\mathbb{P}(v_i = 1 | h) = \sigma \left(a_i + \sum_{j=1}^{M} w_{ij} h_j \right) \quad \text{and} \quad \mathbb{P}(v_i = 0 | h) = 1 - \sigma \left(a_i + \sum_{j=1}^{M} w_{ij} h_j \right).$$

Therefore, every unit communicates at most one bit of information. This is especially important for the hidden units since this feature implements the information bottleneck structure, which acts as a strong regulariser [142]. The hidden layer of the network can learn the low-dimensional probabilistic representation of the dataset if the network is organised and trained as an autoencoder [30].

5.2.3 Sample encoding

Figure 5.2 illustrates the binary representation of an input signal that enters the network through the visible layer. The number of activation units in the visible layer is determined by the number of features we have to encode and the desired precision of their binary representation. For example, if our sample consists of m continuous features and each feature is encoded as an n-digit binary number, the total number of activation units in the visible layer is $m \times n$.

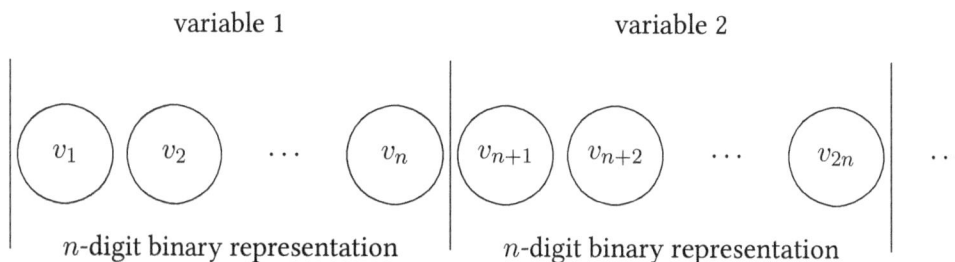

Figure 5.2: Schematic binary encoding of continuous variables.

5.2.4 Boltzmann distribution

The network learns the probability distribution $\mathbb{P}(v, h)$ of the configurations of visible and hidden activation units – the Boltzmann distribution – by trying to reconstruct the inputs from the training dataset (visible unit values) through finding an optimal set of the network weights and biases:

$$\mathbb{P}(v, h) = \frac{1}{Z} e^{-E(v, h)}, \tag{5.2.4}$$

where the energy function reads

$$E(\mathrm{v}, \mathrm{h}) = -\sum_{i=1}^{N} a_i v_i - \sum_{j=1}^{M} b_j h_j - \sum_{i=1}^{N}\sum_{j=1}^{M} w_{ij} v_i h_j. \qquad (5.2.5)$$

Here, Z is the partition function:

$$Z = \sum_{\mathrm{v},\mathrm{h}} e^{-E(\mathrm{v},\mathrm{h})}.$$

However, we are usually interested either in learning the probability distribution of the visible layer configurations if we want to generate new samples that would have the same statistical properties as the original training dataset, or in learning the probability distribution of the hidden layer configurations if we want to build a deep neural network where the RBM layer performs the feature extraction and dimensionality reduction function. The probabilities of the visible (hidden) states are given by summing over all possible hidden (visible) vectors:

$$\mathbb{P}(\mathrm{v}) = \frac{1}{Z}\sum_{\mathrm{h}} e^{-E(\mathrm{v},\mathrm{h})} \quad \text{and} \quad \mathbb{P}(\mathrm{h}) = \frac{1}{Z}\sum_{\mathrm{v}} e^{-E(\mathrm{v},\mathrm{h})}.$$

The most popular training algorithm for RBM, k-step Contrastive Divergence (CD), was proposed by Hinton [141, 142]. The algorithm aims to maximise the log probability of a training vector, i.e., to find such network weights and biases that the "energy" function E is minimised for the samples from the training dataset (a smaller value of energy corresponds to a larger probability of a configuration). The k-step CD algorithm is fully specified in Section 5.3.2, and the interested reader can also find an excellent introduction to the training of RBMs in the work by Fischer and Igel [103].

5.2.5   Extensions of the Bernoulli RBM

The standard Bernoulli RBM setup we considered above restricts the visible layer v to a Bernoulli distribution. In fact, as long as the Hammersley-Clifford theorem holds, we can consider any distribution or any form of energy function. It was shown in [62, 193], for example, that a Bernoulli distribution for the hidden layer combined with a Gaussian distribution for the visible layer are compatible with an energy function of the form

$$E(\mathrm{v},\mathrm{h}) = \sum_{i=1}^{N} \frac{(v_i - a_i)^2}{2\sigma_i^2} - \sum_{j=1}^{M} b_j h_j - \sum_{i=1}^{N}\sum_{j=1}^{M} w_{ij} \frac{v_i h_j}{\sigma_i^2},$$

for some parameters a_i, σ_i, b_j, w_{ij}, $i = 1,\ldots,N$, $j = 1,\ldots,M$. In this case, for any h_j, the conditional probabilities $\mathbb{P}(h_j = 1|\mathrm{v})$ remain of sigmoid form and the conditional distribution of the visible layer is Gaussian as

$$\mathrm{Law}(v_i|\mathrm{h}) = \mathcal{N}\left(a_i + \sum_{j=1}^{M} w_{ij} h_j, \sigma_i^2\right), \quad \text{for each } i = 1,\ldots,N.$$

The RBMs we have considered do not account for time series, i.e., probability structures with temporal dependence. By enlarging the corresponding graph, in particular by adding a conditional layer with directed connections to the classical hidden and visible layers, Taylor [304] showed that such dependence can be accounted for.

An RBM is a neural network represented by a bipartite graph. Its power is derived from operating on stochastic binary activation units. It is a generative model that encodes learned probability distribution in its weights and biases and then generates new samples that are statistically indistinguishable from the samples in the original dataset.

If it is organised as an autoencoder with the bottleneck information structure, an RBM is able to learn the low-dimensional representation of the dataset. This property suggests that an RBM can be used as a feature extraction layer in a machine learning pipeline for certain supervised and unsupervised learning problems.

5.3   Training and Running an RBM

To build a neural network means to specify the network architecture and training algorithm. Having described the RBM architecture in the previous section, we now outline the training routines.

5.3.1   Training an RBM with Boltzmann sampling

The goal of RBM training is to estimate the optimal vector θ of model parameters (weights and biases) so that $\mathbb{P}_\theta(\mathrm{v}) = \mathbb{P}_{\mathrm{data}}(\mathrm{v})$. For a given training sample $\mathrm{v} := (v_1, \ldots, v_N)$, the RBM aims at maximising the log-likelihood function, namely

$$\max_\theta \sum_{i=1}^n \mathfrak{L}(\theta|v_i),$$

where, for any v,

$$\mathfrak{L}(\theta|\mathrm{v}) = \log(\mathbb{P}(\mathrm{v})) = \log\left(\frac{1}{Z}\sum_{\mathrm{h}} e^{-E(\mathrm{v},\mathrm{h})}\right) = \log\left(\sum_{\mathrm{h}} e^{-E(\mathrm{v},\mathrm{h})}\right) - \log\left(\sum_{\mathrm{v},\mathrm{h}} e^{-E(\mathrm{v},\mathrm{h})}\right).$$

The standard optimisation method, as proposed in [141], is a standard gradient ascent method, i.e., starting from an initial guess θ^0, we update it as

$$\theta^{k+1} = \theta^k + \partial_\theta \sum_{i=1}^N \mathfrak{L}(\theta^k|v_i)$$

until we reach good enough convergence. In order to compute it, one first needs to compute the joint probabilities $\mathbb{P}(v_i, h_j)$, which is classically done via Boltzmann (Gibbs) sampling [3], which is possible since we know exactly the conditional distributions.

5.3.2   The Contrastive Divergence algorithm

While training RBMs can be performed with Boltzmann sampling, this is usually prohibitively expensive to run. A more efficient training algorithm, the k-step CD algorithm, was proposed in [142].

Algorithm 2: k-step CD

Result: Weights and biases updates.

Input:

- Training minibatch S;

- Model parameters a_i, b_j, w_{ij} for $i = 1, \ldots, N, j = 1, \ldots, M$ (before update).

Initialisation: for all $i, j : \Delta w_{ij} = \Delta a_i = \Delta b_j = 0$

for $\mathrm{v} \in S$ **do**

    $\mathrm{v}^{(0)} \leftarrow \mathrm{v}$

    for $t = 0, \ldots, k - 1$ **do**

        for $j = 1, \ldots, M$ **do**

            sample Bernoulli random variable $h_j^{(t)} \sim \mathbb{P}(h_j | \mathrm{v}^{(t)})$

        end

        for $i = 1, \ldots, N$ **do**

            sample Bernoulli random variable $v_i^{(t+1)} \sim \mathbb{P}(v_i | \mathrm{h}^{(t)})$

        end

    end

    for $i = 1, \ldots, N, \ j = 1, \ldots, M$ **do**

        $\Delta w_{ij} \leftarrow \Delta w_{ij} + \eta \left(\mathbb{P}(h_j = 1 | \mathrm{v}^{(0)}) v_i^{(0)} - \mathbb{P}(h_j = 1 | \mathrm{v}^{(k)}) v_i^{(k)} \right)$

    end

    for $i = 1, \ldots, N$ **do**

        $\Delta a_i \leftarrow \Delta a_i + \eta \left(v_i^{(0)} - v_i^{(k)} \right)$

    end

    for $j = 1, \ldots, M$ **do**

        $\Delta b_j \leftarrow \Delta b_j + \eta \left(\mathbb{P}(h_j = 1 | \mathrm{v}^{(0)}) - \mathbb{P}(h_j = 1 | \mathrm{v}^{(k)}) \right)$

    end

end

The choice of k balances accuracy and speed. For many practical purposes $k = 1$ is an optimal choice, even though the expectations may be biased in this case. However, the bias tends to be small [53]. The network is trained through the updates of weights and biases, which increase the log probability of a training vector and are given by the following expressions:

$$\Delta w_{ij} = \eta \frac{\partial \log(\mathbb{P}(\mathrm{v}))}{\partial w_{ij}} = \eta \left(\langle v_i h_j \rangle_{\text{data}} - \langle v_i h_j \rangle_{\text{model}} \right), \qquad (5.3.1)$$

$$\Delta a_i = \eta \frac{\partial \log(\mathbb{P}(\mathrm{v}))}{\partial a_i} = \eta \left(\langle v_i \rangle_{\text{data}} - \langle v_i \rangle_{\text{model}} \right), \qquad (5.3.2)$$

$$\Delta b_j = \eta \frac{\partial \log(\mathbb{P}(\mathrm{v}))}{\partial b_i} = \eta \left(\langle h_j \rangle_{\text{data}} - \langle h_j \rangle_{\text{model}} \right), \qquad (5.3.3)$$

where $\langle \cdot \rangle$ denotes expectations under the distribution specified by the subscript and η is the chosen learning rate. Expectations $\langle \cdot \rangle_{\text{data}}$ can be calculated directly from the training dataset, while getting unbiased samples of $\langle \cdot \rangle_{\text{model}}$ requires performing alternating sampling from the model Boltzmann distribution for a long time (this is needed to achieve the state of thermal equilibrium), starting from some randomly initialised state. However, the k-step CD method can be used to approximate $\langle \cdot \rangle_{\text{model}}$ with another, easier-to-calculate expectation, as shown in Algorithm 2.

5.3.3 Generation of synthetic samples

Once fully trained, the network can be used to generate new samples from the learned distribution. For example, the RBM can be used as a market generator that produces new market scenarios in the form of the new synthetic samples drawn from the multivariate distribution of the market risk factors encoded in the network weights and biases.

The first step is the generation of a random input: each visible unit is initialised with a randomly generated binary variable. The second step is performing a large number of forward and backward passes between the visible and the hidden layers, until the system reaches a state of *thermal equilibrium*: a state where the initial random vector is transformed into a sample from the learned distribution. The number of cycles needed to reach the state of thermal equilibrium is problem dependent and is a function of network architecture and network parameters (weights and biases). In some cases, the generation of independent samples requires 10^3–10^4 forward and backward passes through the network [187].

The final step is the readout from the visible layer, which gives us a bitstring, encoding the sample from the target distribution. Figure 5.3 displays the QQ-plots of the samples drawn from the distributions of daily returns for two stock indices: German DAX and Brazilian BOVESPA. Recall that a quantile-quantile (or QQ) plot is a scatter plot created by plotting two sets of quantiles against one another. If both sets come from the same distribution, all points should lie close to the diagonal. The dataset consists of 536 samples – daily index returns observed between 5 January 2009 and 22 February 2011 (UCI Machine Learning Repository [9, 10]). The "Normal" distribution models daily returns as Normally distributed with a mean and variance that match those from the historical dataset. The "RBM" distribution is a dataset of RBM-generated samples that, ideally, should have exactly the same statistical properties as the original historical dataset. If the samples drawn from two distributions have identical quantiles, the QQ-plots will have all points placed on the diagonal and we can conclude that the two distributions are identical. Figure 5.3 shows that this is indeed the case (with reasonably good accuracy) for the samples from the "Data" and "RBM" distributions, while both demonstrate much heavier tails in comparison with the fitted Normal distribution.

The results shown in Figure 5.3 were obtained with an RBM trained on a dataset of daily returns. Each return from the training dataset was converted into a 12-digit binary number. Every digit of the binary number was treated as a separate binary feature (12 features per index; 24 features in total) – this required placing 24 activation units in the visible layer of the RBM network. The number of hidden units was set to 16. Thus, the network was trained as a strongly regularised autoencoder. The generated returns (in binary format) were then converted back into their continuous representation. The model was Bernoulli RBM (`sklearn.neural_network.BernoulliRBM`) from the open source `scikit-learn` package [249] with the following set of parameters:

- n_components = 16 – number of hidden activation units
- learning_rate = 0.0005
- batch_size = 10 – size of the training minibatches
- n_iter = 40000 – number of iterations

The synthetic data generation approach can be formulated as Algorithm 3.

Algorithm 3: Synthetic Data Generation

1: The construction of the binary representation of the original dataset:

    a) A continuous feature can be converted into an equivalent binary representation with the required precision.

    b) An integer feature $x \in \{x_1, \ldots, x_n\}$ can be translated into an N-digit binary number through the standard procedure, where
$$2^{N-1} \leq \max_{1 \leq j \leq n}(x_j) - \min_{1 \leq j \leq n}(x_j) < 2^N.$$

    c) A categorical feature can be binarised either through the one-hot encoding method or following the same procedure as for the integer numbers since categorical values can be enumerated.

    d) The same applies to class labels, both integer and categorical.

2: The training of an RBM on the binary representation of the original dataset with the help of a 1-step CD algorithm.

3: The generation of the required number of new synthetic samples in binary format.

4: For each synthetic data sample: the conversion of the generated binary features into the corresponding categorical, integer, and continuous representations.

5: The generated synthetic dataset is ready to be used for the training of various classifiers and regressors.

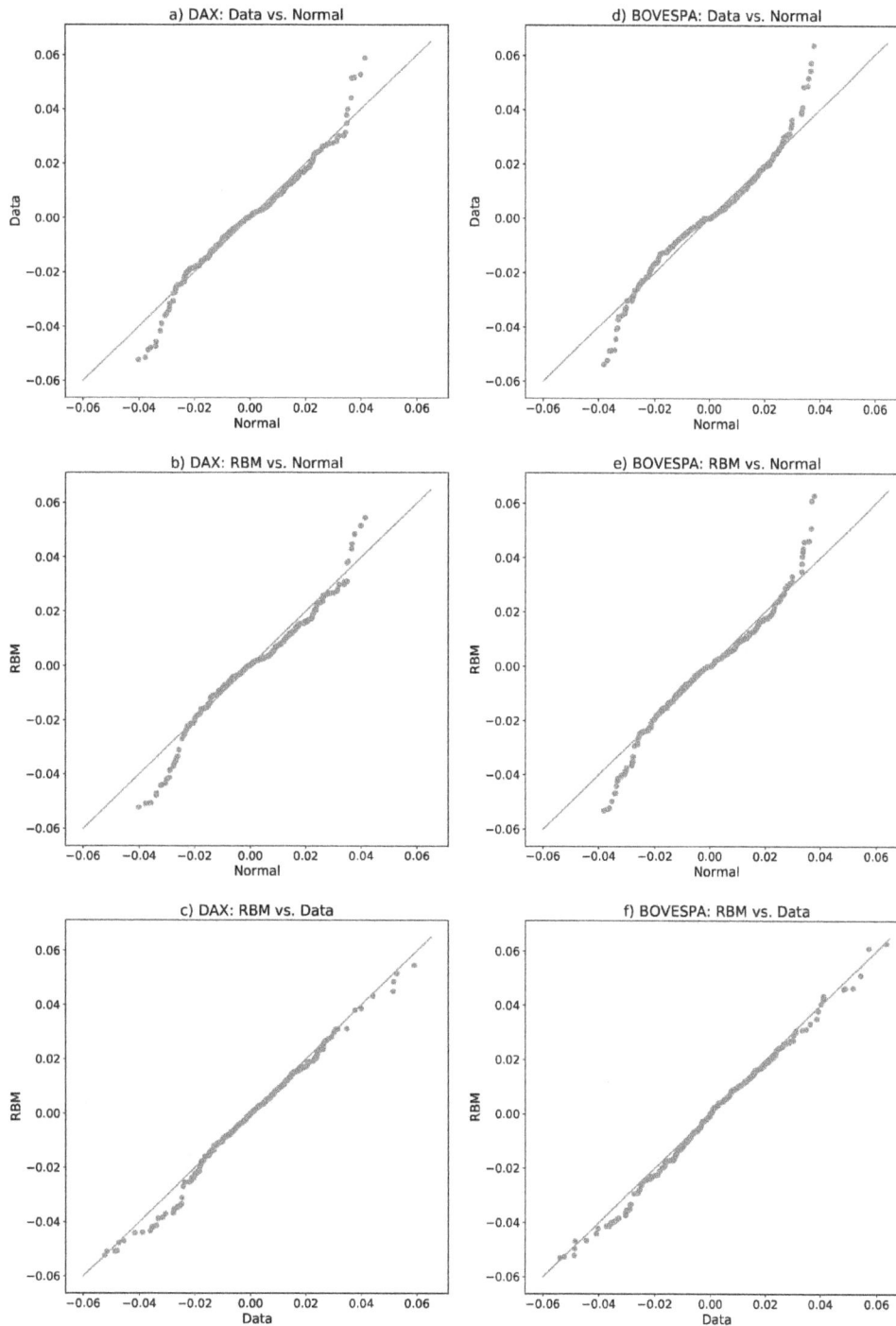

Figure 5.3: QQ-plots of the generated and historical returns. a)-c) DAX. d)-f) BOVESPA. The RBM learns the heavy-tailed empirical distribution of stock index returns.

Kondratyev and Schwarz [187] proposed an RBM-based market generator and investigated its properties on a dataset of daily spot FX log-returns. The time series of four currency pairs' log-returns covered a 20-year time interval (1999–2019), which allowed the RBM to learn the dependence structure of the multivariate distribution and successfully reconstruct linear and rank correlations as well as joint tail behaviour. Also, it was shown that an RBM can be used to perform conditional sampling (e.g., from low-volatility/high-volatility regimes) and achieve the desired degree of autocorrelation by varying the thermalisation parameter. Other productive applications of RBM-based synthetic data generators are data anonymisation, fighting overfitting, and the detection of outliers as demonstrated by Kondratyev, Schwarz, and Horvath [188].

> In addition to operating on stochastic binary activation units, the RBM gains extra resistance to overfitting through the autoencoder architecture and being trained with stochastic gradient ascent. This allows RBMs to learn complex multivariate probability distributions from relatively small datasets while avoiding overfitting.

5.4   Quantum Annealing and Boltzmann Sampling

The application of quantum annealing to Boltzmann sampling is based on the direct correspondence between the RBM energy function given by (5.2.5) and the Hamiltonian in quantum annealing. Recall from Chapter 2 that quantum annealing is based on the principles of adiabatic evolution from the initial state at $t = 0$ given by a Hamiltonian $\mathcal{H}_0$ to a final state at $t = T$ given by a Hamiltonian $\mathcal{H}_F$, such that the system Hamiltonian at time $t \in [0, T]$ is given by

$$\mathcal{H}(t) = r(t)\mathcal{H}_0 + (1 - r(t))\mathcal{H}_F, \tag{5.4.1}$$

where $r(t)$ decreases from 1 to 0 as t goes from 0 to T. An ideal adiabatic evolution scenario envisages the system always staying in the ground state of $\mathcal{H}(t)$: if the system starts in the ground state of $\mathcal{H}_0$ and the evolution proceeds slowly enough to satisfy the conditions of the quantum adiabatic theorem (Chapter 2), then the system will end up in the ground state of $\mathcal{H}_F$.

In practice, existing quantum annealing hardware does not strictly satisfy the conditions of the quantum adiabatic theorem. Quantum annealers operate at very low temperatures of about 15mK [90], but some residual thermal noise is still present. There is also some amount of cross-talk between the qubits, and the chains of physical qubits that represent logical qubits can be broken. Cross-talk is the effect of a desired action on one or more qubits unintentionally affecting one or more other qubits. In some cases, cross-talk is the major source of computational errors. This poses serious issues for quantum annealers solving optimisation problems where the main objective is to find an exact ground state. But some residual amount of thermal and electromagnetic noise is desirable if we want to use a quantum annealer as a sampler.

5.4.1 Boltzmann sampling

The quantum annealer as a sampling engine is based on the central proposal [4] that the distribution of excited states can be modelled as a Boltzmann distribution:

$$\mathbb{P}(\mathbf{x}) = \frac{1}{Z} \exp\left(-\beta \mathcal{H}_F(\mathbf{x})\right), \tag{5.4.2}$$

where β is some parameter (which can be seen as an effective inverse temperature) and Z is the partition function:

$$Z = \sum_{\mathbf{x}} \exp\left(-\beta \mathcal{H}_F(\mathbf{x})\right). \tag{5.4.3}$$

If we define the binary vector $\mathbf{x}$ to be the concatenation of the visible node vector $\mathbf{v}$ and the hidden node vector $\mathbf{h}$:

$$\mathbf{x} := (v_1, v_2, \ldots, v_N, h_1, h_2, \ldots, h_M),$$

then, by comparing (5.2.4) and (5.4.2), we can establish a direct correspondence between the energy function E and the Hamiltonian $\mathcal{H}_F$. Therefore, we can suggest an alternative way of calculating the expectations $\langle\cdot\rangle_{\text{model}}$ formulated as in the following algorithm [4]:

Algorithm 4: Boltzmann Sampling

1: Use the RBM energy function E as the final Hamiltonian $\mathcal{H}_F$.

2: Run quantum annealing K times and collect the readout statistics for $v_i(k)$ and $h_j(k)$, $i = 1, \ldots, N$, $j = 1, \ldots, M$, $k = 1, \ldots, K$.

3: Calculate the unbiased expectations:

$$\langle v_i h_j \rangle_{\text{model}} := \frac{1}{K} \sum_{k=1}^{K} v_i(k) h_j(k),$$

$$\langle v_i \rangle_{\text{model}} := \frac{1}{K} \sum_{k=1}^{K} v_i(k),$$

$$\langle h_j \rangle_{\text{model}} := \frac{1}{K} \sum_{k=1}^{K} h_j(k).$$

There are two main motivations for using quantum annealing to perform Boltzmann sampling as described in Algorithm 4. First, it bypasses the need for running the CD algorithm (Algorithm 2), which only provides approximations to the expectations $\langle\cdot\rangle_{\text{model}}$ (even though these approximations can be sufficiently accurate). Second, the anneal time needed to generate a new sample from the Boltzmann distribution is of the order of $\sim$1 microsecond regardless of the graph size. This is not the case with the classical RBM, where it is often necessary to perform thousands of forward and backward passes through the network before a new independent sample from the Boltzmann distribution encoded in the network weights and biases can be read out [187]. For large RBM graphs, it can easily take tens of milliseconds on standard hardware. Thus, we have two avenues of exploring the potential quantum advantage offered by quantum annealing for Boltzmann sampling: accuracy and speedup.

5.4.2 Mapping

The first step in performing Boltzmann sampling on a quantum annealer is the mapping of the RBM onto the quantum annealing hardware graph. We start with writing an expression for the RBM energy function E in the following form:

$$E(\mathrm{v}, \mathrm{h}) = E(\mathrm{x}) = \beta \mathrm{x}^T Q \mathrm{x}. \tag{5.4.4}$$

Here, Q is the $(N + M) \times (N + M)$ matrix whose elements are RBM weights and biases:

$$Q = \frac{1}{\beta} \left[\begin{array}{cccc|cccc} a_1 & 0 & \dots & 0 & w_{11} & w_{12} & \dots & w_{1M} \\ 0 & a_2 & \dots & 0 & w_{21} & w_{22} & \dots & w_{2M} \\ \vdots & \vdots & \ddots & \vdots & \vdots & \vdots & \ddots & \vdots \\ 0 & 0 & \dots & a_N & w_{N1} & w_{N2} & \dots & w_{NM} \\ 0 & 0 & \dots & 0 & b_1 & 0 & \dots & 0 \\ 0 & 0 & \dots & 0 & 0 & b_2 & \dots & 0 \\ \vdots & \vdots & \ddots & \vdots & \vdots & \vdots & \ddots & \vdots \\ 0 & 0 & \dots & 0 & 0 & 0 & \dots & b_M \end{array} \right].$$

Quantum annealers operate on spin variables $\{-1, +1\}$ instead of binary variables $\{0, 1\}$. The vector of binary variables x can be transformed into the vector of spin variables s using

$$\mathrm{x} \longrightarrow \mathrm{s} = 2\mathrm{x} - 1,$$

and we obtain the following expression for the RBM energy:

$$E = -\sum_{i=1}^{N} g_i s_i - \sum_{j=N+1}^{N+M} g_j s_j - \sum_{i=1}^{N} \sum_{j=N+1}^{N+M} J_{ij} s_i s_j - \mathrm{const} = E_{\mathrm{Ising}} - \mathrm{const}, \tag{5.4.5}$$

where, for $i = 1, \ldots, N$ and $j = N + 1, \ldots, N + M$,

$$g_i := \frac{a_i}{2} + \frac{1}{4} \sum_{j=N+1}^{N+M} w_{ij}, \quad g_j := \frac{b_j}{2} + \frac{1}{4} \sum_{i=1}^{N} w_{ij}, \quad J_{ij} := \frac{1}{4} w_{ij},$$

and $(s_i)_{i=1,\ldots,N}$ are spin variables corresponding to the visible nodes and $(s_j)_{j=N+1,\ldots,N+M}$ are spin variables corresponding to the hidden nodes.

We can ignore the constant term in the RBM energy expression (5.4.5) since the same factor will appear in the numerator and denominator of $\mathbb{P}(\mathrm{v}, \mathrm{h})$. Thus, we have

$$\langle v_i h_j \rangle_{\mathrm{model}}^{E_{\mathrm{Ising}}} = \langle v_i h_j \rangle_{\mathrm{model}}^{E}.$$

To express the Ising Hamiltonian using a quantum mechanical description of spins, we replace the spin variables with their respective Pauli operators:

$$\mathcal{H}_{\mathrm{Ising}} = -\sum_{i=1}^{N} g_i \sigma_z^i - \sum_{j=N+1}^{N+M} g_j \sigma_z^j - \sum_{i=1}^{N} \sum_{j=N+1}^{N+M} J_{ij} \sigma_z^i \sigma_z^j, \qquad (5.4.6)$$

with σ_z^i being the usual Pauli matrix representation for an Ising quantum spin. With the initial Hamiltonian given by

$$\mathcal{H}_0 = \sum_{i=1}^{N+M} \sigma_x^i,$$

the time-dependent Hamiltonian (5.4.1) takes the form

$$\mathcal{H}(t) = r(t)\mathcal{H}_0 + (1 - r(t))\mathcal{H}_{\mathrm{Ising}}.$$

5.4.3   Hardware embedding and parameter optimisation

In the standard programming practices of existing quantum annealers, each spin variable s_i should ideally be assigned to a specific chip element, a superconducting flux qubit, modelled by a quantum two-level system that could represent the quantum Hamiltonian

$$\mathcal{H}_{\text{local}} = \sum_i g_i \sigma_z^i.$$

While each qubit supports the programming of the g_i terms, the J_{ij} parameters can then be implemented energetically through inductive elements, meant to represent

$$\mathcal{H}_{\text{couplers}} = \sum_{ij} J_{ij} \sigma_z^i \sigma_z^j,$$

if and only if the required circuitry exists between qubits i and j, which cannot be manu-factured too far apart in the spatial layout of the processor due to engineering considera-tions [320]. In other words, $J_{ij} = 0$ unless $(i, j) \in G$, where G is a particular quantum annealing graph (for example, *Chimera* or *Pegasus* graphs in the case of D-Wave quantum annealers).

It would be straightforward to embed the final Hamiltonian (5.4.6) on the quantum chip had all the physical qubits been connected to each other. Unfortunately, this is not the case. The existing quantum annealers have rather limited qubit connectivity. For example, in the case of the *Chimera* (*Pegasus*) graph, a physical qubit is connected with a maximum of six (fifteen) other physical qubits.

To get around this restriction, the standard procedure is to employ the minor-embedding compilation technique for fully connected graphs. By means of this procedure, we obtain another Ising form, where qubits are arranged in ordered 1D chains (forming the *logical* qubits that represent the spin variables) interlaced on the quantum annealer graph:

$$\mathcal{H}_{\text{Ising}} = -\sum_{i=1}^{N} |J_F| \left[\sum_{c=1}^{N_c-1} \sigma_z^{ic} \sigma_z^{i(c+1)} \right] - \sum_{j=N+1}^{N+M} |J_F| \left[\sum_{c=1}^{N_c-1} \sigma_z^{jc} \sigma_z^{j(c+1)} \right] \qquad (5.4.7)$$

$$-\sum_{i=1}^{N} \frac{g_i}{N_c} \left[\sum_{c=1}^{N_c} \sigma_z^{ic} \right] - \sum_{j=N+1}^{N+M} \frac{g_j}{N_c} \left[\sum_{c=1}^{N_c} \sigma_z^{jc} \right] \qquad (5.4.8)$$

$$-\sum_{i=1}^{N}\sum_{j=N+1}^{N+M} J_{ij}\left[\sum_{c_i,c_j=1}^{N_c} \delta_{ij}^{G}(c_i, c_j)\sigma_z^{ic_i}\sigma_z^{jc_j}\right]. \qquad (5.4.9)$$

In (5.4.7), we explicitly isolate the encoding of the *logical* quantum variable: the classical binary variable s_i is associated with N_c Ising spins σ_z^{ic}, ferromagnetically coupled directly by strength J_F, forming an ordered 1D chain subgraph of G. The value of J_F should be strong enough to correlate the value of the magnetisation of each individual spin if measured in the computational basis ($\langle\sigma_z^{ic}\rangle = \langle\sigma_z^{i(c+1)}\rangle$).

In (5.4.8) and (5.4.9), we encode the Ising Hamiltonian (5.4.6) through our extended set of variables: the local field g_i is evenly distributed across all qubits belonging to the logical chain i, and each coupler J_{ij} is active only between one specific pair of qubits ($\sigma_z^{ic_i^\star}, \sigma_z^{jc_j^\star}$), which is specified by the adjacency check function $\delta_{ij}^{G}(c_i, c_j)$, which assumes a unit value only if ($c_i = c_i^\star$) and ($c_j = c_j^\star$), and is zero otherwise.

Given this particular embedding scheme, we can turn our attention to finding an optimal value for the parameter β in (5.4.4), which can only be done experimentally. Since the final Hamiltonian is programmed on the quantum annealer using dimensionless coefficients, the parameter β cannot be expressed in the usual form $1/kT$, where k is the Boltzmann constant and T the effective temperature. Instead, it should be viewed as an empirical parameter that depends on the network architecture, the embedding scheme, and the physical characteristics of the quantum annealer (such as the operating temperature, the anneal time, the energy scale of the superconducting flux qubit system, etc.).

The experimental approach of estimating β consists of the following five steps [4]:

1: Construct an RBM.

2: Map the RBM to a final Hamiltonian assuming a particular value of β (Alg. 4-Step 1).

3: Run quantum annealing (Alg. 4-Step 2).

4: Compute the model expectations using the quantum samples (Alg. 4-Step 3).

5: Compare the resulting expectations with the "correct" benchmark values (e.g., obtained with the classical CD algorithm).

This process is repeated for different choices of β. The value of β that gives the best fit can then be used for the given RBM architecture. As noted in [4], even with the optimal settings for β, the estimates of the model expectations will still have some error. However, in comparison to the noise associated with the Boltzmann sampling in the CD algorithm, this may be sufficient to estimate the gradients in (5.3.1), (5.3.2), and (5.3.3).

5.4.4 Generative models

The main application of the Boltzmann sampling we've considered so far is in providing an unbiased estimate of the model expectations as specified in Algorithm 4. Once fully trained with the help of quantum annealing, an RBM can be used in a conventional classical way to generate new synthetic samples from the learned probability distribution. In this case, quantum annealing is only used as a subroutine in the hybrid quantum-classical training protocol.

However, it is possible to use a quantum annealer as a generator in its own right. Rather than assisting in training the classical RBM, a quantum annealer can output the binary representation of the continuous samples as per the distribution encoded in the final Hamiltonian (5.4.6). The Quantum Variational Autoencoder [175] is another example of a QBM that can be trained end to end by maximising a well-defined cost function: a quantum lower bound to a variational approximation of the log-likelihood.

> Boltzmann sampling is the key element of RBM training and the generation of new samples. Quantum annealing can provide orders of magnitude speedup by replacing classical Boltzmann sampling with quantum sampling.

5.5   Deep Boltzmann Machine

Deep Boltzmann Machines (DBMs) can be constructed from several RBMs where the hidden layer of the first RBM becomes the visible layer of the second, and so on, as shown in Figure 5.4.

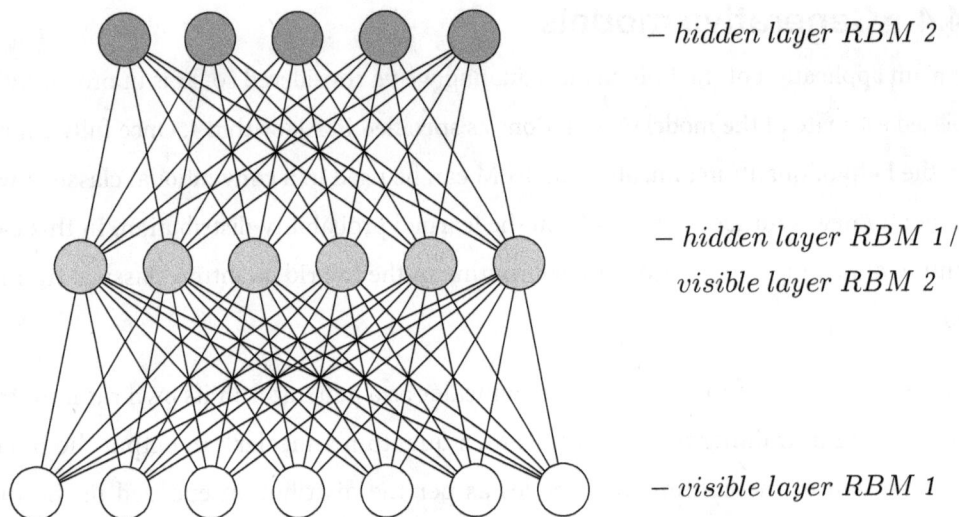

— *hidden layer RBM 2*

— *hidden layer RBM 1 /*
  visible layer RBM 2

— *visible layer RBM 1*

Figure 5.4: Schematic representation of a DBM.

A DBM can be trained layer by layer, one RBM at a time. This will result in a powerful generative model capable of learning complex multivariate distributions and dependence structures. However, the generative training of the DBM can be used as the first step towards building a discriminative model if the training dataset samples are labelled. In this case, all DBM weights and biases found with the help of either CD or quantum Boltzmann sampling algorithms are seen as initial values of the weights and biases of the corresponding feedforward neural network. The discriminative model will consist of all the layers of the original DBM with an extra output layer performing the assignment of the class labels. The discriminative model can be fine-tuned through the standard backpropagation of the error algorithm.

5.5.1 Training DBMs with quantum annealing

The generative training of DBMs can be seen as a pre-training of the discriminative model. Figure 5.5 provides a schematic illustration of the hybrid quantum-classical training process.

(1) Generative training of RBMs with quantum Boltzmann sampling

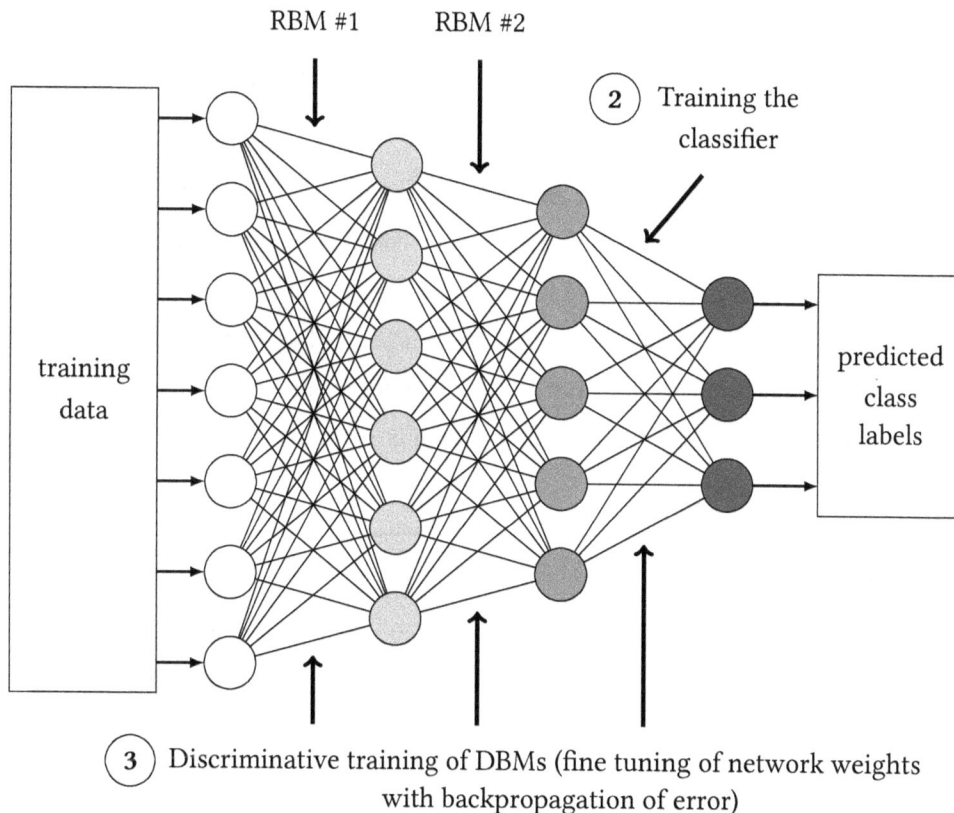

Figure 5.5: Generative and discriminative training of a DBM.

In the DBM training scheme shown in Figure 5.5, only Step 1 relies on quantum annealing. Steps 2 and 3 are completely classical. Step 3 is optional: without it, we have a standard machine learning "pipeline" where one or several RBMs (Step 1) perform "feature extraction" by building a low-dimensional representation of the samples in the dataset, thus helping the discriminative model (Step 2) to achieve better classification results.

5.5.2   A DBM pipeline example

The pipeline approach can be illustrated using the popular "King+Rook vs. King+Pawn" dataset from the UCI Machine Learning Repository [286, 287]. The task is to classify the end game positions with the black pawn one move from queening and the white side (King+Rook) to move. The possible outcomes are "white can win" (Class 1) and "white cannot win" (Class 0). The board is described by 36 categorical attributes that can be encoded as 38 binary variables. The dataset consists of 3,196 samples (white can win in 52% of all cases in the dataset).

The `scikit-learn` package provides all the necessary components for building the classical part of a DBM pipeline. The pipeline itself can be constructed with the help of `sklearn.-pipeline.make_pipeline`. The DBM is constructed from two RBMs implemented with the help of `sklearn.neural_network.BernoulliRBM`. RBM #1 has 38 nodes in the visible layer and 30 nodes in the hidden layer; RBM #2 has 30 nodes in the visible layer and 20 nodes in the hidden layer. The exact pipeline configuration is as follows (all other parameters were set at their default values):

RBM #1	RBM #2	MLP Classifier
n_components = 30	n_components = 20	hidden_layer_sizes = (20)
learning_rate = 0.00025	learning_rate = 0.00025	activation = 'tanh'
batch_size = 10	batch_size = 10	solver = 'adam'
n_iter = 100000	n_iter = 100000	alpha = 0.1
		max_iter = 5000

Table 5.1: Configuration of the DBM pipeline for the "King+Rook vs. King+Pawn" classification problem.

Thus, both RBMs are trained as autoencoders: the DBM translates each 38-feature sample into its 20-feature low-dimensional representation. These new "extracted" features, ideally, should have higher predicting power in comparison with the original features, assuming that both RBMs learned the main characteristics and dependence structure of the dataset and stripped away the noise or the less important characteristics. The discriminator is `sklearn.neural_network.MLPClassifier` with 20 tanh activation units in its single hidden layer.

With this setting, the DBM achieves the following out-of-sample classification results (with the dataset split 70:30 into the training and testing datasets using `sklearn.model_-selection.train_test_split`):

- Classification accuracy: 95.2%

This compares favourably with, for example, an ensemble learning classifier such as random forest (`sklearn.ensemble.RandomForestClassifier`). The random forest classifier with the number of estimators set at 1,000 and the maximum depth set equal to 5 has the following out-of-sample classification results:

- Classification accuracy: 94.9%

The architecture of DBMs allows them to be trained as either generative or discriminative models. In both cases, Boltzmann sampling can play an important role in improving their performance by providing quantum speedup and higher accuracy.

Summary

In this chapter, we learned about energy-based models – a special class of powerful generative models. We learned how to build, train, and run RBMs in order to generate synthetic samples that are statistically indistinguishable from the original training dataset.

We familiarised ourselves with the Boltzmann sampling and Contrastive Divergence algorithms. Boltzmann sampling can be efficiently performed on NISQ-era quantum annealers that may improve the quality of the model and achieve orders of magnitude of speedup in generating new samples.

We learned how to combine individual RBMs together to construct a DBM. Quantum annealing can be productively applied to the pre-training of a DBM before it is fine-tuned as a deep feedforward neural network classifier.

Finally, we explored the possibility of using RBMs and DBMs as the first model in the machine learning pipeline for denoising and feature extraction.

In the next chapter, we will shift our attention to gate model quantum computing. We will start with the concept of a classical binary digit (bit) and classical logic gates before introducing their quantum counterparts: the quantum binary digit (qubit) and one-qubit/multi-qubit quantum logic gates and quantum circuits.

PART II

GATE MODEL QUANTUM COMPUTING

6

Qubits and Quantum Logic Gates

A computation can be broadly defined as a transformation of one memory state into another. Put slightly differently, a computation is a function that transforms information [305]. In the case of classical digital computing, the fundamental memory unit is a *binary digit* (bit) of information. Functions that operate on bits of information are called *logic gates*. Logic gates are Boolean functions that can be combined into *circuits* capable of performing addition and multiplication, as well as more complex operations. In logic gates, the number of output bits does not have to be the same as the number of input bits.

A computation may seem to be an abstract mathematical concept but it always requires some physical system in order to be executed. It does not matter what this physical system is: billiard balls, electric switches, transistors, or anything else – the computation is substrate independent. However, it is always some physical process that changes the state of the system in a controlled way.

Classical digital computing requires some physical implementation of two distinct deterministic states (usually denoted as 0 and 1) and a set of gates that perform controlled transitions between them. In the following sections, we will see how classical digital computation can be implemented, what set of basis operations is required, and how the logic of classical computation can be extended to more general logic of quantum computation, of which classical computing is just a special case.

6.1 Binary Digit (Bit) and Logic Gates

In this section, we briefly review classical logic gates and their universality in order to draw a parallel later to quantum gates.

6.1.1 Logic gates

A logic gate is an implementation of a Boolean function, a logical operation performed on one or more binary inputs that produces a single binary output. Logic gates are represented by their *truth tables*. A truth table has one column for each input variable, and one final column showing all of the possible results of the logical operation that the table represents. Each row of the truth table contains one possible configuration (a single bit or a bitstring) of the input variables, and the result of the operation for those values.

Figures and Tables 6.1, 6.2, 6.3, and 6.4 are schematic circuit representations of the AND, OR, NAND (not AND), and XOR (exclusive OR) logic gates as well as their corresponding truth tables.

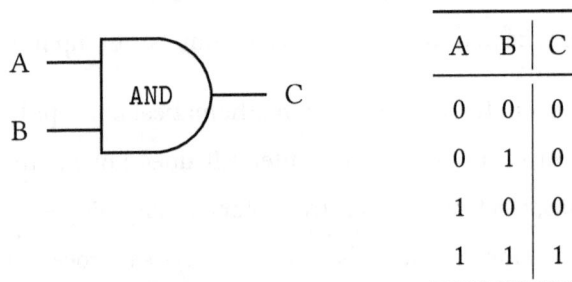

A	B	C
0	0	0
0	1	0
1	0	0
1	1	1

Figure 6.1: AND gate diagram and truth table.

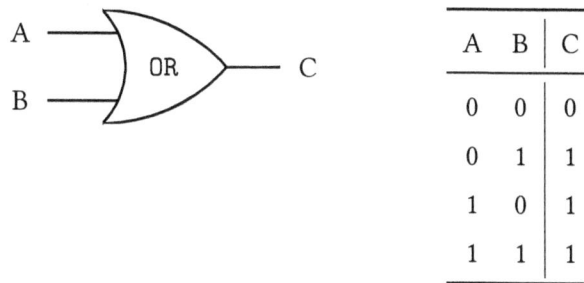

Figure 6.2: OR *gate diagram and truth table.*

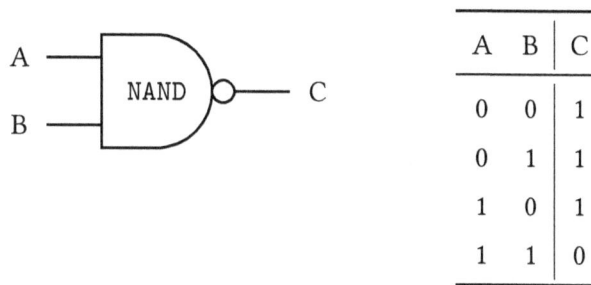

Figure 6.3: NAND *gate diagram and truth table.*

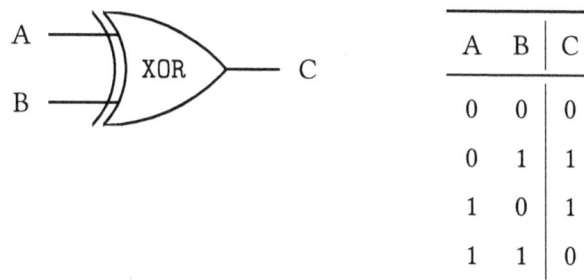

Figure 6.4: XOR *gate diagram and truth table.*

6.1.2   NAND as a universal logic gate

Logic gates can be combined into circuits where the output of one is the input of another. This allows us i) to implement more complex operators than basic Boolean functions and ii) to implement all necessary Boolean functions using only a small number of easy-to-build logic gates. For example, all Boolean functions can be constructed using only a NAND gate and a fan-out operation (several inputs connected to the same output). This makes NAND a *universal* gate in classical computing. Figure 6.5 illustrates this by presenting the decomposition of four basic logic gates into circuits consisting only of NAND gates.

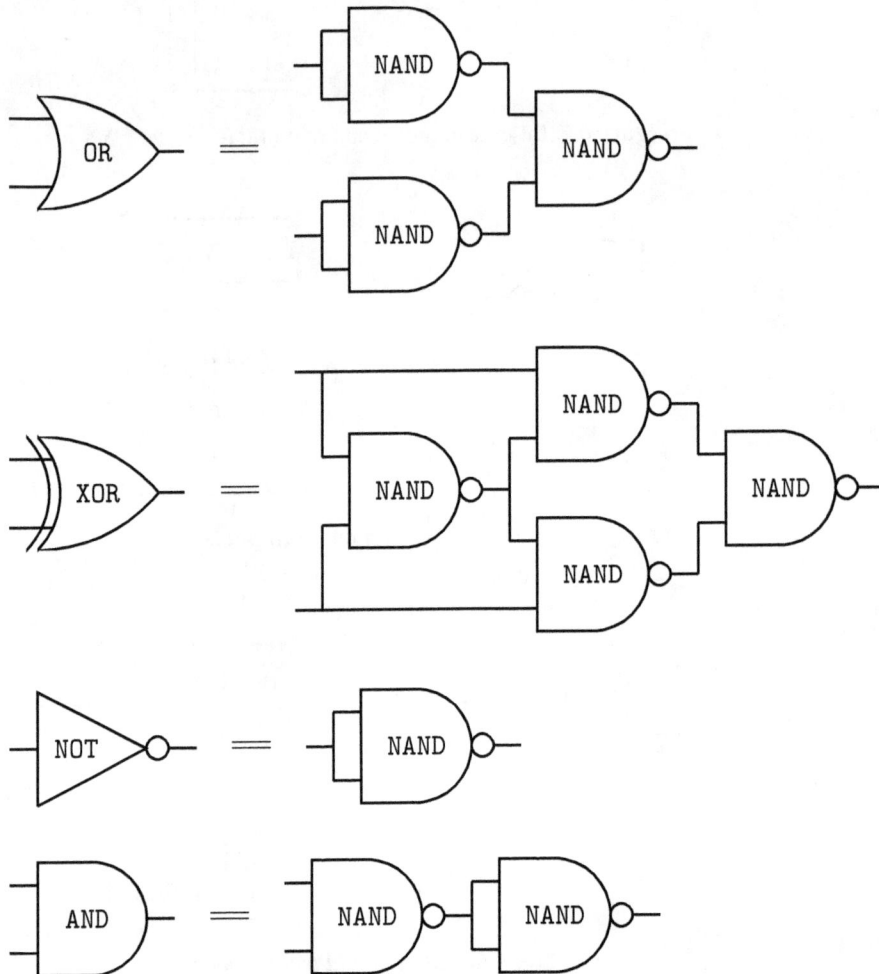

Figure 6.5: Examples of logic gates' decomposition into NAND gates and fan-out operations.

6.1.3   Building an addition operator from the NAND gates

Figure 6.6 shows how NAND and XOR gates can be combined into a circuit that implements the basic addition operator. As we know, the XOR gate itself can be constructed from the combination of NAND gates. The addition operator takes three 1-bit binary numbers as inputs and outputs two 1-bit binary numbers that can be read as a 2-bit bitstring (a 2-bit binary number). This 2-bit binary number can be translated into its integer number representation – an integer number between 0 and 3, as shown in the truth table (Table 6.1).

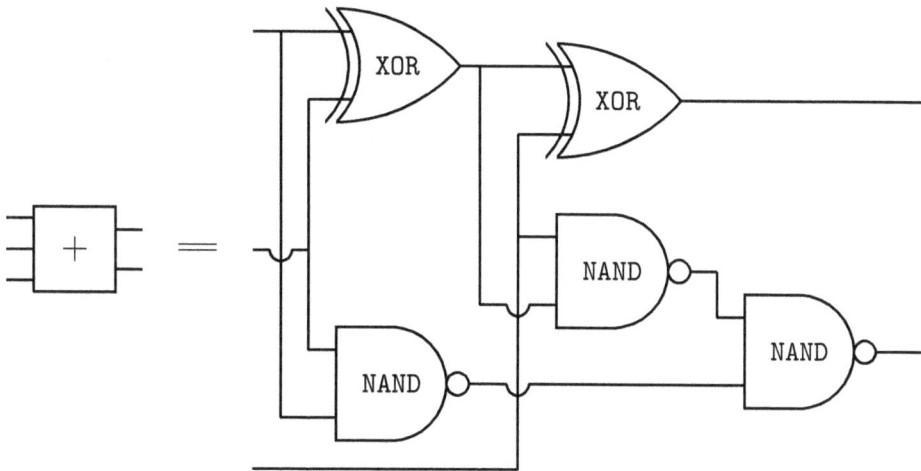

Figure 6.6: Addition operator: the input is three 1-bit binary numbers and the output is a single 2-bit binary number.

input 1	input 2	input 3	output 1	output 2	binary	integer
0	0	0	0	0	00	0
0	0	1	0	1	01	1
0	1	0	0	1	01	1
1	0	0	0	1	01	1
0	1	1	1	0	10	2
1	0	1	1	0	10	2
1	1	0	1	0	10	2
1	1	1	1	1	11	3

Table 6.1: Addition operator truth table.

> Computation is a transformation of one memory state into another. Functions that perform such transformations are called logic gates. Logic gates are fully specified by their truth tables. A universal logic gate is one from which all other Boolean functions can be constructed. We only need to find an efficient physical realisation of a universal gate in order to perform computations of arbitrary complexity.

6.2 Physical Realisations of Classical Bits and Logic Gates

We have so far defined bits and classical logic gates from a theoretical computer science point of view. We now provide an overview of the most efficient hardware techniques used to effectively implement such operations.

6.2.1 Implementation of the NAND gate

The NAND gate (together with the fan-out operator) is a universal gate in classical digital computing. Therefore, it should be sufficient to find a practical physical implementation of the NAND Boolean function in order to build a universal computer. Figure 6.7 displays several possible realisations of the NAND gate using different technologies, from electrical switches to semiconductors.

Relay Logic: Switches are interpreted as bits with $0 =$ open and $1 =$ closed. When switches A and B are both closed, an electromagnet opens switch C. If either or both of switches A and B are open, the circuit is broken and an electromagnet cannot open switch C.

Resistor-Transistor (RT) Logic: Voltages are interpreted as bits with $0 =$ zero volts and $1 = 3$ volts. When wires A and B are both at $+3$ volts, the two transistors conduct electricity and wire C drops to zero volts. If either or both of inputs A and B are zero volts, the corresponding transistors do not conduct and output C stays at $+3$ volts.

Complementary Metal-Oxide-Semiconductor (CMOS) Logic: Similar to RT logic, voltages are interpreted as bits with $0 =$ zero volts and $1 = 3$ volts. The PMOS transistor is open when the input is 1 ($+3$ volts) and closed when the input is 0 (zero volts). NMOS is the logical opposite of PMOS. The PMOS circuit is placed between the voltage and the output. The NMOS circuit is placed between the output and the ground.

If both of the A and B inputs are high, both the NMOS transistors will conduct, neither of the PMOS transistors will conduct, and a conductive path will be established between the output, C, and the ground, thus bringing the output low. If both of the A and B inputs are low, then neither of the NMOS transistors will conduct, while both of the PMOS transistors will conduct, establishing a conductive path between the output and the voltage source, bringing the output high. If either of the A or B inputs is low, one of the NMOS transistors will not conduct, one of the PMOS transistors will, and a conductive path will be established between the output and the voltage source, bringing the output high. Therefore, the circuit implements the NAND gate as the only configuration of the two inputs that results in a low output is when both are high.

a) Relay Logic   b) Resistor-Transistor Logic   c) CMOS Logic

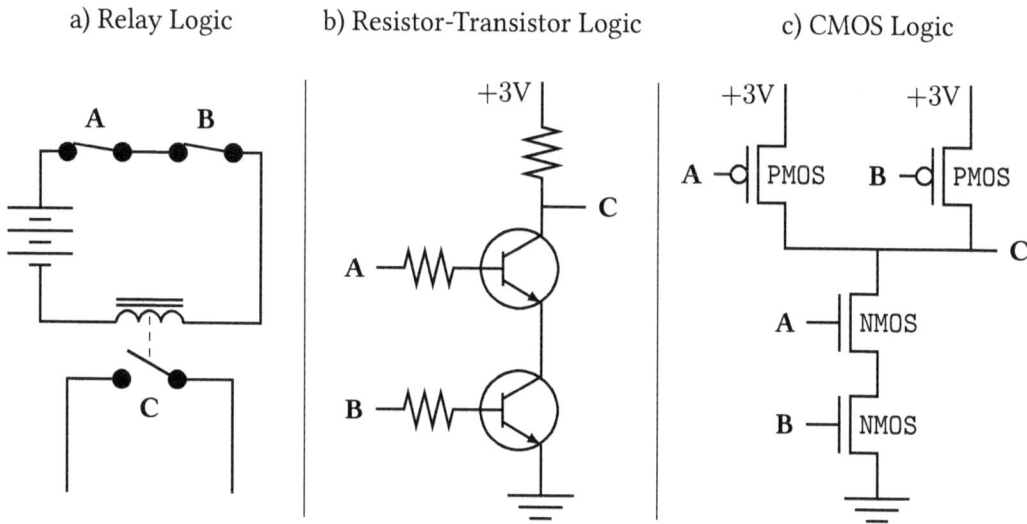

Figure 6.7: Physical realisations of the NAND gate.

6.2.2   Implementation of the RAM memory cell

Random Access Memory (RAM) is used to store instructions and data currently used by the CPU. It is called *volatile memory* in the sense that it is wiped out when the computer is switched off. RAM may consist of many billions of elementary *memory cells*, with each memory cell being able to store one bit of information.

Now that we know how to build a universal logic gate, we can try to design a circuit that would implement the elementary memory unit we need to build RAM. For example, Figure 6.8 shows how the memory cell can be built from four NAND logic gates.

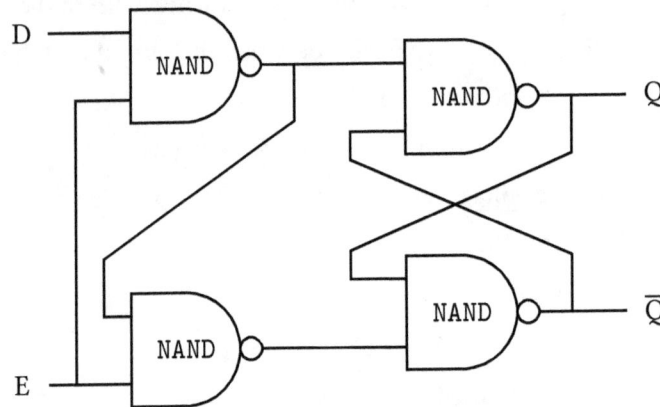

Figure 6.8: Construction of the elementary memory cell from NAND gates.

D	E	Q	$\overline{Q}$
0	1	0	1
1	1	1	0
0	0	Q	$\overline{Q}$
1	0	Q	$\overline{Q}$

Table 6.2: Memory cell truth table.

The circuit in Figure 6.8 has two input pins, D (Data) and E (Enabler), and two output pins, Q and $\overline{Q}$ (NOT Q).

The truth table (Table 6.2) of the memory cell circuit explains how it works:

- When the enabler input E is set to 1, the output Q can be set to the data input D.
- When the enabler input E is set to 0, the output Q cannot be changed – it retains its value.

These are the key features that allow the circuit to serve as a memory cell.

> Computation is substrate independent. Any physical system that can exist in two discrete, stable states with controlled transitions between them can be used to implement gate model digital computing. At the same time, some of the implementations are more efficient (faster, cheaper, and more reliable) than others.

These classical logic gates provide a natural framework to understand their quantum formulations, which we'll investigate now.

6.3 Quantum Binary Digit (Qubit) and Quantum Logic Gates

Quantum bits and quantum logic gates are the quantum computing counterparts of classical bits and logic gates. While they share common features, the quantum aspects yield a multitude of specific properties, which are the subject of this section.

6.3.1 Computation according to the laws of quantum mechanics

Classical logic gates operating on bits implement Boolean functions, forming the basis of digital classical computing. As we have seen, there are many possible physical implementations of a classical bit – a system that has two distinct, stable states with controlled transitions between them. What can we say about such a system from the quantum mechanical point of view?

As we know from Chapter 1, any such system may exist in a superposition of states, and the state of a qubit $|\psi\rangle$ is described by the expression

$$|\psi\rangle = \alpha\,|0\rangle + \beta\,|1\rangle,$$

where α and β are complex numbers satisfying

$$|\alpha|^2 + |\beta|^2 = 1. \tag{6.3.1}$$

The coefficients α and β are *probability amplitudes*. Any attempt to *measure* the state $|\psi\rangle$ results in getting $|0\rangle$ with probability $|\alpha|^2$, and $|1\rangle$ with probability $|\beta|^2$. The measurement consists in coupling the quantum system to the environment, which collapses the superposition. After the measurement, the system is in the measured state and further measurements on the same basis will always yield the same result.

Since the qubit state $|\psi\rangle$ is described by two complex probability amplitudes satisfying (6.3.1), we can say that the state of a qubit is a unit vector in the two-dimensional complex vector space. In other words, the state $|\psi\rangle$ can be written as the vector

$$\begin{bmatrix} \alpha \\ \beta \end{bmatrix} = \alpha \begin{bmatrix} 1 \\ 0 \end{bmatrix} + \beta \begin{bmatrix} 0 \\ 1 \end{bmatrix}.$$

This means that the basis states $|0\rangle$ and $|1\rangle$ are represented by the standard orthonormal basis vectors

$$|0\rangle := \begin{bmatrix} 1 \\ 0 \end{bmatrix}, \quad |1\rangle := \begin{bmatrix} 0 \\ 1 \end{bmatrix}.$$

The standard orthonormal basis $|0\rangle$ and $|1\rangle$ is not the only possible choice of the basis vectors. Any pair of *linearly independent* unit vectors $|u\rangle$ and $|v\rangle$ from the complex two-dimensional vector space can serve as a basis:

$$\alpha\,|0\rangle + \beta\,|1\rangle = \alpha'\,|u\rangle + \beta'\,|v\rangle.$$

For example, we can use the Hadamard basis $\{|+\rangle, |-\rangle\}$ defined by

$$|+\rangle := \frac{1}{\sqrt{2}}|0\rangle + \frac{1}{\sqrt{2}}|1\rangle = \begin{bmatrix} \frac{1}{\sqrt{2}} \\ \frac{1}{\sqrt{2}} \end{bmatrix} \quad \text{and} \quad |-\rangle := \frac{1}{\sqrt{2}}|0\rangle - \frac{1}{\sqrt{2}}|1\rangle = \begin{bmatrix} \frac{1}{\sqrt{2}} \\ -\frac{1}{\sqrt{2}} \end{bmatrix}.$$

The basis is determined by the measurement process or the physical realisation of the quantum computer.

It is important to specify the choice of the basis. For example, the vector

$$\begin{bmatrix} \frac{1}{\sqrt{2}} \\ \frac{1}{\sqrt{2}} \end{bmatrix}$$

measured in the standard orthonormal basis (the *computational basis*) gives outcomes $|0\rangle$ and $|1\rangle$ with equal probability $1/2$. Measured in the Hadamard basis, it gives the outcome $|+\rangle$ with probability 1.

The state of a two-qubit system can be represented by a unit vector in the four-dimensional complex vector space. In this case, the standard orthonormal basis consists of four orthonormal unit vectors

$$|00\rangle := \begin{bmatrix} 1 \\ 0 \\ 0 \\ 0 \end{bmatrix}, \quad |01\rangle := \begin{bmatrix} 0 \\ 1 \\ 0 \\ 0 \end{bmatrix}, \quad |10\rangle := \begin{bmatrix} 0 \\ 0 \\ 1 \\ 0 \end{bmatrix}, \quad |11\rangle := \begin{bmatrix} 0 \\ 0 \\ 0 \\ 1 \end{bmatrix}, \tag{6.3.2}$$

and the system state is described by four probability amplitudes:

$$|\psi\rangle = \alpha\,|00\rangle + \beta\,|01\rangle + \gamma\,|10\rangle + \delta\,|11\rangle,$$

with $\alpha, \beta, \gamma, \delta \in \mathbb{C}$ such that $|\alpha|^2 + |\beta|^2 + |\gamma|^2 + |\delta|^2 = 1$. The basis vectors (6.3.2) of the two-qubit states are constructed as *tensor products* of the individual qubit basis vectors:

$$|00\rangle = |0\rangle \otimes |0\rangle = \begin{bmatrix} 1 \cdot \begin{bmatrix} 1 \\ 0 \end{bmatrix} \\ 0 \cdot \begin{bmatrix} 1 \\ 0 \end{bmatrix} \end{bmatrix} = \begin{bmatrix} 1 \\ 0 \\ 0 \\ 0 \end{bmatrix}, \quad |01\rangle = |0\rangle \otimes |1\rangle = \begin{bmatrix} 1 \cdot \begin{bmatrix} 0 \\ 1 \end{bmatrix} \\ 0 \cdot \begin{bmatrix} 0 \\ 1 \end{bmatrix} \end{bmatrix} = \begin{bmatrix} 0 \\ 1 \\ 0 \\ 0 \end{bmatrix},$$

$$|10\rangle = |1\rangle \otimes |0\rangle = \begin{bmatrix} 0 \cdot \begin{bmatrix} 1 \\ 0 \end{bmatrix} \\ 1 \cdot \begin{bmatrix} 1 \\ 0 \end{bmatrix} \end{bmatrix} = \begin{bmatrix} 0 \\ 0 \\ 1 \\ 0 \end{bmatrix}, \quad |11\rangle = |1\rangle \otimes |1\rangle = \begin{bmatrix} 0 \cdot \begin{bmatrix} 0 \\ 1 \end{bmatrix} \\ 1 \cdot \begin{bmatrix} 0 \\ 1 \end{bmatrix} \end{bmatrix} = \begin{bmatrix} 0 \\ 0 \\ 0 \\ 1 \end{bmatrix}.$$

Generally, the n-qubit system can exist in any superposition of the 2^n basis states and requires 2^n probability amplitudes to be fully specified.

Computation is a transformation of the memory state. The qubit states are transformed by an application of quantum logic gates. Quantum logic gates are unitary linear operators that are represented by unitary matrices. The action of a quantum logic gate on a specific quantum state is found by multiplying the unitary matrix representing the gate by the vector representing the state. The result is the new quantum state.

6.3.2   Qubit

It is convenient to visualise the state of a qubit as a point on the unit sphere, named the Bloch sphere after physicist Felix Bloch. Every point on the Bloch sphere is uniquely specified by two angles, $\theta \in [0, \pi]$ and $\phi \in [0, 2\pi]$, as shown in Figure 6.9.

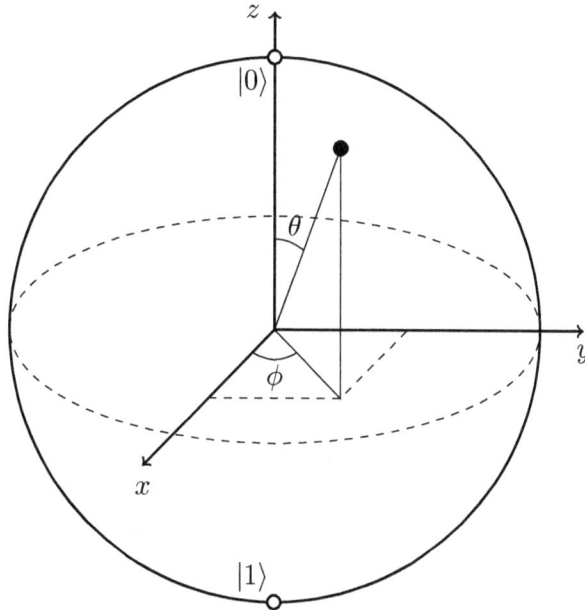

Figure 6.9: Quantum state $|\psi\rangle$ on the Bloch sphere.

With the mapping

$$\alpha = \cos\left(\frac{\theta}{2}\right), \quad \beta = e^{i\phi}\sin\left(\frac{\theta}{2}\right),$$

we obtain the canonical representation of the qubit state:

$$|\psi\rangle = \alpha\,|0\rangle + \beta\,|1\rangle = \begin{bmatrix} \cos\left(\dfrac{\theta}{2}\right) \\ e^{i\phi}\sin\left(\dfrac{\theta}{2}\right) \end{bmatrix}.$$

A transformation of the qubit state can be visualised as a transition from one point on the Bloch sphere to another. Therefore, the unitary matrix (quantum logic gate) that performs this transformation can be seen as a rotation operator and we can speak about *rotation* as a synonym of a gate operation and *rotation angles* as gate parameters.

6.3.3   One-qubit quantum logic gates

Unlike classical computing where we can only define two logic gates operating on a single bit (the identity gate and the NOT gate), quantum computing has infinitely many single qubit logic gates: any unitary 2×2 matrix (rotation) is a quantum logic gate. Some of these gates are more important (or easier to implement) than others. Below, we provide detailed descriptions of some of them, starting with the identity I and the Pauli matrices X, Y, and Z. The action of the I gate is obvious – it leaves the state of the qubit unchanged; Pauli matrices perform rotation of the qubit state by π radians around the x, y, and z axes, respectively:

$$I = \begin{bmatrix} 1 & 0 \\ 0 & 1 \end{bmatrix}, \quad X = \begin{bmatrix} 0 & 1 \\ 1 & 0 \end{bmatrix}, \quad Y = \begin{bmatrix} 0 & -i \\ i & 0 \end{bmatrix}, \quad Z = \begin{bmatrix} 1 & 0 \\ 0 & -1 \end{bmatrix}.$$

By performing simple algebraic operations, we can easily verify that the X gate flips the bit and that the Z gate flips the phase:

$$\text{X gate:} \quad \begin{bmatrix} 0 & 1 \\ 1 & 0 \end{bmatrix} \begin{bmatrix} 1 \\ 0 \end{bmatrix} = \begin{bmatrix} 0 \\ 1 \end{bmatrix}, \quad \begin{bmatrix} 0 & 1 \\ 1 & 0 \end{bmatrix} \begin{bmatrix} 0 \\ 1 \end{bmatrix} = \begin{bmatrix} 1 \\ 0 \end{bmatrix}.$$

$$\text{Z gate:} \quad \begin{bmatrix} 1 & 0 \\ 0 & -1 \end{bmatrix} \begin{bmatrix} 1 \\ 0 \end{bmatrix} = \begin{bmatrix} 1 \\ 0 \end{bmatrix}, \quad \begin{bmatrix} 1 & 0 \\ 0 & -1 \end{bmatrix} \begin{bmatrix} 0 \\ 1 \end{bmatrix} = - \begin{bmatrix} 0 \\ 1 \end{bmatrix}.$$

These operations can be visualised with the help of the following graphical representation of the quantum gates:

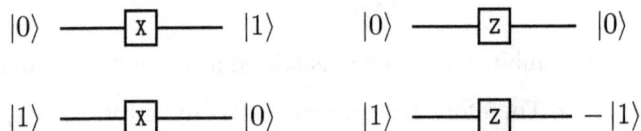

Figure 6.10: Graphical representation of the X and Z gates.

Here, horizontal lines represent *quantum registers* and boxes represent *quantum gates*. Together, quantum registers and quantum gates form the graphical representation of *quantum circuits* – the sequence of quantum gates that transform the quantum state, thus implementing quantum computation. The quantum circuits are read from left to right: the initial quantum state is shown at the left end of the quantum circuit and the final state is shown at the right end. Often, the last operator on the quantum register is the *measurement operator*. Following measurement (in the computational basis), a qubit is transformed into a classical bit and its value is a known binary number.

Since the X gate flips the state of the qubit, it is also called the NOT gate. Similarly, the Z gate that flips the phase of the qubit state is known as the PHASE gate.

We can draw a direct analogy between the NOT gate in classical computing and the NOT gate in quantum computing but there are quantum gates that perform operations that do not exist in classical computing. One such example is the $\sqrt{\text{NOT}}$ gate (represented by the matrix M introduced in Section 1.2.1). In classical computing, we do not have a function that, when applied twice, would flip the bit. But such a function exists in quantum computing:

$$\sqrt{\text{NOT}} \equiv \sqrt{\text{X}} = \frac{1}{2} \begin{bmatrix} 1+i & 1-i \\ 1-i & 1+i \end{bmatrix}.$$

We know that one of the main sources of power in quantum computing is the ability of a qubit to exist in a superposition of basis states. But how can we put a qubit that was initialised as $|0\rangle$ (or $|1\rangle$) in a superposition of states $|0\rangle$ and $|1\rangle$? The answer is the Hadamard gate, H, which creates an equal superposition of states $|0\rangle$ and $|1\rangle$ when applied to either state $|0\rangle$ or state $|1\rangle$:

$$H = \frac{1}{\sqrt{2}} \begin{bmatrix} 1 & 1 \\ 1 & -1 \end{bmatrix},$$

$$H \begin{bmatrix} 1 \\ 0 \end{bmatrix} = \frac{1}{\sqrt{2}} \begin{bmatrix} 1 \\ 0 \end{bmatrix} + \frac{1}{\sqrt{2}} \begin{bmatrix} 0 \\ 1 \end{bmatrix} \quad \text{and} \quad H \begin{bmatrix} 0 \\ 1 \end{bmatrix} = \frac{1}{\sqrt{2}} \begin{bmatrix} 1 \\ 0 \end{bmatrix} - \frac{1}{\sqrt{2}} \begin{bmatrix} 0 \\ 1 \end{bmatrix}.$$

$$|0\rangle \;\;\text{—}\boxed{\text{H}}\text{—}\;\; \tfrac{1}{\sqrt{2}}|0\rangle + \tfrac{1}{\sqrt{2}}|1\rangle \qquad\qquad |1\rangle \;\;\text{—}\boxed{\text{H}}\text{—}\;\; \tfrac{1}{\sqrt{2}}|0\rangle - \tfrac{1}{\sqrt{2}}|1\rangle$$

Figure 6.11: Graphical representation of the Hadamard H *gate.*

Interestingly, the Hadamard H gate is its own inverse, so that the second application of the Hadamard gate reverses the action of the first (mathematically, $H^2 = I$, or $H = H^{-1}$):

$$|0\rangle \;\;\text{—}\boxed{\text{H}}\text{—}\boxed{\text{H}}\text{—}\;\; |0\rangle \qquad\qquad |1\rangle \;\;\text{—}\boxed{\text{H}}\text{—}\boxed{\text{H}}\text{—}\;\; |1\rangle$$

Figure 6.12: Hadamard H *gate applied twice.*

Some other useful one-qubit gates are the *phase shift* gates, where the phase is shifted by $\pi/2$ and $\pi/4$ rather than by π, as is the case for the Z (PHASE) gate:

$$S = \begin{bmatrix} 1 & 0 \\ 0 & e^{i\pi/2} \end{bmatrix} \quad \text{and} \quad T = \begin{bmatrix} 1 & 0 \\ 0 & e^{i\pi/4} \end{bmatrix}.$$

Finally, it is necessary to mention the *adjustable* one-qubit gates that perform rotation of the qubit state around a specific axis by an arbitrary angle θ. For any given gate G, define

$$R_G(\theta) := \exp\left(-\frac{1}{2}i\theta G\right).$$

Using Lemma 1 covered in Chapter 1, we can then immediately compute R_X, R_Y, and R_Z as

$$R_X(\theta) = \begin{bmatrix} \cos\left(\frac{\theta}{2}\right) & -i\sin\left(\frac{\theta}{2}\right) \\ -i\sin\left(\frac{\theta}{2}\right) & \cos\left(\frac{\theta}{2}\right) \end{bmatrix},$$

$$R_Y(\theta) = \begin{bmatrix} \cos\left(\frac{\theta}{2}\right) & -\sin\left(\frac{\theta}{2}\right) \\ \sin\left(\frac{\theta}{2}\right) & \cos\left(\frac{\theta}{2}\right) \end{bmatrix}, \qquad (6.3.3)$$

$$R_Z(\theta) = \begin{bmatrix} e^{-i\theta/2} & 0 \\ 0 & e^{i\theta/2} \end{bmatrix}.$$

The adjustable gates play a very important role in Parameterised Quantum Circuits (PQC), which we will consider in the following chapters of this book.

6.3.4 Two-qubit quantum logic gates

Similar to one-qubit gates specified by unitary 2×2 matrices, we can construct any number of multi-qubit gates. The n-qubit gates would be represented by $2^n \times 2^n$ unitary matrices. Since multi-qubit gates act on several qubits at the same time, they can be used to *entangle* them – that is, make their states depend on each other. We also have the possibility to create conditional operators, where an operator is applied to a target qubit only if a control qubit is in state $|1\rangle$. Such gates are called *controlled* gates, and we consider some of them next.

Controlled gates are shown in the quantum circuit as a straight line connecting two quantum registers. One quantum register represents the control qubit, and is indicated by the dot placed at the end of the line connecting the quantum registers. Another quantum register represents the target qubit: the desired conditional operator is put on this register. Figure 6.13 illustrates this by displaying a `Controlled Y` (CY) gate. Here, q_1 is the quantum register representing the control qubit, q_2 is the quantum register representing the target qubit, and the operator applied to the target qubit is Y.

$$\text{CY} = \begin{bmatrix} 1 & 0 & 0 & 0 \\ 0 & 1 & 0 & 0 \\ 0 & 0 & 0 & -i \\ 0 & 0 & i & 0 \end{bmatrix}.$$

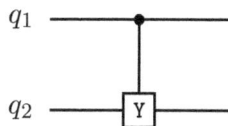

Figure 6.13: CY gate.

The `Controlled NOT` gate, usually denoted as `CNOT` or `CX`, is another example of a two-qubit controlled gate. It consists of applying the Pauli `X` gate to the target qubit if the control qubit is in state $|1\rangle$ and is given by the following unitary matrix:

$$\text{CNOT} \equiv \text{CX} = \begin{bmatrix} 1 & 0 & 0 & 0 \\ 0 & 1 & 0 & 0 \\ 0 & 0 & 0 & 1 \\ 0 & 0 & 1 & 0 \end{bmatrix}.$$

This gate is often represented in the quantum circuit by an `XOR` logical symbol (circled plus) placed on the target qubit quantum register since its truth table (for the target qubit) coincides with the truth table of the `XOR` logic gate.

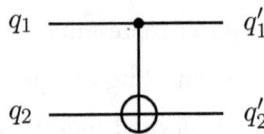

Figure 6.14: CX (CNOT) gate.

q_1	q_2	q_1'	q_2'
0	0	0	0
0	1	0	1
1	0	1	1
1	1	1	0

Table 6.3: Truth table for CX (CNOT) gate.

Seen differently, note that we in fact have the equality

$$\text{CX} |q_1 q_2\rangle = |q_1\rangle |q_1 \oplus q_2\rangle,$$

for any $q_1, q_2 \in \{0, 1\}$, where $\oplus$ denotes addition modulo 2.

The CZ gate is a Pauli Z (phase flip) applied to the target qubit conditional on the control qubit being in state $|1\rangle$ and is given by the following unitary matrix:

$$\text{CPHASE} \equiv \text{CZ} = \begin{bmatrix} 1 & 0 & 0 & 0 \\ 0 & 1 & 0 & 0 \\ 0 & 0 & 1 & 0 \\ 0 & 0 & 0 & -1 \end{bmatrix}.$$

Interestingly, for CZ, it does not really matter which qubit is the target qubit and which is the control qubit – the result is the same:

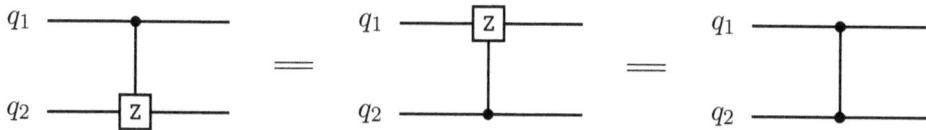

Figure 6.15: CZ (CPHASE) gates.

The SWAP gate swaps the states of two qubits. The $\sqrt{\text{SWAP}}$ gate is universal in the sense that any multi-qubit gate can be constructed from only $\sqrt{\text{SWAP}}$ and single qubit gates.

$$\text{SWAP} = \begin{bmatrix} 1 & 0 & 0 & 0 \\ 0 & 0 & 1 & 0 \\ 0 & 1 & 0 & 0 \\ 0 & 0 & 0 & 1 \end{bmatrix}, \quad \sqrt{\text{SWAP}} = \begin{bmatrix} 1 & 0 & 0 & 0 \\ 0 & \frac{1+i}{2} & \frac{1-i}{2} & 0 \\ 0 & \frac{1-i}{2} & \frac{1+i}{2} & 0 \\ 0 & 0 & 0 & 1 \end{bmatrix}.$$

Very often, the choice of the set of universal gates from which all other gates can be constructed is dictated by the characteristics of the physical system used to perform quantum computation. The $\sqrt{\text{SWAP}}$ is a *native gate* in the systems that exploit exchange interactions [236]. Related gates such as iSWAP and $\sqrt{\text{iSWAP}}$ are natural gates in the systems with Ising-like interactions [265]:

$$\text{iSWAP} = \begin{bmatrix} 1 & 0 & 0 & 0 \\ 0 & 0 & i & 0 \\ 0 & i & 0 & 0 \\ 0 & 0 & 0 & 1 \end{bmatrix}, \quad \sqrt{\text{iSWAP}} = \begin{bmatrix} 1 & 0 & 0 & 0 \\ 0 & \frac{1}{\sqrt{2}} & \frac{i}{\sqrt{2}} & 0 \\ 0 & \frac{i}{\sqrt{2}} & \frac{1}{\sqrt{2}} & 0 \\ 0 & 0 & 0 & 1 \end{bmatrix}.$$

An example of an adjustable two-qubit gate is the XY gate, which is a rotation by some angle θ between the $|01\rangle$ and $|10\rangle$ states:

$$\text{XY}(\theta) = \begin{bmatrix} 1 & 0 & 0 & 0 \\ 0 & \cos\left(\frac{\theta}{2}\right) & i\sin\left(\frac{\theta}{2}\right) & 0 \\ 0 & i\sin\left(\frac{\theta}{2}\right) & \cos\left(\frac{\theta}{2}\right) & 0 \\ 0 & 0 & 0 & 1 \end{bmatrix}.$$

Note that $\text{XY}(\pi) = \text{iSWAP}$ and $\text{XY}(\pi/2) = \sqrt{\text{iSWAP}}$. Together with CZ, the iSWAP gate plays an important role in the construction of quantum circuits since any two-qubit gate can be expressed with, at most, three CZ or three iSWAP gates [2].

6.3.5   The Toffoli gate

The classical Toffoli gate, invented by Tommaso Toffoli [309], is a three-bit logic gate, which is universal in classical computing. In quantum computing, it is a three-qubit Controlled Controlled NOT (CCNOT) gate that is represented by the following quantum circuit, where the qubit C is the target qubit and the qubits A and B are the control qubits:

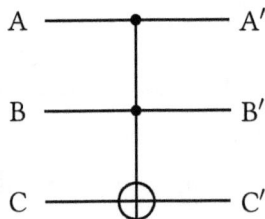

Figure 6.16: Toffoli (CCNOT) gate.

The classical Toffoli gate is given by the following truth table:

A	B	C	A'	B'	C'
0	0	0	0	0	0
1	0	0	1	0	0
0	1	0	0	1	0
1	1	0	1	1	1
0	0	1	0	0	1
1	0	1	1	0	1
0	1	1	0	1	1
1	1	1	1	1	0

Table 6.4: Truth table for the Toffoli gate.

The quantum Toffoli gate is represented by the unitary matrix:

$$
\text{CCNOT} =
\begin{bmatrix}
1 & 0 & 0 & 0 & 0 & 0 & 0 & 0 \\
0 & 1 & 0 & 0 & 0 & 0 & 0 & 0 \\
0 & 0 & 1 & 0 & 0 & 0 & 0 & 0 \\
0 & 0 & 0 & 1 & 0 & 0 & 0 & 0 \\
0 & 0 & 0 & 0 & 1 & 0 & 0 & 0 \\
0 & 0 & 0 & 0 & 0 & 1 & 0 & 0 \\
0 & 0 & 0 & 0 & 0 & 0 & 0 & 1 \\
0 & 0 & 0 & 0 & 0 & 0 & 1 & 0
\end{bmatrix}.
$$

It is clear from the Toffoli gate truth table that it also implements the AND and NAND gates. With $C = 0$, it can be viewed as the AND gate:

$$\text{if } C = 0 : \quad C' = A \text{ AND } B.$$

And with C = 1, it can be viewed as the NAND gate:

$$\text{if } C = 1: \quad C' = A \text{ NAND } B.$$

The Toffoli gate can be decomposed into a quantum circuit consisting of CNOT and one-qubit gates:

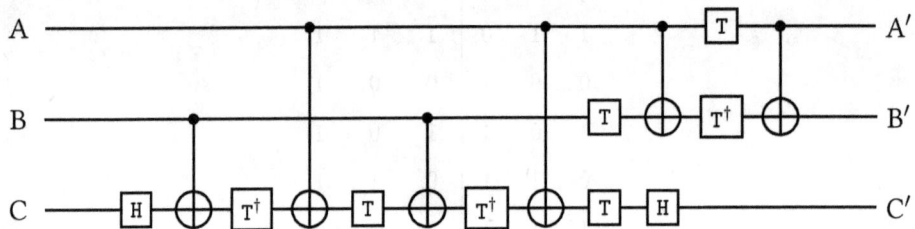

Figure 6.17: Decomposition of the Toffoli (CCNOT) gate. The "dagger" superscript after the gate symbol such as T† *indicates the* adjoint *operator (see Section 1.1.1).*

The fact that the Toffoli circuit allows us to implement the NAND gate, which is universal in classical computing, demonstrates the fact that quantum computing can perform all operations that are possible on classical computers. In other words, quantum computers can simulate classical computers. At the same time, we have seen examples of quantum operations that do not have their analogues in classical computing. In the most general case, classical simulation of an n-qubit quantum system would require the ability to store 2^n probability amplitudes – an impossible task for n larger than several hundred as there would not be enough matter in the visible universe to implement such classical memory. Consequently, quantum computation is more general than classical computation. Computation, as a concept, is really quantum computation. Classical computation is just a special case of quantum one [32].

Quantum computing offers a wider range of logic gates than classical computing.

The Toffoli gate demonstrates that quantum computers can perform all operations implementable on classical computers. At the same time, an attempt to simulate quantum computing classically will immediately run into memory issues.

As discussed, quantum gates correspond to unitary matrices, which have the property of being invertible. Since quantum circuits are fundamentally classical and tensor products of such matrices, they can easily be inverted, yielding the concept of *reversible computing*, which we focus on in the next section.

6.4   Reversible Computing

The importance of the Toffoli gate goes beyond mere universality. It is a universal *reversible* logic gate, meaning that it can serve as a basis for reversible computing. Here, we should note that *all* quantum logic gates that are represented by unitary matrices are reversible. So, what does reversible computing mean?

Reversible computing is a model of computation where the computational process is time-reversible. It also means that no information is lost through the computation process and we can always reconstruct the initial state. The ability to physically realise reversible computing is hugely important due to the deep physical link between loss of information and generation of heat.

According to the principle formulated by Landauer [198], in order for a computational process to be physically reversible, it must also be logically reversible. Fundamentally, this is due to the fact that the act of computation can only be performed by some physical system and is subject to the physical laws of thermodynamics.

The loss of information leads to the increase in information entropy. Similarly, the increase in thermodynamic entropy leads to the generation of heat. In both cases, we are moving from the more ordered state to the less ordered state, which is an irreversible process.

This can be illustrated by the definitions of entropy (as a measure of disorder) in both statistical mechanics and information theory. Entropy in statistical mechanics is given by

$$S = -k_B \sum_i p_i \log(p_i),$$

where k_B is the Boltzmann constant and p_i is the probability of the microstate i taken from the equilibrium ensemble (macroscopic thermodynamic state), while entropy in information theory is given by

$$H = -\sum_i p_i \log_2(p_i),$$

where p_i is the probability of the message i taken from the message space. A very high probability of a particular microstate and a very high probability of a specific message indicate highly ordered systems with low entropy. The entropy is maximised (and information is minimised) when the microstates/messages are uniformly distributed.

Any probability distribution can be approximated arbitrarily closely by some thermodynamic system [234]. If h is information (bits) per particle, then for N particles, the entropy measured in the *natural unit of information* (1 bit = $\log(2)$ nat) is given by the expression

$$S = -k_B \log(2) N h.$$

In energy units, $k_B T \log(2)$ of heat is generated for each bit of information lost. Here, T is the temperature of the heat sink (in Kelvins). For example, if we take $T = 300\text{K}$ (approx. 27C), then the minimum possible amount of energy required to erase one bit of information is 2.87 zJ (zeptojoule).

In practical terms, it means that every *logically irreversible* operation (e.g., NAND or XOR gates) must be accompanied by the corresponding entropy increase and generation of heat. As the energy efficiency of the computation process becomes progressively more important, efforts to develop reversible computing are increasing since it may prove difficult for traditional technology based on the laws of classical physics to progress very far beyond current levels of energy efficiency if reversible computing principles are not used [106].

> In contrast with classical computing, all quantum computing operations are reversible (except measurement). This means that the quantum advantage is likely to be demonstrated not only in quantum speedup and the expressive power of quantum circuits but also in achieving superior energy efficiency.

We have so far seen forward and backward (reversible) quantum operations. However, what fundamentally distinguishes quantum computing from classical computing is the concept of entanglement.

6.5 Entanglement

The key aspect of quantum computing is entanglement, which allows for quantum states to encode more information than the sum of their individual components. We explain this in detail here and provide examples for two-qubit systems.

6.5.1 Quantum entanglement and why it matters

An n-qubit system can exist in any superposition of the 2^n basis states:

$$\sum_{i=0}^{2^n-1} c_i \left|i\right\rangle = c_0 \left|00\ldots00\right\rangle + c_1 \left|00\ldots01\right\rangle + \ldots + c_{2^n-1} \left|11\ldots11\right\rangle,$$

with

$$\sum_{i=0}^{2^n-1} |c_i|^2 = 1.$$

If such a state can be represented as a tensor product of individual qubit states, then the qubit states are *not entangled*. For example, it is easy to check that

$$\frac{1}{4\sqrt{2}} \left(\sqrt{3}\left|000\right\rangle + \left|001\right\rangle + 3\left|010\right\rangle + \sqrt{3}\left|011\right\rangle + \sqrt{3}\left|100\right\rangle + \left|101\right\rangle + 3\left|110\right\rangle + \sqrt{3}\left|111\right\rangle \right)$$

$$= \left(\frac{1}{\sqrt{2}}\left|0\right\rangle + \frac{1}{\sqrt{2}}\left|1\right\rangle \right) \otimes \left(\frac{1}{2}\left|0\right\rangle + \frac{\sqrt{3}}{2}\left|1\right\rangle \right) \otimes \left(\frac{\sqrt{3}}{2}\left|0\right\rangle + \frac{1}{2}\left|1\right\rangle \right), \qquad (6.5.1)$$

so that the quantum state is not entangled (only in superposition). An *entangled* state cannot be represented as a tensor product of individual qubit states.

For example, the two-qubit state

$$\frac{1}{\sqrt{2}}\,|00\rangle + \frac{1}{\sqrt{2}}\,|11\rangle \tag{6.5.2}$$

does not allow a tensor product decomposition. Namely, for any $a, b, c, d \in \mathbb{C}$ such that $|a|^2 + |b|^2 = |c|^2 + |d|^2 = 1$, we have

$$\frac{1}{\sqrt{2}}\,|00\rangle + \frac{1}{\sqrt{2}}\,|11\rangle \neq (a\,|0\rangle + b\,|1\rangle) \otimes (c\,|0\rangle + d\,|1\rangle).$$

We notice that we need 2^n probability amplitudes to describe the state on the left side of (6.5.1) while we only need $2n$ probability amplitudes to describe the state on the right side of (6.5.1). The number of probability amplitudes needed to fully describe the state of a system is directly related to the amount of information the system can store. Entanglement allows us to encode a significantly larger amount of information than is possible with individual independent qubits. One can say that most of the information encoded in the state of a quantum mechanical system is stored non-locally in the correlations between the qubit states. This non-locality of information is one of the major distinguishing features of quantum computing over classical computing and is essential for a number of applications.

What happens if we measure the entangled qubits? In (6.5.2), both qubits are in the state of equal superposition, i.e., if we measure the first qubit, we will get both 0 and 1 with probability 1/2. If, instead, we measure the second qubit, we will also get 0 and 1 with equal probability. However, the situation is completely different if we measure the second qubit after the first has already been measured. In this case, the state of the second qubit is fully determined by the act of measuring the first qubit and there is no longer any uncertainty about its value: if the first qubit was measured as 0, the second qubit is also in state 0, and if the first qubit was measured as 1, the second qubit is also in state 1. In other words, measuring one qubit collapses the superposition and has an immediate effect on the other.

6.5.2 Entangling qubit states with two-qubit gates

Qubit states can be entangled with the help of two-qubit gates. The two-qubit state given by (6.5.2) is known as one of the four maximally entangled Bell states. It can be constructed from the unentangled state $|00\rangle$:

$$|00\rangle = (1 \cdot |0\rangle + 0 \cdot |1\rangle) \otimes (1 \cdot |0\rangle + 0 \cdot |1\rangle)$$

by applying the *Bell circuit* consisting of H and CNOT gates:

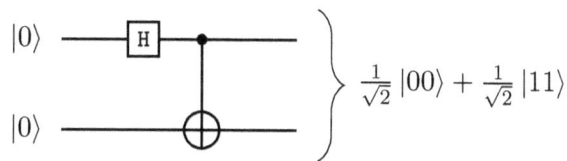

Figure 6.18: Bell circuit.

Running this circuit on the unentagled states $|01\rangle$, $|10\rangle$, and $|11\rangle$ will result in the construction of the other three Bell states:

$$|01\rangle \rightarrow \frac{1}{\sqrt{2}}(|01\rangle + |10\rangle),$$

$$|10\rangle \rightarrow \frac{1}{\sqrt{2}}(|00\rangle - |11\rangle),$$

$$|11\rangle \rightarrow \frac{1}{\sqrt{2}}(|01\rangle - |10\rangle).$$

Entanglement can be achieved with other two-qubit gates as well. Depending on the hardware implementation, it may be a SWAP, a CPHASE, or some other fixed two-qubit gate, or it can be an adjustable two-qubit gate such as $XY(\theta)$.

Entanglement allows us to store most of the information in the correlations between the states rather than in the states of individual qubits.

> Entanglement is one of the main sources of the expressive power of quantum circuits that underpins our search for the quantum advantage.

We saw that entanglement is a distinctive feature of quantum computing. We now see how it comes into play when analysing the quantum equivalents of classical logic gate decompositions studied in Section 6.1.

6.6   Quantum Gate Decompositions

The most widely used NISQ computing technologies are trapped ions and superconducting qubits. In both cases, one-qubit gates are much faster than two-qubit gates (by an order of magnitude). Additionally, one-qubit gates have much higher fidelity [46, 178]. This means that we can treat one-qubit gates as computationally inexpensive and should not worry too much about their quantities. At the same time, we have to be economical with two-qubit gates: out of two equivalent circuits, the one with the smaller number of two-qubit gates would generally perform better. Therefore, we should be aware of the two-qubit gates that are native to any particular system – gates that can be implemented naturally using standard hardware control techniques. More complex gates can be decomposed into a subcircuit of the native gates but an even better solution would be to specify the algorithm that takes advantage of the native gates and bypasses the need of having non-native two-qubit gates. For example, Rigetti's Aspen system [299] is based on the superconducting qubits with two native two-qubit gates CZ and XY – constructing a circuit based on these gates rather than, for example, SWAP gates would achieve better performance.

However, it is not always practical or desirable to make an algorithm hardware-dependent. And since the choice of the native gates is inevitably limited, it is useful to keep in mind several basic decompositions. The following relationships can be verified by direct calculations and play an important role in quantum circuit construction:

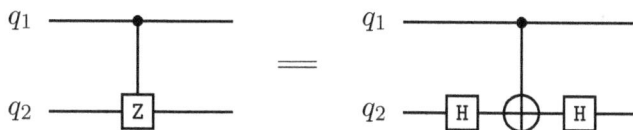

Figure 6.19: CZ *gate decomposition into* CX *and Hadamard gates.*

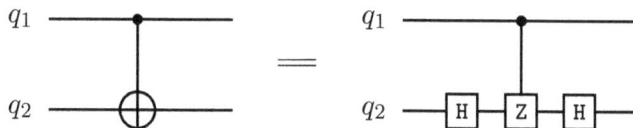

Figure 6.20: CX *gate decomposition into* CZ *and Hadamard gates.*

Given the limited connectivity of NISQ devices (nearest neighbours for most qubits), the SWAP gate that swaps the states of the qubits is especially useful and its efficient implementation using available native gates is very important. The SWAP gate can be represented by a subcircuit consisting of three CX gates:

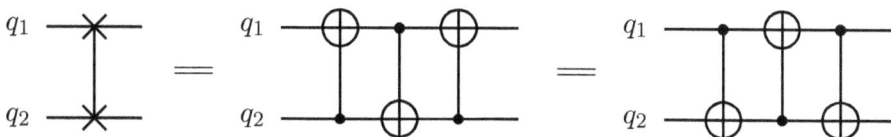

Figure 6.21: SWAP *gate decomposition into three* CX *gates.*

Taking into account the relationship between the CX and CZ gates in Figure 6.20, the SWAP gate can also be decomposed into a subcircuit of three CZ gates and a handful of one-qubit Hadamard gates:

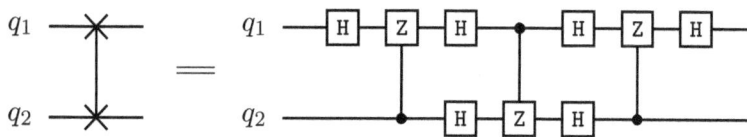

Figure 6.22: SWAP *gate decomposition into three* CZ *and six Hadamard gates.*

Alternatively, the SWAP gate can be implemented with the help of three iSWAP gates:

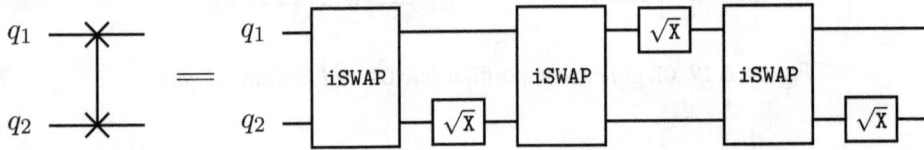

Figure 6.23: SWAP *gate decomposition into three* iSWAP *gates (*iSWAP $= XY(\pi)$*) and three* $\sqrt{X}$ *gates.*

It is easy to verify by direct calculations that the circuit on the right side of Figure 6.23 performs the following transformations:

$$
|0\rangle \otimes |1\rangle \equiv \begin{bmatrix} 0 \\ 1 \\ 0 \\ 0 \end{bmatrix} \longrightarrow \begin{bmatrix} 0 \\ 0 \\ i \\ 0 \end{bmatrix} = \exp\left(i\frac{\pi}{2}\right) \begin{bmatrix} 0 \\ 0 \\ 1 \\ 0 \end{bmatrix} \equiv \exp\left(i\frac{\pi}{2}\right) |1\rangle \otimes |0\rangle,
$$

$$
|1\rangle \otimes |0\rangle \equiv \begin{bmatrix} 0 \\ 0 \\ 1 \\ 0 \end{bmatrix} \longrightarrow \begin{bmatrix} 0 \\ i \\ 0 \\ 0 \end{bmatrix} = \exp\left(i\frac{\pi}{2}\right) \begin{bmatrix} 0 \\ 1 \\ 0 \\ 0 \end{bmatrix} \equiv \exp\left(i\frac{\pi}{2}\right) |0\rangle \otimes |1\rangle.
$$

The coefficient $\exp(i\pi/2)$ is a *global phase* and can be ignored. We can do this because a global phase is not *observable*: measuring the states $|\psi\rangle$ and $\exp(i\phi)|\psi\rangle$ will yield the same result (i.e., the same states with the same probabilities) for any $\phi \in \mathbb{R}$. Said differently, two states differing only by a global phase represent the same physical system.

Finally, we mention the iSWAP representation of the CNOT gate. To do this, we need two iSWAP and several one-qubit gates, as shown in Figure 6.24.

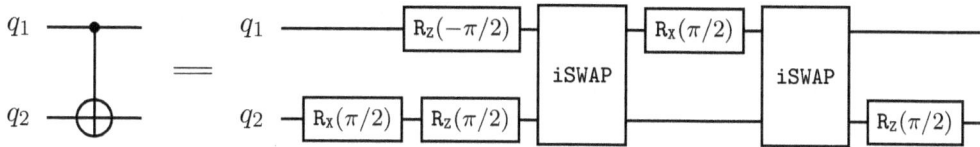

Figure 6.24: CNOT gate decomposition into two iSWAP *gates* (iSWAP $=$ XY(π)) *and several one-qubit rotation gates.*

The CNOT gate applies the NOT gate to the target qubit if the control qubit is in state $|1\rangle$ while leaving the control qubit state unchanged. This is exactly what we see when we apply the circuit shown on the right side of Figure 6.24 to the states $|10\rangle$ and $|11\rangle$:

$$|1\rangle \otimes |0\rangle \equiv \begin{bmatrix} 0 \\ 0 \\ 1 \\ 0 \end{bmatrix} \longrightarrow \begin{bmatrix} 0 \\ 0 \\ 0 \\ \frac{1-i}{\sqrt{2}} \end{bmatrix} = \exp\left(-i\frac{\pi}{4}\right) \begin{bmatrix} 0 \\ 0 \\ 0 \\ 1 \end{bmatrix} \equiv \exp\left(-i\frac{\pi}{4}\right) |1\rangle \otimes |1\rangle,$$

$$|1\rangle \otimes |1\rangle \equiv \begin{bmatrix} 0 \\ 0 \\ 0 \\ 1 \end{bmatrix} \longrightarrow \begin{bmatrix} 0 \\ 0 \\ \frac{1-i}{\sqrt{2}} \\ 0 \end{bmatrix} = \exp\left(-i\frac{\pi}{4}\right) \begin{bmatrix} 0 \\ 0 \\ 1 \\ 0 \end{bmatrix} \equiv \exp\left(-i\frac{\pi}{4}\right) |1\rangle \otimes |0\rangle.$$

Here, the accumulated unobservable global phase is $\exp\left(-i\pi/4\right)$.

Decomposition of non-native two-qubit gates into the subcircuits consisting of the native two-qubit gates and high fidelity one-qubit gates allows us to build hardware-independent quantum algorithms.

Mimicking the previous setup for classical logic gates, we now investigate how qubits and quantum logic gates can be effectively (physically) realised.

6.7 Physical Realisations of Qubits and Quantum Gates

Now that the theoretical framework for quantum bits and quantum gates has been set, it is important to understand how these can actually be realised from a hardware point of view.

6.7.1 The DiVincenzo criteria

The modern approach to building quantum computing hardware was marked by the set of requirements for the physical implementation of quantum computation proposed in 2000 by DiVincenzo [87]. These requirements, now known as the "DiVincenzo criteria", are as follows:

1. A scalable physical system with well characterised qubits. A qubit being "well characterised" means the following:

- Its physical parameters should be accurately known, including the internal Hamiltonian of the qubit, which determines the qubit energy eigenstates. Typically, the ground state is taken as $|0\rangle$ and the first excited state is taken as $|1\rangle$.
- The presence of and couplings to other states of the qubit.
- The couplings to external fields, needed to manipulate the state of the qubit.
- The interactions with other qubits, needed to implement multi-qubit gates.

2. The ability to initialise the state of the system to a simple fiducial state, such as an all-zero state. This requirement arises from the clear need to initialise quantum registers to a known value before the start of the computation. Another motivation for this requirement is the fact that quantum error correction requires a continuous, fresh supply of qubits in a low-entropy state ($|0\rangle$ state). The need for a continuous supply of 0s, rather than just an initial supply, is a non-trivial problem that may rule out some of the potentially promising qubit implementations.

3. Long relevant decoherence times, much longer than the gate operation time. Decoherence can be viewed as the loss of information from a quantum system into the environment. Coupling with the environment leads to entanglement between the system and environment and the transfer of quantum information to the surroundings. As a result, the system dynamics is no longer unitary and the computation becomes irreversible (though the combined system plus environment evolves in a unitary fashion). This means that the quantum computer behaves as a classical machine. Therefore, it is important to preserve coherence for a sufficiently long time to ensure that the uniquely quantum features of this style of computation have a chance to come into play. The term "relevant" emphasises that a physical system that realises a qubit can have many decoherence times pertaining to different degrees of freedom but many of these can be irrelevant to the functioning of this system as a qubit.

4. A universal set of quantum gates. In all the physical implementations, only particular sorts of Hamiltonians can be turned on and off. In most cases, we are limited by only two-body (two-qubit) interactions. This poses a problem for a quantum computation specified with multi-qubit unitary transformations. Fortunately, these can always be re-expressed in terms of sequences of one- and two-qubit gates, and the two-qubit gates can be of just one type, which is "native" to a particular implementation (e.g., CNOT, CPHASE, or XY).

5. A qubit-specific measurement capability. This is a straightforward requirement for the efficient quantum computing process: the result of a computation must be read out, and this requires the ability to measure specific qubits.

There are many possible realisations of quantum computers satisfying the DiVincenzo criteria. The fundamental building blocks of quantum computers – qubits – can be constructed from electrons, photons, trapped ions, neutral atoms, superconducting circuits, to name just a few possibilities. Essentially, any quantum mechanical system that can exist in a superposition of two distinct states with controlled transitions between them can serve as a physical realisation of a qubit. This can be the spin of an electron ("up", "down") or the polarisation of a photon ("vertical", "horizontal"). In this section, we start with considering how the DiVincenzo criteria can be satisfied by superconducting qubits.

6.7.2   Superconducting qubits

Qubits constructed from tiny superconducting circuits are strong candidates for the scalable physical realisation of the principles of digital quantum computing. In a normal conductor, the charge carriers are individual electrons. Electrons are spin-$\frac{1}{2}$ elementary particles (fermions) satisfying the Pauli principle: no two fermions can simultaneously occupy the same state. In a superconducting circuit, the basic charge carriers are pairs of electrons (known as Cooper pairs), which are bosons (the total spin of a Cooper pair is an integer number) and can occupy the same quantum energy level. This effect is known as the Bose-Einstein condensate. The condensate wave function allows designing and measuring macroscopic quantum effects. The parameters of the superconducting circuits may be designed by setting the classical values of the electrical elements that compose them, for example, adjusting the capacitance and inductance.

This gives us a concrete idea of how to build macroscopic qubits with desired quantum properties. Let us first have a look at the system known as Quantum Harmonic Oscillator (QHO), shown schematically in Figure 6.25. Before explaining the physical aspects, let us have a look at its mathematical justification.

From classical to quantum harmonic oscillator

Consider a simple harmonic oscillator, namely, a spring on a flat frictionless surface, attached on one side to an unmovable object and on the other side to a movable one (say, a weight). In the equilibrium state, in the resting position, nothing moves. After applying some force, say, by pulling (or pushing) the movable object, the spring starts oscillating due to its restoring force F_R. Hooke's law states that this force is proportional to the extension, namely,

$$F_R(x(t)) = -kx(t),$$

starting from $x(0) = x_0 \in \mathbb{R}$, where $x(t)$ denotes the position of the spring at time t and k is the spring constant. Newton's second law of motion also states that

$$F_R(x(t)) = ma(t),$$

for $t \geq 0$, where $a(t)$ denotes the acceleration at time t, and m is the mass of the spring. Since $a(t) = \ddot{x}(t)$, combining the two equations yields, for each $t \geq 0$,

$$\ddot{x}(t) = -\frac{k}{m}x(t),$$

starting from $x(0) = x_0$, which is the equation of motion for the simple oscillator. It is a simple one-dimensional, second order, linear, ordinary differential equation, which can be solved simply as

$$x(t) = x_0 \cos(\omega t) + \frac{v_0}{\omega}\sin(\omega t), \quad \text{for all } t \geq 0, \tag{6.7.1}$$

where

$$\omega := \sqrt{\frac{k}{m}} \quad \text{and} \quad v_0 := \dot{x}(0)$$

are, respectively, the natural frequency of the oscillator and the speed. Trigonometric manipulations show that (6.7.1) can equivalently be written as

$$x(t) = \alpha \cos(\omega t - \varphi), \quad \text{for all } t \geq 0, \tag{6.7.2}$$

where

$$\alpha := \sqrt{x_0^2 + \frac{v_0^2}{\omega^2}} \quad \text{and} \quad \tan(\varphi) := \frac{v_0}{\omega x_0}.$$

Recall now that the potential energy $\mathfrak{V}$ is the energy stored in the oscillator when it is extended or compressed, i.e., (considering $x = 0$ as the equilibrium state),

$$\mathfrak{V}(x,t) = -\int_0^{x(t)} F_R(z)\mathrm{d}z = \frac{k}{2}x(t)^2.$$

The total energy of the system is then the sum of the kinetic and the potential energies:

$$\begin{aligned}
\mathfrak{E}_{\text{total}}(t) &= \mathfrak{E}_{\text{kinetic}}(t) + \mathfrak{V}(x(t)) \\
&= \frac{m}{2}v(t)^2 + \frac{k}{2}x(t)^2.
\end{aligned}$$

Using the explicit solution (6.7.2), with $\omega := \sqrt{\frac{k}{m}}$, we obtain:

$$
\begin{aligned}
\mathfrak{E}_{\text{total}}(t) &= \frac{m}{2}\left(\frac{\mathrm{d}}{\mathrm{d}t}\Big(\alpha\cos(\omega t - \varphi)\Big)\right)^2 + \frac{k}{2}\Big(\alpha\cos(\omega t - \varphi)\Big)^2 \\
&= \frac{m\alpha^2\omega^2}{2}\sin(\omega t - \varphi)^2 + \frac{m\alpha^2\omega^2}{2}\cos(\omega t - \varphi)^2 \\
&= \frac{m\alpha^2\omega^2}{2} = \frac{k\alpha^2}{2}.
\end{aligned}
$$

In this classical setting, we see that the total energy of the system can take a continuum of values. The quantum counterpart is fundamentally different and we shall see below that it is in fact quantised (giving rise to the "quantum" theory), as originally proposed by Bohr in 1913, and later detailed by Schrödinger and Heisenberg in 1926. Recall now the general form of the time-dependent Schrödinger equation describing the evolution of a system over time:

$$
i\hbar\frac{\mathrm{d}\Psi(x,t)}{\mathrm{d}t} = \mathcal{H}\Psi(x,t), \tag{6.7.3}
$$

where $\mathcal{H}$ represents the Hamiltonian of the system. Since the latter (representing the energy of the system) is the sum of the kinetic energy and the potential energy, we have

$$
\mathcal{H} = \mathfrak{E}_{\text{kinetic}} + \mathfrak{E}_{\text{potential}} = \frac{p^2}{2m} + \mathfrak{V} = -\frac{\hbar^2}{2m}\frac{\mathrm{d}^2}{\mathrm{d}x^2} + \mathfrak{V},
$$

where m is the mass of the particle, $\hbar$ is the usual Planck constant, $\mathfrak{V}$ is the potential representing the environment, and p is the momentum operator:

$$
p = -i\hbar\frac{\mathrm{d}}{\mathrm{d}x}.
$$

Plugging this Hamiltonian into (6.7.3) yields

$$
i\hbar\frac{\mathrm{d}\Psi(x,t)}{\mathrm{d}t} = \left(-\frac{\hbar^2}{2m}\frac{\mathrm{d}^2}{\mathrm{d}x^2} + \mathfrak{V}(x)\right)\Psi(x,t).
$$

Since the potential $\mathfrak{V}$ does not depend on time, separation of variables, with $\Psi(x,t) = \psi(x)u(t)$, gives

$$i\hbar\psi(x)u'(t) = \left(-\frac{\hbar^2}{2m}\psi''(x) + \mathfrak{V}(x)\psi(x)\right)u(t),$$

or else

$$i\hbar\frac{u'(t)}{u(t)} = \frac{-\frac{\hbar^2}{2m}\psi''(x) + \mathfrak{V}(x)\psi(x)}{\psi(x)}.$$

Since both sides depend on a different variable, they must be equal to a constant, say E, and we thus obtain the ordinary differential equation

$$i\hbar\frac{u'(t)}{u(t)} = E,$$

as well as the eigenvalue equation

$$\mathcal{H}\psi(x) = -\frac{\hbar^2}{2m}\psi''(x) + \mathfrak{V}(x)\psi(x) = E\psi(x). \qquad (6.7.4)$$

The first one immediately admits the solution, with normalisation $u(0) = 1$,

$$u(t) = \exp\left(-\frac{iEt}{\hbar}\right).$$

The eigenvalue equation (6.7.4) can be solved, for example, by spectral method. In fact, it can be proved (and we refer the interested reader to [282, Section 3.1] for details) that the operator $\mathcal{H}$ here admits a finite spectrum, with the set of (eigenvalues, eigenfunctions) $\{(E_n, \psi_n)\}_{n\geq 0}$ given by

$$\begin{cases} E_n = \left(n+\frac{1}{2}\right)\hbar\omega, \\ \psi_n(x) = \frac{1}{\sqrt{2^n n!}}\left(\frac{m\omega}{\pi\hbar}\right)^{1/4}\exp\left(-\frac{m\omega x^2}{2\hbar}\right)H_n\left(\sqrt{\frac{m\omega}{\hbar}}x\right), \end{cases} \qquad (6.7.5)$$

for each $n \geq 0$, $x \in \mathbb{R}$, where H_n denotes the n-th physicists' Hermite polynomial

$$H_n(z) := (-1)^n e^{z^2} \frac{d^n}{dz^n} \left(e^{-z^2} \right).$$

Physical representation of the QHO

A schematic representation of the QHO is shown in Figure 6.25. In this system, energy oscillates between electrical energy in the capacitor C and magnetic energy in the inductor L.

a) QHO circuit   b) QHO energy levels

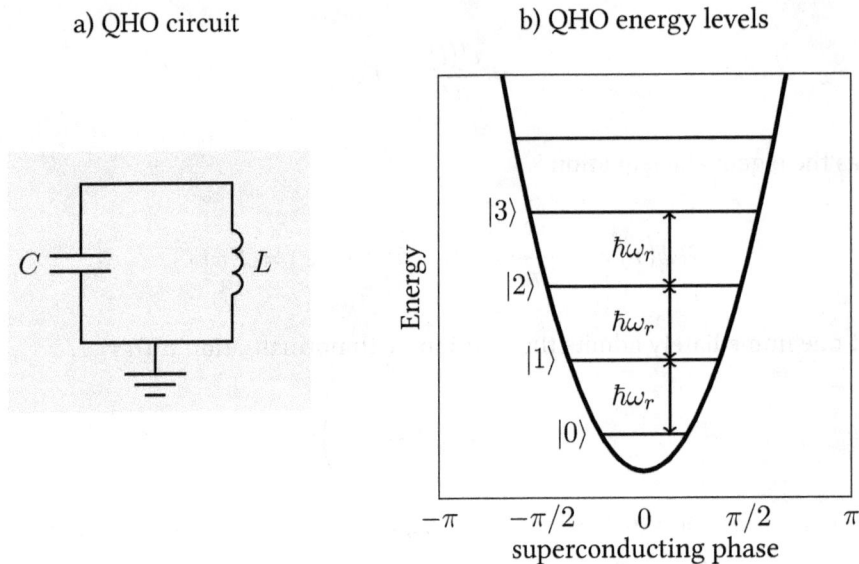

Figure 6.25: Quantum Harmonic Oscillator (QHO).

The Hamiltonian of this system is identical to the one describing a particle in a one-dimensional quadratic potential [191]. The solution to the eigenvalue problem above (see (6.7.5)) gives an infinite series of eigenstates $(|n\rangle)_{n \geq 0}$, whose corresponding eigenenergies, $(E_n)_{n \geq 0}$, are all equidistantly spaced, as can be seem from (6.7.5):

$$E_{n+1} - E_n = \hbar \omega_r, \quad \text{for all } n \geq 0,$$

where the resonant frequency, ω_r, is given by the Thomson formula [308]

$$\omega_r = \frac{1}{\sqrt{LC}}.$$

Our first task is to define a computational subspace consisting of only two energy states, $|0\rangle$ and $|1\rangle$, usually the lowest two energy eigenstates in between which transitions can be driven without also exciting other levels in the system. However, we cannot use the lowest two energy eigenstates of the QHO for this purpose since the quantum logic gate operations depend on frequency selectivity. The equidistant level-spacing of the QHO makes this impossible.

Therefore, we need to add anharmonicity (or non-linearity) into the system. We require the transition frequencies ω_{01} between eigenstates $|0\rangle$ and $|1\rangle$ and ω_{12} between eigenstates $|1\rangle$ and $|2\rangle$ to be sufficiently different in order to be individually addressable. The required non-linearity can be introduced by replacing the inductor L with the Josephson junction module J, as shown schematically in Figure 6.26.

a) QAO circuit   b) QAO energy levels

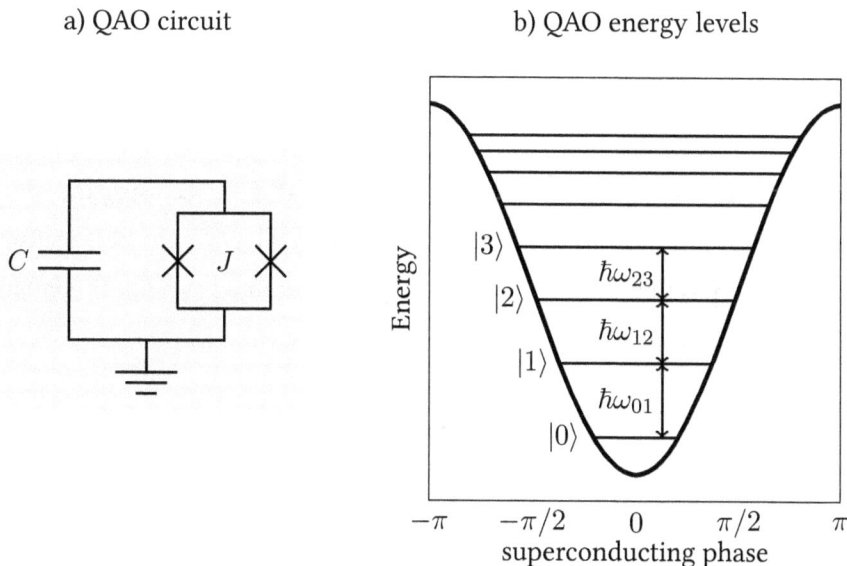

Figure 6.26: *Qubit implemented as Quantum Anharmonic Oscillator (QAO). The two lowest energy eigenstates* $|0\rangle$ *and* $|1\rangle$ *form the qubit's computational space.*

The Josephson junction is the key element that transforms a superconducting circuit into a qubit. The description of the Josephson effect (the quantum tunnelling of the Cooper pairs) is outside the scope of this book but interested readers are encouraged to learn more about it from the excellent Feynman's Lectures on Physics [101].

After introducing the Josephson module to the circuit (the electric circuit symbol for the Josephson junction is an "X"), the potential energy no longer has the parabolic form (as a function of the superconducting phase), but rather takes a cosinusoidal form, which makes the energy spectrum non-equidistant. Now we can identify the two lowest energy eigenstates as a qubit computational subspace [191, 219].

A superconducting loop with two Josephson junctions in either arm is very sensitive to the magnetic flux enclosed [126]. In the following, we shall use a more compact symbol for the Josephson junction subcircuit:

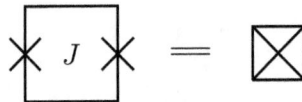

Figure 6.27: Josephson junction module subcircuit.

Remark: It may also be possible to form a computational subspace using the three lowest energy eigenstates: $|0\rangle$, $|1\rangle$, and $|2\rangle$. In this case, we would have a physical realisation of a *qutrit*, whose superposition state vector, $|\psi\rangle$, can be represented as a linear combination of the three orthonormal basis states:

$$|\psi\rangle = \alpha |0\rangle + \beta |1\rangle + \gamma |2\rangle,$$

where $\alpha, \beta, \gamma \in \mathbb{C}$ are probability amplitudes such that $|\alpha|^2 + |\beta|^2 + |\gamma|^2 = 1$. Qutrits increase the amount of information encoded in a single element, enable techniques that decrease readout errors [215], and reduce the cost of decomposing three-qubit gates into basic two-qubit components [140].

Controlling and measuring superconducting qubits

Having satisfied the first DiVincenzo requirement (a well-defined qubit), we have to demonstrate how superconducting qubits can be controlled, coupled together to build scalable systems, and measured. We start with the control and measurement of the superconducting qubit states.

The capacitive coupling between a resonator (or a feedline) and the superconducting qubit allows for microwave control to implement single-qubit rotations as well as certain two-qubit gates [191]. Figure 6.28 provides a schematic representation of the superconducting qubit coupled to a microwave source (also referred to as a *qubit drive*). The qubit is controlled by the pulses of microwave radiation. The control parameters are the frequency, the phase, and the duration of the pulses.

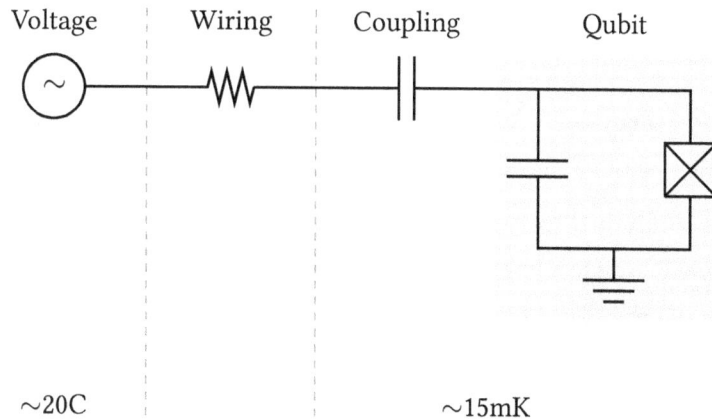

Figure 6.28: Qubit capacitively coupled to the feedline.

Figure 6.29 shows the qubit that is capacitively coupled to a microwave resonator (nonlinear) whose frequency is shifted by the qubit state [277]. This frequency shift is exploited for reading the qubit state using the dispersive readout method. When sending a microwave pulse to the resonator, the phase of the reflected (or transmitted) signal conveys information on the qubit state.

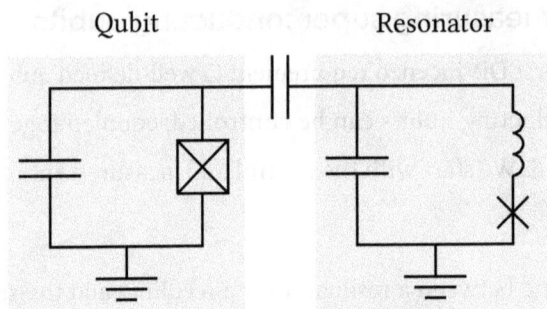

Figure 6.29: Qubit readout circuit that features a non-linear resonator with a Josephson junction.

Entanglement with superconducting qubits

For the implementation of multi-qubit gates (and, therefore, entanglement), qubits must be connected. The connectivity of superconducting qubits is realised via capacitive coupling – either directly or with the help of a coupler, as shown schematically in Figure 6.30, where capacitive coupling is achieved via a coupler in the form of a linear resonator [191].

The fixed-frequency superconducting qubits typically feature longer coherence times and are less sensitive to flux noise. The two-qubit gate developed for these qubits is the cross-resonance gate CR. In the schematic circuit diagram of two fixed-frequency superconducting qubits coupled through a linear resonator (Figure 6.30), the CR gate is realised when qubit 1 is driven at the frequency of qubit 2.

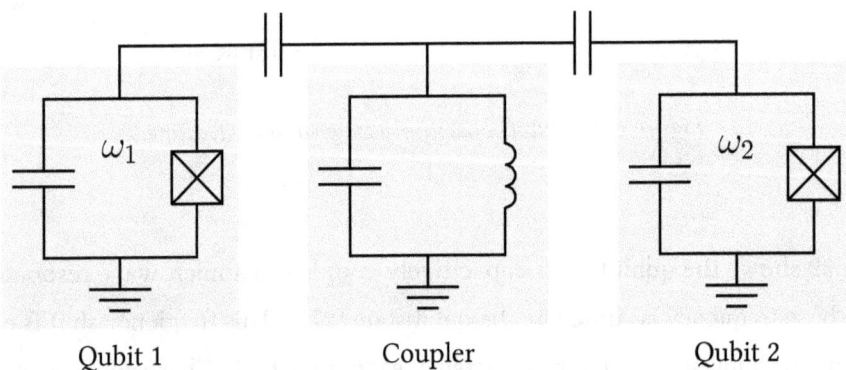

Figure 6.30: Capacitive coupling via coupler (linear resonator). Qubit 1 and Qubit 2 are fixed-frequency qubits with frequencies ω_1 and ω_2, respectively.

The unitary matrix representation of the $\mathtt{CR}(\theta)$ gate is given by the following:

$$\mathtt{CR}(\theta) = \mathtt{R_{ZX}}(\theta) = \exp\left(-\frac{1}{2}i\theta\mathtt{ZX}\right)$$

$$= \begin{bmatrix} \cos\left(\frac{\theta}{2}\right) & -i\sin\left(\frac{\theta}{2}\right) & 0 & 0 \\ -i\sin\left(\frac{\theta}{2}\right) & \cos\left(\frac{\theta}{2}\right) & 0 & 0 \\ 0 & 0 & \cos\left(\frac{\theta}{2}\right) & i\sin\left(\frac{\theta}{2}\right) \\ 0 & 0 & i\sin\left(\frac{\theta}{2}\right) & \cos\left(\frac{\theta}{2}\right) \end{bmatrix}, \qquad (6.7.6)$$

where the effective rotation angle θ is a function of the physical characteristics of the qubits, the coupler, and the driving microwave pulse.

Due to the form of (6.7.6), the $\mathtt{CR}(\theta)$ gate can also be denoted as the $\mathtt{ZX}(\theta)$ gate. This also tells us how we can use the cross-resonance gate to generate a $\mathtt{CNOT}$ gate (up to a global phase $\exp(-i\pi/4)$) in combination with only one-qubit gates:

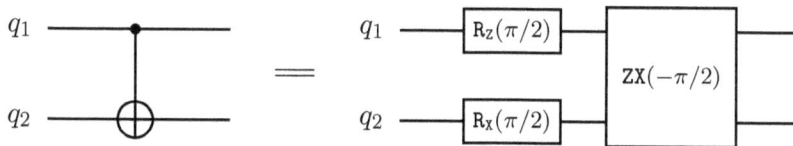

Figure 6.31: *CNOT gate decomposition into* ZX, Rx, *and* Rz *gates.*

6.7.3 Photonic qubits

At the time of writing, it is not clear which qubit construction technology will become the industry standard, if any. There are a lot of exciting experiments and technological breakthroughs ahead of us. The superconducting qubits clearly satisfy the DiVincenzo criteria but there are many other interesting solutions at various stages of development.

One of the possibilities is to encode qubits in photons. Single photons are largely free of noise and can be easily manipulated to realise one-qubit gates. A qubit can be encoded in any of multiple photons' degrees of freedom: temporal, path, and polarisation. One-qubit gates can be implemented using *birefringent waveplates* and conversion between polarisation and path encoding can be achieved using a *polarising beam splitter* [240], where

$|0\rangle$ or $|1\rangle$ represents a photon in the upper or lower path, respectively (see Figure 6.33).

As we know, in order to entangle qubits, we need to find a suitable physical implementation of two-qubit gates. Let us have a look at the possible realisation of the CNOT gate using photonic qubits. First of all, we notice that the CNOT gate can be expressed in terms of the CPHASE gate, which can be naturally implemented on the photonics hardware:

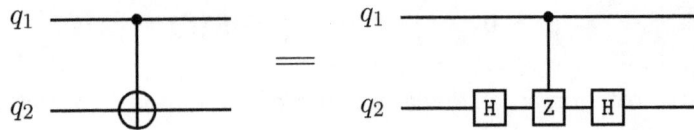

Figure 6.32: CNOT gate decomposition into CPHASE and H gates.

When the control qubit is in state $|0\rangle$, the two H gates cancel each other, and when it is in state $|1\rangle$, the combination of the gates acts as a NOT gate.

Figure 6.33 displays a schematic representation of the possible photonic implementation of the CNOT gate [240]. The two paths used to encode the target qubit are mixed at a 50% reflecting beam splitter (BS) that performs the Hadamard operation. If the phase shift is not applied, the second beam splitter (second Hadamard gate) undoes the first by returning the target qubit to the same state it started in. This is an example of classical interference. If a π phase shift is applied, the target qubit is flipped.

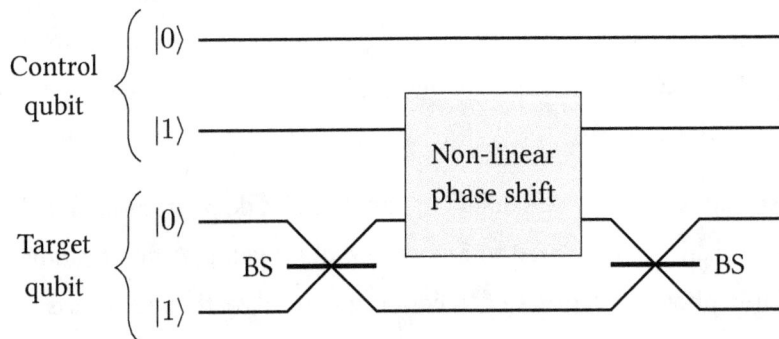

Figure 6.33: Photonic implementation of the CNOT gate.

When the control qubit is in state $|0\rangle$, the π-phase shift is not applied, while it is applied when the control qubit is in state $|1\rangle$. A CNOT gate must implement this phase shift when the control qubit is in the $|1\rangle$ path, otherwise not.

Although the proposed realisation of the CNOT gate is simple in principle, it is a hard practical problem to find a material with optical non-linearity strong enough to implement the conditional phase shift [240]. However, it is possible to achieve a CNOT gate with the help of single photon sources, single photon detectors, and linear optical circuits consisting of beam splitters, as was proposed by Knill, Laflamme, and Milburn [179]. The first integrated photonic CNOT gate for polarisation-encoded qubits was demonstrated in [74].

6.7.4 Trapped ion qubits

Another promising approach to building large-scale quantum computers is based on the trapped ion technology [68]. Ions (positively charged atoms that lost an electron) are *trapped* in the electromagnetic field potential, which fixes their positions in space. The quantum chip is cooled and placed in the vacuum chamber. The ions themselves are cooled and made almost motionless by the laser beams that drain their energy through the rapid absorption-emission of photons (ions emit photons of slightly higher frequency than absorbed photons, thus losing their kinetic energy).

The most widely used technique is a *linear trap*, shown in Figure 6.34. The two states of the i-th qubit can be identified with the internal states of the corresponding ion: a ground state $|g\rangle_i$ and an excited state $|e\rangle_i$. The trapped ions do not sit perfectly still but can oscillate around their equilibrium positions. Figure 6.34 depicts a situation where N ions are confined in a linear trap and interact with different laser beams in standing wave configurations [67].

The confinement of the motion along the x, y, and z axes can be described by a harmonic potential of frequencies $\omega_x \ll \omega_y$, and ω_z, respectively. Additionally, the Coulomb repulsion between the positively charged ions provides the coupling of the motion of the ions along the x axis. The collective motion (excitation) along the x axis, if present, behaves as a quasiparticle called a *phonon*. We denote the state of the Centre-of-Mass (CM) mode of N

ions moving in the x direction as $|0\rangle$ (no phonon) or $|1\rangle$ (one phonon).

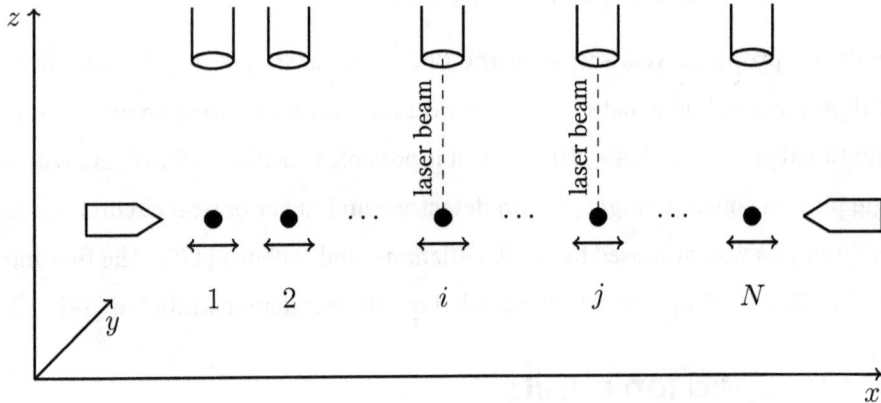

Figure 6.34: Schematic representation of the linear ion trap.

By applying the laser beam at the right frequency, it is possible to exclusively excite either a single ion or the CM mode. Addressing a single ion (and, thus, implementing a one-qubit gate) is straightforward. Let us see how we can implement a multi-qubit gate needed to create entanglement.

The following protocol that implements a two-qubit gate was proposed by Cirac and Zoller [68]. First, we note that the excited state $|e\rangle_i$ is not unique and depends on the polarisation of the laser beam applied to ion i. If we have two possible polarisations, which we denote as $q = 0$ and $q = 1$, then the corresponding excited states are denoted as $|e_0\rangle_i$ and $|e_1\rangle_i$. The computational basis is $\{|g\rangle_i, |e_0\rangle_i\}$.

The protocol reads as follows:

1: Apply a π laser pulse with polarisation $q = 0$ to excite the i-th ion. The π laser pulse has the meaning of a laser pulse applied for time π/ω, where ω is a characteristic frequency of the trapped ion system.

2: Direct the laser to the j-th ion and turn it on for a time of a 2π pulse with polarisation $q = 1$.

3: Direct the laser back to the i-th ion and turn it on for a time of a π pulse with polarisation $q = 0$.

The effect of this procedure is to change the sign of the state only when both ions are initially excited, as shown in Table 6.5.

Initial state	State after step 1	State after step 2	State after step 3
$|g\rangle_i |g\rangle_j |0\rangle$	$|g\rangle_i |g\rangle_j |0\rangle$	$|g\rangle_i |g\rangle_j |0\rangle$	$|g\rangle_i |g\rangle_j |0\rangle$
$|g\rangle_i |e_0\rangle_j |0\rangle$	$|g\rangle_i |e_0\rangle_j |0\rangle$	$|g\rangle_i |e_0\rangle_j |0\rangle$	$|g\rangle_i |e_0\rangle_j |0\rangle$
$|e_0\rangle_i |g\rangle_j |0\rangle$	$-\mathrm{i}|g\rangle_i |g\rangle_j |1\rangle$	$\mathrm{i}|g\rangle_i |g\rangle_j |1\rangle$	$|e_0\rangle_i |g\rangle_j |0\rangle$
$|e_0\rangle_i |e_0\rangle_j |0\rangle$	$-\mathrm{i}|g\rangle_i |e_0\rangle_j |1\rangle$	$-\mathrm{i}|g\rangle_i |e_0\rangle_j |1\rangle$	$-|e_0\rangle_i |e_0\rangle_j |0\rangle$

Table 6.5: Two-qubit gate (CPHASE) with the trapped ion qubits.

We note that the state of the CM mode is restored to the initial state $|0\rangle$ (no phonon) after the process. The protocol realises the CPHASE two-qubit gate.

Trapped ion-based qubits are characterised by a longer coherence time (how long the quantum state survives) and higher fidelity (accuracy of the gate operations) in comparison with the superconducting qubits. On the flip side, superconducting qubits enjoy orders of magnitude shorter gate times.

Similar to many possible physical realisations of a classical bit, there exist many competing quantum computing technologies. It is too early to say which one will become the ultimate winner in the long run, or whether multiple technologies will co-exist by occupying their respective niches.

We now have all the tools, theoretical and physical, to actually build quantum circuits. In the next section, we discuss the quantum hardware and quantum simulators that would allow us to do it.

6.8   Quantum Hardware and Simulators

The current state-of-the-art quantum computing technology demonstrates impressive qubit fidelity and coherence time:

- Qubits made of superconducting circuits (coherence time: $\sim 10\mu s$) [158, 178]

one-qubit gate		two-qubit gate	
Gate time:	$\sim 10^{-2}\mu s$	Gate time:	$\sim 10^{-2}$–$10^{-1}\mu s$
Fidelity:	99.9%	Fidelity:	99.9%

Table 6.6: Superconducting qubits.

- Qubits made of trapped ions (coherence time: $> 10^7 \mu s$) [46]

one-qubit gate		two-qubit gate	
Gate time:	~ 1–$10\mu s$	Gate time:	$\sim 10\mu s$
Fidelity:	99.9999%	Fidelity:	99.9%

Table 6.7: Trapped ion qubits.

Even more importantly, the pace of technological improvements remains very fast. Recent experiments conducted by MIT researchers on superconducting qubits [301] demonstrated a possibility to sharply reduce errors in two-qubit gates, bringing two-qubit gate fidelity for CZ and iSWAP gates to near 99.9%. Interestingly, this improvement in two-qubit gate fidelity has been achieved through the introduction of tunable couplers (Figure 6.30 schematically shows a two-qubit interaction via a coupler). To eliminate the error-generating qubit-qubit interactions, the higher energy levels of the coupler were used to cancel out the problematic interactions. As was schematically shown in Figure 6.26, the higher energy levels are usually ignored, although they have a non-negligible contribution. Better control and design of the coupler is key to tailoring the qubit-qubit interaction as needed.

This is a big step toward implementing *error correction*: additional qubits can be added to improve the robustness of quantum computation. Qubit errors can be actively addressed by adding redundancy. However, in order for the hardware redundancy to be practical, higher qubit fidelity is required. Different error correction protocols require different fidelity thresholds, and 99.9% two-qubit gate fidelity is not a bad place to start.

However, we are still some years away from sufficiently fault-tolerant quantum computers. This is why it is useful (and even necessary) to experiment with quantum simulators – classical computers operating according to the logic of quantum computing. There is nothing preventing classical digital computers from operating according to the laws of quantum computing except for prohibitive memory requirements. The state of an n-qubit quantum system can be stored in classical memory as 2^n probability amplitudes. This prevents most classical computers from performing quantum computing operations on more than 35–40 qubits. But it is perfectly feasible to run quantum computing programs on up to 25–30 quantum registers.

Although quantum simulators can only operate on a relatively small number of qubits, they are *ideal* quantum computers that do not suffer from any type of quantum hardware imperfections. This makes them invaluable in testing principles and small-scale versions of quantum algorithms. They can be used for the proof-of-concept and to help develop new ideas in a situation where actual quantum hardware is still too noisy and not readily available.

There are many open source quantum simulators (and even specialised quantum computing programming languages). In this book, we investigate the performance of various quantum algorithms using `Qiskit` [260] – an open source Python package that implements the logic of quantum computing in an intuitive and user-friendly way. `Qiskit` also owes its popularity to a well-written and highly educational textbook with many well-thought-out examples that make it a pleasure to learn the principles of quantum computing. The `Qiskit` package and related tutorials are available at

```
https://www.ibm.com/quantum/qiskit
```

and

```
https://learning.quantum.ibm.com/
```

The field of quantum programming is growing fast and several languages or software development kits are now competing. Quantum instruction sets such as `Quil`, `cQASM`, `OpenQASM`, and `Blackbird` allow us to translate high-level algorithms into physical instructions run on quantum computers. They are used in quantum software development kits (QSDKs) to represent quantum circuits. The most important QSDKs at the time of writing are as follows:

- `Ocean` (D-Wave)
- `PyQuil` (Rigetti)
- `Qiskit` (IBM)
- `Cirq` (Google)
- `Quantum Development Kit` (Microsoft)
- `Braket SDK` (Amazon)
- `ProjectQ` (ETH Zurich)
- `PennyLane` (Xanadu)

Parallel to these QSDKs, quantum programming languages have been developed, both of the imperative type (step-by-step instructions) such as `QCL`, `QMASM`, and `Silq` and of the functional type such as `QML`, `Quantum Lambda Calculus`, `QFC`, `QPL`, and `Q♯`.

> We have observed an exceptionally fast pace of quantum computing hardware development over the last several years, with multiple technological breakthroughs. Additionally, the progress on the quantum software development side assists in relaxing requirements for the physical qubits needed to build fault-tolerant quantum computers.

Summary

In this chapter, we covered and contrasted the basic elements of classical and quantum computing. We started with the concept of a fundamental memory unit (bit) and functions that transform the memory states (logic gates). We also provided examples of the possible physical realisations of the logic gates and a memory cell – this highlights the fundamental dualism of both classical and quantum computing: computation is substrate independent but its practical realisation requires the existence of a suitable physical system.

Then, we introduced the concept of the qubit and its canonical mathematical representation. Visualisation of a qubit with the help of the Bloch sphere allows for the natural representation of single qubit quantum gates as rotation operators.

Next, we studied two-qubit gates and their matrix representation. We learned how to assemble one-qubit and multi-qubit gates into quantum circuits – a good example is the Bell circuit, which creates a maximally entangled state of two qubits from the completely unentangled initial states. We also touched on the important topics of reversible computing and cutting-edge quantum hardware.

In the next chapter, we will introduce a particular type of quantum circuit, so-called Parameterised Quantum Circuits, that provide great flexibility for applications. We will also explore various data encoding schemes – the mapping of samples from the classical datasets into the corresponding quantum states.

7

Parameterised Quantum Circuits and Data Encoding

Having built the quantum hardware, how can we use it to the maximum effect given its scale, connectivity, and fidelity rate? This question can be best answered if we split it into two parts. First, what problems are in principle solvable on NISQ computers? Second, how do we encode classical data into quantum states?

The rest of this book focuses on the first part: problems and models that can be formulated in a way that doesn't require a massive number of qubits and that are, at least to some extent, noise tolerant. The first step in this direction is the concept of the Parameterised Quantum Circuit (PQC) as a generic quantum machine learning model.

The second part – data encoding – is equally important and relies on several practical methods described in this chapter. This is an active area of research where we can expect most of the progress to come from the quantum software side.

7.1 Parameterised Quantum Circuits

We have seen how to combine quantum gates to form arbitrarily wide and deep quantum circuits. A quantum circuit transforms an initial quantum state, $|\psi\rangle$, into a final quantum state $|\psi'\rangle$ by applying a sequence of unitary operators:

$$|\psi'\rangle = U_m(\theta_m) \ldots U_2(\theta_2) U_1(\theta_1) |\psi\rangle \,.$$

Here, $(U_i)_{i=1,\ldots,m}$ and $(\theta_i)_{i=1,\ldots,m}$ denote, respectively, the individual gates and the associated vectors of gate parameters. Some gates may be *fixed* (e.g., a two-qubit CNOT gate viewed as a controlled rotation of the target qubit state around the x axis by a fixed angle $\theta = \pi$), while some gates may be *adjustable* (e.g., a one-qubit $R_X(\theta)$ gate that rotates the qubit state around the x axis by an arbitrary angle $\theta \in [-\pi, \pi]$).

Once the final quantum state $|\psi'\rangle$ is constructed, the individual qubits can be measured. After measurement, the qubit states stay the same in the basis in which they were measured, which we always assume to be the standard computational basis unless explicitly specified otherwise. Therefore, the final output of running the quantum circuit and then measuring the qubits (not necessarily all qubits have to be measured) is a classical bitstring.

What we just described is a parameterised quantum circuit, schematically shown in Figure 7.1. The PQC can be used in many different ways. First of all, let us note that the PQC can be *trained*. Training the PQC has the meaning of finding an optimal set of adjustable parameters (the vectors $\theta_1, \ldots, \theta_m$ above, for example) given the overall PQC *ansatz* (architecture). The meaning of "optimal" is problem dependent but generally means a configuration of adjustable parameters that ensures maximum closeness of the final quantum state $|\psi'\rangle$ to some desired target quantum state that corresponds to a particular probability distribution we aim to encode.

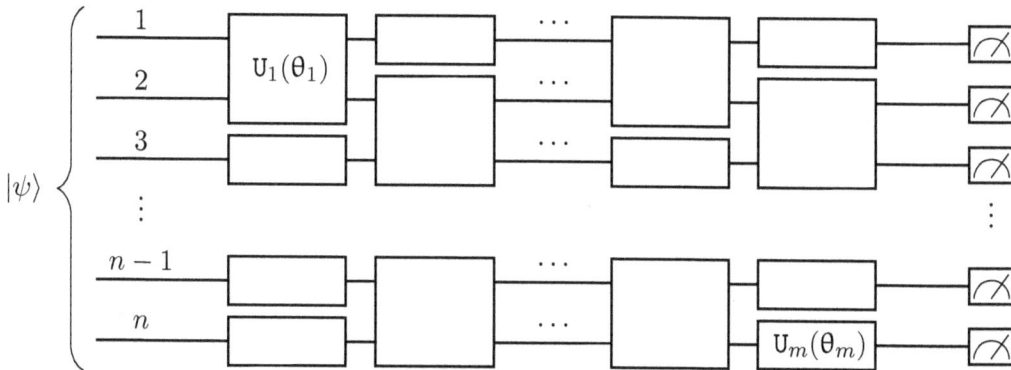

Figure 7.1: Schematic representation of a Parameterised Quantum Circuit.

In the following chapters, we will see how PQCs can be used as discriminative ML models (Chapter 8), as generative ML models (Chapter 9), and as optimisers (Chapters 10 and 11):

- In the case of the Variational Quantum Eigensolver, a PQC is used to construct the final quantum state $|\psi'\rangle$ that is close to the eigenstate of the problem Hamiltonian that corresponds to the smallest eigenvalue (the ground state energy that is linked to the minimum of the cost function).

- When we build a quantum discriminative model – a Quantum Neural Network trained as a classifier – we are interested in measuring only a handful of qubits (or even just a single qubit). This should give us the binary representation of the "class label" for the given sample. The input (initial quantum state $|\psi\rangle$) encodes the sample we want to classify.

- When our objective is to build a quantum generative model – the Quantum Circuit Born Machine – we measure all qubits. This gives us a bitstring that is a generated sample from the probability distribution encoded in the final quantum state $|\psi'\rangle$ constructed by the PQC. The initial state is initialised as $|0\rangle^{\otimes n}$.

PQCs are invariably trained using hybrid quantum-classical protocols. The hybrid approach is shown schematically in Figure 7.2 and consists of three components: the user, the classical computer, and the quantum computer [28].

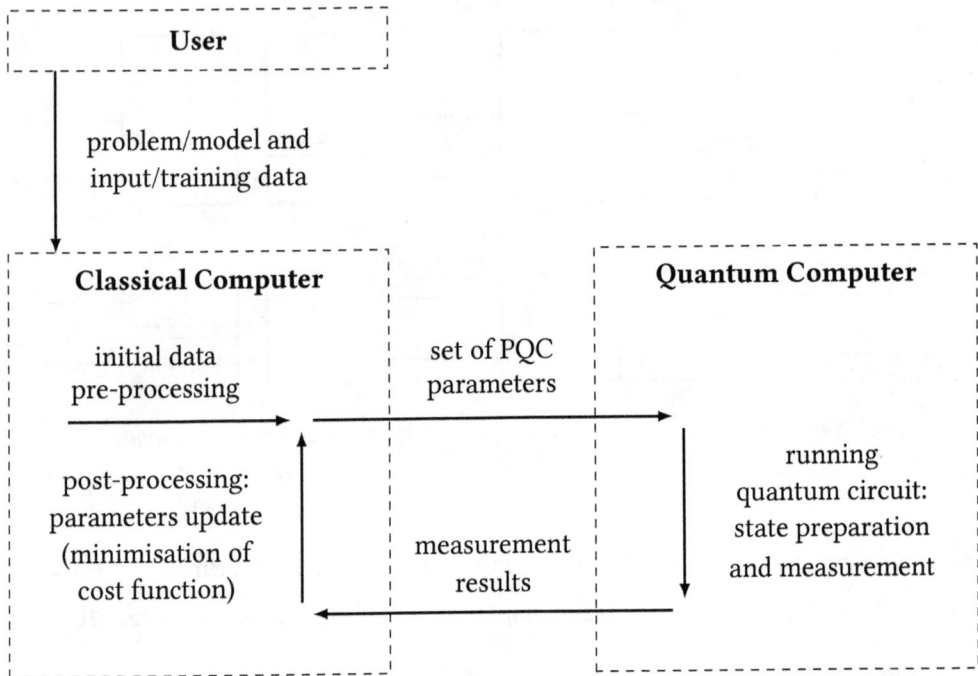

Figure 7.2: Training PQCs – schematic process.

The user provides the model for the problem; the classical computer pre-processes the data and produces the initial set of parameters for the PQC; the quantum computer runs the PQC by preparing the quantum state as prescribed by the PQC and by performing measurements. Measurement outcomes are then post-processed by the classical computer, which updates the model parameters as per the chosen training algorithm (backpropagation of error with gradient descent, non-differentiable learning method, etc.) The overall algorithm is run in a closed loop between the classical and quantum computers.

PQC is one of the most successful frameworks for applying NISQ computers to solve non-trivial real-world problems. It follows the paradigm of a hybrid quantum-classical computational protocol and can be used to experiment with the wide range of quantum machine learning models.

7.2 Angle Encoding

Let us go back to the Bloch sphere (Figure 7.3) that visualises the canonical representation of the qubit state – a unit vector in the two-dimensional complex vector space:

$$|\psi\rangle = \begin{bmatrix} \cos\left(\dfrac{\theta}{2}\right) \\ e^{i\phi} \sin\left(\dfrac{\theta}{2}\right) \end{bmatrix}.$$

The angles $\theta \in [0, \pi]$ and $\phi \in [0, 2\pi]$ uniquely determine the position of the qubit on the unit sphere. Since we need two continuous variables to specify the qubit state, a single qubit can encode two real-valued features.

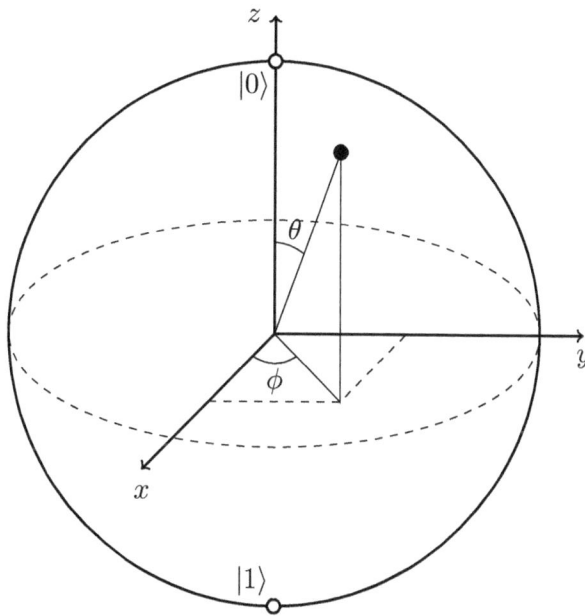

Figure 7.3: Quantum state $|\psi\rangle$ on the Bloch sphere.

7.2.1 The basic encoding scheme

We can illustrate this with the following schematic example: let us assume that we have an 8-feature dataset consisting of N samples and that all features $X_1, \ldots, X_8$ are real-valued and such that their extremal values, $X_i^{\min}$ and $X_i^{\max}$, $i = 1, \ldots, 8$, can be computed. Then, for every sample $j = 1, \ldots, N$ from the dataset, we can establish a one-to-one mapping between the values of the features X_i^j and the corresponding rotation angles θ_i^j:

$$\theta_i^j = \frac{X_i^j - X_i^{\min}}{X_i^{\max} - X_i^{\min}} \pi, \tag{7.2.1}$$

where $X_i^{\min} := \min_j X_i^j$ and $X_i^{\max} := \max_j X_i^j$. The rotation angles θ_i^j generalise angles θ and ϕ in Figure 7.3.

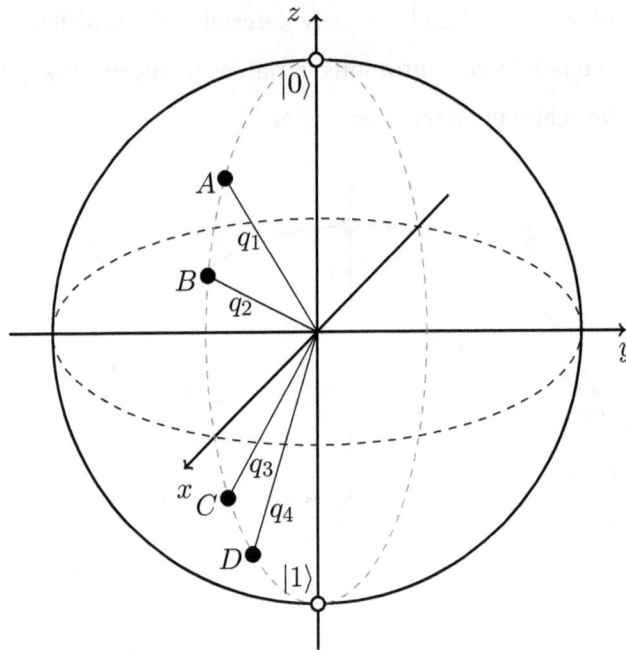

Figure 7.4: Feature encoding – rotations around the y axis. The states of qubits $q_1, \ldots, q_4$ are shown on the same Bloch sphere. The initial state of all qubits is $|0\rangle$ and the end states after rotations around the y axis by angles $\theta_1, \ldots, \theta_4$ are denoted as A, B, C, and D.

The 8-feature sample can be encoded in a 4-qubit state (unentangled). For example, starting with four quantum registers initialised as $|0\rangle$ in the computational basis, we can first perform rotations around the y axis: rotation by θ_1 for qubit 1, rotation by θ_2 for qubit 2, and so on. This is shown schematically in Figure 7.4, where qubits move from their initial state $|0\rangle$ to states A, B, C, and D.

7.2.2 Encoding two features per quantum register

After that, we encode the remaining features by performing rotations around the z axis: rotation by θ_5 for qubit 1, rotation by θ_6 for qubit 2, and so on as shown in Figure 7.5.

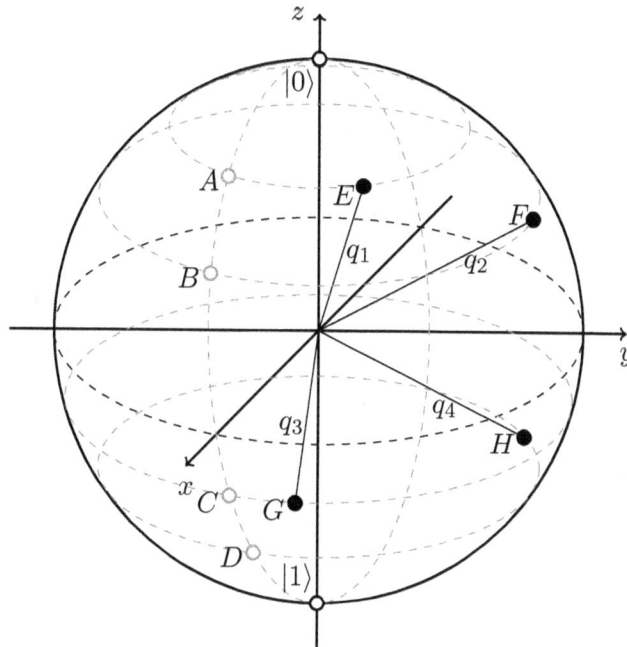

Figure 7.5: Feature encoding – rotations around the z axis. The initial states of qubits $q_1, \ldots, q_4$ are A, B, C, D. After rotation around the z axis by angles $\theta_5, \ldots, \theta_8$, the final qubit states are E, F, G, H.

The qubit states move from A to E, from B to F, from C to G, and from D to H. The corresponding quantum circuit looks as follows:

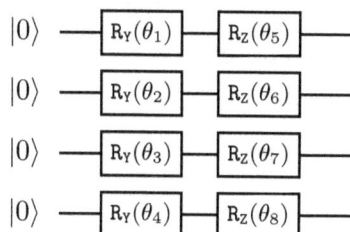

Figure 7.6: 4-qubit quantum circuit for 8-feature sample encoding.

7.2.3   Mapping a classical data sample into a quantum state

We can summarise the angle encoding scheme as follows. For the classical data sample $X^j := (X_1^j, \ldots, X_K^j) \in \mathbb{R}^K$, $j = 1, \ldots, N$, angle encoding works by constructing the map

$$X^j \longmapsto \bigotimes_{i=1}^{K} \left(\cos \left(\frac{\theta_i^j}{2} \right) |0\rangle + \sin \left(\frac{\theta_i^j}{2} \right) |1\rangle \right),$$

where angles $(\theta_i^j)_{i=1,\ldots,K;j=1,\ldots,N}$ are given by the expression (7.2.1). This scheme only requires one rotation gate for each qubit, hence it encodes as many features as the number of qubits. However, we know that a single quantum register can encode two real variables. The following scheme maps the classical sample into the quantum state with the help of an extra phase gate:

$$X^j \longmapsto \bigotimes_{i=1}^{K} \left(\cos \left(\frac{\theta_{2i-1}^j}{2} \right) |0\rangle + \exp \left(i\theta_{2i} \right) \sin \left(\frac{\theta_{2i-1}^j}{2} \right) |1\rangle \right).$$

This scheme allows us to encode $2n$ features with n qubits.

> n quantum registers have capacity to encode $2n$ continuous features with just two layers of one-qubit gates.

7.3   Amplitude Encoding

So far, we have not utilised the information encoding possibilities provided by entanglement, although, in principle, most of the information in large quantum systems can be stored in correlations. In the case of our 8-feature dataset example considered in the previous section, we can reduce the number of necessary qubits to just three if we use entanglement. The first six rotation angles $\theta_1, \ldots, \theta_6$ can still be used for the single-qubit rotations R_Y, R_Z. The last two, θ_7 and θ_8, can be used for controlled rotations that entangle qubits 1 and 2 and qubits 2 and 3, as shown in Figure 7.7:

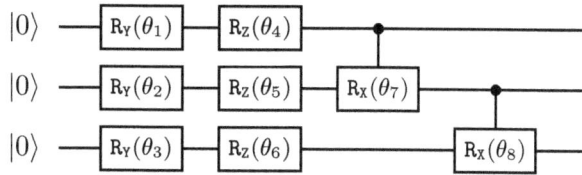

Figure 7.7: 3-qubit quantum circuit for 8-feature sample encoding.

In principle, since the n-qubit state can be uniquely described by specifying 2^n probability amplitudes, we only need n qubits to encode 2^n features. However, this superdense encoding is not always practical or desirable. The *amplitude encoding* was used in the seminal work by Harrow, Hassidim, and Lloyd [133], but obtaining the amplitude encoding is a non-trivial task for problems of realistic size, and this, usually, is the main bottleneck for many quantum algorithms [199].

The amplitude encoding can be formalised in the following way. Consider an N-dimensional (with $N = 2^n$) data point $\mathrm{x} := (x_1, \ldots, x_N) \in \mathbb{C}^N$. We can associate quantum amplitudes to the coordinates as

$$|\psi_{\mathrm{x}}\rangle = \frac{1}{\|\mathrm{x}\|} \sum_{i=1}^{N} x_i |i\rangle ,$$

where $\|\mathrm{x}\| := \sum_{i=1}^{N} |x_i|^2$ is the normalisation factor. We can therefore encode the dataset $\mathrm{D} := (\mathrm{x}^1, \ldots, \mathrm{x}^M)$ consisting of M points in $\mathbb{R}^N$ as

$$|\mathrm{D}\rangle = \frac{1}{C_{\mathrm{D}}} \sum_{i=1}^{2^p} \overline{x}_i |i\rangle ,$$

for some integer p, where

$$\overline{\mathrm{x}} = (\overline{x}_i)_{i=1,\ldots,2^p} = (x_1^1, \ldots, x_N^1, x_1^2, \ldots, x_N^2, \ldots, x_1^M, \ldots, x_N^M) \in \mathbb{R}^{MN}$$

is the concatenation of all the data points and C_{D} is a normalisation constant. Here, the constraint is hence that $2^p \geq MN$, namely $p \geq \log_2(MN)$. Note that there may again be some sparsity in the case where $2^p > MN$.

The clear advantage is that it can store 2^n features with only n qubits, but it unfortunately has a depth of $\mathcal{O}(2n)$ and is hence hard to build.

> With the amplitude encoding, n quantum registers have capacity to encode 2^n continuous features. However, this requires construction of the deep quantum circuits with the circuit depth growing as $\mathcal{O}(2n)$. This solution may not be practical for NISQ computers when dealing with a large number of features.

7.4   Binary Inputs into Basis States

Consider a real number $x \in \mathbb{R}$ approximated with the binary representation

$$x \approx \widehat{x} = (x_\mathrm{i}, x_{\mathrm{i}-1}, \cdots, x_{-\eth}) := (-1)^{x_\mathrm{i}} \left(\sum_{j=0}^{\mathrm{i}-1} x_j 2^j + \sum_{j=1}^{\eth} x_{-j} 2^{-j} \right)$$

$$\longmapsto |x_\mathrm{i} x_{\mathrm{i}-1} \cdots x_{-\eth}\rangle =: |x\rangle,$$

for some non-negative integers i and $\eth$, where $x_\mathrm{i} \in \{0, 1\}$ accounts for the sign of x and $(x_j)_{j=0,\ldots,\mathrm{i}-1} \in \{0,1\}^\mathrm{i}$ and $(x_j)_{j=-1,\ldots,-\eth} \in \{0,1\}^\eth$ represent, respectively, the integer and decimal parts of x. Consider now a vector $\mathrm{x} := (x^1, \ldots, x^N) \in \mathbb{R}^N$. We can concatenate all the binary approximations $\widehat{x}^1, \ldots, \widehat{x}^N$ into one vector

$$\left(x_\mathrm{i}^1, x_{\mathrm{i}-1}^1, \cdots, x_{-\eth}^1, \cdots, x_\mathrm{i}^N, x_{\mathrm{i}-1}^N, \cdots, x_{-\eth}^N \right) \in \{0,1\}^{(1+\mathrm{i}+\eth)N}$$

to obtain a quantum state representation with $(1 + \mathrm{i} + \eth)N$ qubits of the form

$$|x_\mathrm{i}^1 x_{\mathrm{i}-1}^1 \cdots x_{-\eth}^1 \cdots x_\mathrm{i}^N, x_{\mathrm{i}-1}^N \cdots x_{-\eth}^N\rangle.$$

Since the vector thus obtained only contains 0 and 1, starting from the quantum state $|0\rangle^{\otimes(1+\mathrm{i}+\eth)N}$, we only need to apply the NOT gate X whenever the corresponding value is equal to 1, so that the encoding circuit simply reads

$$|0\rangle^{\otimes(1+i+\partial)N} \longmapsto \bigotimes_{l=1}^{N} \bigotimes_{k=-\partial}^{i} X^{x_k^l} |0\rangle^{\otimes(1+i+\partial)N}.$$

While the algorithm is straightforward and only requires the use of the single-qubit quantum gate X, it requires a large number of qubits and is in general not efficient in practice. Indeed, for a given dimension N, there are 2^N possible basis states. If a dataset contains only M points, with M being much smaller than N, the quantum representation will therefore be sparse.

Example: Consider a dataset $D = (x^1, x^2)$ with $x^1, x^2 \in [-2, 2]$, each approximated with four qubits:

$$x^1 \approx (-1)^{x_1^1}(2^0 x_0^1 + 2^{-1}x_{-1}^1 + 2^{-2}x_{-2}^1) = (-1)^{x_1^1}\left(x_0^1 + \frac{1}{2}x_{-1}^1 + \frac{1}{4}x_{-2}^1\right)$$

and

$$x^2 \approx (-1)^{x_1^2}(2^0 x_0^2 + 2^{-1}x_{-1}^2 + 2^{-2}x_{-2}^2) = (-1)^{x_1^2}\left(x_0^2 + \frac{1}{2}x_{-1}^2 + \frac{1}{4}x_{-2}^2\right),$$

with $x_k^i \in \{0,1\}$ for each $i = 1, 2$ and $k = -2, -1, 0, 1$. Their quantum embeddings therefore read $|x_1^1 x_0^1 x_{-1}^1 x_{-2}^1\rangle$ and $|x_1^2 x_0^2 x_{-1}^2 x_{-2}^2\rangle$, and the quantum circuit to encode the dataset therefore takes the form

$$|0\rangle^{\otimes 8} \longmapsto \left[\left(X^{x_1^1} \otimes X^{x_0^1} \otimes X^{x_{-1}^1} \otimes X^{x_{-2}^1}\right) \otimes \left(X^{x_1^2} \otimes X^{x_0^2} \otimes X^{x_{-1}^2} \otimes X^{x_{-2}^2}\right)\right] |0\rangle^{\otimes 8}.$$

7.5 Superposition Encoding

As developed in [312, 319], it is possible to build such a superposition of data in time that is linear in the number of points and features. We consider again a dataset $D := (\mathrm{x}^1, \ldots, \mathrm{x}^M)$, with $\mathrm{x}^k := (x_1^k, \ldots, x_n^k) \in \{0,1\}^n$ for each $k = 1, \ldots, M$. We use a quantum system of the form

$$|\psi_0\rangle := |0\rangle^{\otimes n} |00\rangle |0\rangle^{\otimes n},$$

where the left-most part with n qubits is called the loading register while the right-most one (also with n qubits) is the storage register. The middle one is an ancilla register that will be used to control manipulations between the loading and storage registers. The encoding algorithm works recursively. We first apply an Hadamard gate to the second ancilla qubit and store the first data point x^1 into the storage register. Since

$$\left(\bigotimes_{i=1}^{n} \mathrm{X}^{x_i^1} \right) |0\rangle^{\otimes n} = |x_1^1 \cdots x_n^1\rangle = |\mathrm{x}^1\rangle,$$

this can be achieved (after the Hadamard operation) by applying the unitary operator

$$\mathrm{I}^{\otimes n} \otimes \mathrm{I} \otimes \mathrm{H} \otimes \left(\bigotimes_{i=1}^{n} \mathrm{X}^{x_i^1} \right)$$

controlled with the second ancilla qubit, and the resulting quantum state reads

$$|\psi_1\rangle := \frac{|0\rangle^{\otimes n} |00\rangle |0\rangle^{\otimes n}}{\sqrt{2}} + \frac{|0\rangle^{\otimes n} |01\rangle |\mathrm{x}^1\rangle}{\sqrt{2}}.$$

This can easily be turned (see the proof of Lemma 7 below) into

$$|\psi_1\rangle = \frac{|0\rangle^{\otimes n} |00\rangle |\mathrm{x}^1\rangle}{\sqrt{2}} + \frac{|0\rangle^{\otimes n} |01\rangle |0\rangle^{\otimes n}}{\sqrt{2}}.$$

After m steps, we arrive at a quantum state of the form

$$|\psi_m\rangle := \frac{1}{\sqrt{M}} \sum_{k=1}^{m} |0\rangle^{\otimes n} |00\rangle |\mathrm{x}^k\rangle + \sqrt{\frac{M-m}{M}} |0\rangle^{\otimes n} |01\rangle |0\rangle^{\otimes n}. \tag{7.5.1}$$

The following lemma guarantees the validity of the algorithm:

Lemma 7. *There exists a unitary operator* U *such that*

$$\mathrm{U}|\psi_m\rangle = \frac{1}{\sqrt{M}} \sum_{k=1}^{m+1} |0\rangle^{\otimes n} |00\rangle |\mathrm{x}^k\rangle + \sqrt{\frac{M-(m+1)}{M}} |0\rangle^{\otimes n} |01\rangle |0\rangle^{\otimes n} =: |\psi_{m+1}\rangle.$$

Proof. The proof is constructive and shows precisely what the operator U looks like.

1: Construct the successive maps

$$|\psi_m\rangle = \frac{1}{\sqrt{M}} \sum_{k=1}^{m} |0\rangle^{\otimes n} |00\rangle |x^k\rangle + \sqrt{\frac{M-m}{M}} |0\rangle^{\otimes n} |01\rangle |0\rangle^{\otimes n},$$

$$\longmapsto \frac{1}{\sqrt{M}} \sum_{k=1}^{m} |x^{m+1}\rangle |00\rangle |x^k\rangle + \sqrt{\frac{M-m}{M}} |x^{m+1}\rangle |01\rangle |0\rangle^{\otimes n},$$

$$\longmapsto \frac{1}{\sqrt{M}} \sum_{k=1}^{m} |x^{m+1}\rangle |00\rangle |x^k\rangle + \sqrt{\frac{M-m}{M}} |x^{m+1}\rangle |01\rangle |x^{m+1}\rangle,$$

$$\longmapsto \frac{1}{\sqrt{M}} \sum_{k=1}^{m} |x^{m+1}\rangle |00\rangle |x^k\rangle + \sqrt{\frac{M-m}{M}} |x^{m+1}\rangle |11\rangle |x^{m+1}\rangle =: |\widetilde{\psi}_m\rangle.$$

The first one is easily achieved by applying the operator $\left(\bigotimes_{i=1}^{n} X^{x_i^{m+1}}\right) \otimes I^{\otimes 2} \otimes I^{\otimes n}$ to $|\psi_m\rangle$. The second step is realised with controlled gates using the second qubit of the ancilla register as a control. The last one is trivial with a CNOT gate on the first ancilla qubit using the second ancilla qubit as the control.

2: Now define the unitary gate

$$\widetilde{U} := \frac{1}{\sqrt{M-m}} \begin{bmatrix} \sqrt{M-m-1} & 1 \\ -1 & \sqrt{M-m-1} \end{bmatrix},$$

and note that its controlled (by the first ancilla qubit a_1) version $_{a_1}\widetilde{U}$ acts as

$$_{a_1}\widetilde{U} |00\rangle = |00\rangle,$$

$$_{a_1}\widetilde{U} |11\rangle = \frac{1}{\sqrt{M-m}} |1\rangle \otimes \left(|0\rangle + \sqrt{M-m-1}\,|1\rangle\right) = \frac{|10\rangle + \sqrt{M-m-1}\,|11\rangle}{\sqrt{M-m}}.$$

Applying it to the ancilla register of $|\widetilde{\psi}_m\rangle$ in Step 1 (and leaving all other qubits unchanged) yields

$$= \mathbf{I}^{\otimes n} \otimes {}_{a_1}\widetilde{\mathbf{U}} \otimes \mathbf{I}^{\otimes n} |\widetilde{\psi}_m\rangle$$

$$= \frac{1}{\sqrt{M}} \sum_{k=1}^{m} |\mathbf{x}^{m+1}\rangle |00\rangle |\mathbf{x}^k\rangle$$

$$+ \sqrt{\frac{M-m}{M}} |\mathbf{x}^{m+1}\rangle \left\{ \frac{|10\rangle + \sqrt{M-m-1}\,|11\rangle}{\sqrt{M-m}} \right\} |\mathbf{x}^{m+1}\rangle$$

$$= \frac{1}{\sqrt{M}} \sum_{k=1}^{m} |\mathbf{x}^{m+1}\rangle |00\rangle |\mathbf{x}^k\rangle$$

$$+ \frac{1}{\sqrt{M}} |\mathbf{x}^{m+1}\rangle \left\{ |10\rangle + \sqrt{M-m-1}\,|11\rangle \right\} |\mathbf{x}^{m+1}\rangle .$$

We then flip the first ancilla qubit to 0 in the $|10\rangle$ case (easily achievable with SWAP and CNOT gates) and, regrouping the same ancilla terms together, we obtain

$$\frac{1}{\sqrt{M}} \sum_{k=1}^{m+1} |\mathbf{x}^{m+1}\rangle |00\rangle |\mathbf{x}^k\rangle + \sqrt{\frac{M-(m+1)}{M}} |\mathbf{x}^{m+1}\rangle |11\rangle |\mathbf{x}^{m+1}\rangle .$$

Resetting the registers as in (7.5.1) to obtain

$$\frac{1}{\sqrt{M}} \sum_{k=1}^{m+1} |0\rangle^{\otimes n} |00\rangle |\mathbf{x}^k\rangle + \sqrt{\frac{M-(m+1)}{M}} |0\rangle^{\otimes n} |01\rangle |0\rangle^{\otimes n}$$

finishes the proof of the lemma.

$$\square$$

7.6   Hamiltonian Simulation

Hamiltonian encoding, popular in quantum machine learning, is inspired by the Schrödinger equation (1.2.1), which reads

$$i\hbar \frac{\mathrm{d}|\psi(t)\rangle}{\mathrm{d}t} = \mathcal{H}|\psi(t)\rangle,$$

for some Hamiltonian $\mathcal{H}$, where $\hbar$ is the Planck constant, and subject to some boundary condition at $t = 0$. The solution to the equation reads

$$|\psi(t)\rangle = \exp\left(-\frac{i\mathcal{H}t}{\hbar}\right)|\psi(0)\rangle.$$

The idea of Hamiltonian encoding is to encode the initial data into the Hamiltonian $\mathcal{H}$. Consider a cloud of points $X \in \mathcal{M}_{n,n}(\mathbb{C})$. If X is Hermitian, we can define the Hamiltonian matrix $\mathcal{H}_X := X$, otherwise the augmented version

$$\mathcal{H}_X := \begin{bmatrix} \mathbf{0}_{n,n} & X \\ X^\dagger & \mathbf{0}_{n,n} \end{bmatrix}$$

is Hermitian by construction.

Our aim is, for a given precision level ε, to find a state $|\widetilde{\psi}\rangle$ (or an algorithm generating this state) such that

$$\left\||\widetilde{\psi}\rangle - |\psi(t)\rangle\right\| \leq \varepsilon,$$

for some given norm $\|\cdot\|$, where $|\psi(t)\rangle$ solves the Schrödinger equation.

An important bottleneck of this method is the computation of $\exp(\mathcal{H}_X t)$ (where we ignore the ratio $-i/\hbar$ for simplicity). Suppose that the Hamiltonian $\mathcal{H}_X$ can be written as a sum

$$\mathcal{H}_X = \sum_{i=1}^{p} \mathcal{H}_i$$

of easy-to-compute Hamiltonians $(\mathcal{H}_i)_{i=1,\dots,p}$. If the latter do not commute, then the identity

$$\exp\left(\sum_{i=1}^{p} \mathcal{H}_i t\right) = \prod_{i=1}^{p} e^{\mathcal{H}_i t}$$

does not hold, but the first-order Suzuki-Trotter [235, 311] formula (used by Lloyd in [206])

$$\exp\left(\sum_{i=1}^{p}\mathcal{H}_i t\right) = \prod_{i=1}^{p} e^{\mathcal{H}_i t} + \mathcal{O}(t^2),$$

for t small enough, allows us to bypass this issue.

Indeed, even if t is not so small, we may pick $\delta > 0$ small enough and use the factorisation

$$\exp\left(\sum_{i=1}^{p}\mathcal{H}_i t\right) = \left[\exp\left(\sum_{i=1}^{p}\mathcal{H}_i \delta\right)\right]^{t/\delta} = \left[\prod_{i=1}^{p} e^{\mathcal{H}_i \delta} + \mathcal{O}(\delta^2)\right]^{t/\delta},$$

which has a small error (albeit with the caveat that the operation needs to be computed many times). In general, any n-qubit Hamiltonian $\mathcal{H}$ can be decomposed in at most 4^n elementary Hamiltonians (of the Pauli form) as

$$\mathcal{H} = \frac{1}{2^n}\sum_{i_1,\ldots,i_n \in \{\text{I,X,Y,Z}\}} \text{Tr}\left(\bigotimes_{k=1}^{n}\sigma_{i_k}\mathcal{H}\right)\bigotimes_{k=1}^{n}\sigma_{i_k},$$

where σ_{i_k} is a Pauli operator. Of course, 4^n appeals to Pauli operators may be too large in general, but local features of the Hamiltonian (such as sparse [34] or diluted or degree-reduced [8] Hamiltonians) help reduce the complexity.

Remark: An alternative approach, especially for the QML problems analysed in the next chapter, is to encode the data using Quantum Random Access Memory (QRAM), essentially with the bucket-brigade algorithm developed in [114] (see also [15, 148]), and we refer the interested reader to [65] for a good summary of the current state-of-the-art algorithms.

Encoding classical data into a quantum computer has seen many advances recently and several competing techniques are now available depending on the problem under investigation.

Summary

In this chapter, we introduced the concept of a parameterised quantum circuit as a generic QML model. PQCs can be trained and used as discriminative and generative QML models as well as optimisers. They can also be used to encode classical data samples into the corresponding quantum states.

We considered several popular data encoding methods. Arguably, the simplest and easiest to implement is the angle encoding algorithm – we shall use this approach in the next chapter. Other methods also have their strong points, although they tend to be either more demanding in terms of the hardware capabilities or better suited for some niche applications.

In the next chapter, we apply what we have learned so far to the task of building the quantum neural network trained as a classifier and compare its performance on the binary classification problem with standard classical machine learning models.

8

Quantum Neural Network

Quantum neural networks [100] are parameterised quantum circuits that can be trained as either generative or discriminative machine learning models in direct analogy with their classical counterparts. In this chapter, we will consider parameterised quantum circuits trained as classifiers. In the most general case, a classifier is a function that takes an N-dimensional input and returns one of M possible class values. The classifier can be trained on a dataset of samples with known class labels by adjusting the configurable model parameters in such a way as to minimise the classification error. Once the classifier is fully trained, it can be exposed to new unseen samples for which correct class labels are unknown. Therefore, it is critically important to avoid overfitting to the training dataset and ensure that the classifier generalises well to the new data.

There are many similarities between quantum and classical neural networks. In both cases, the key element is the forward propagation of the signal (input), which is transformed by the network activation functions. Both quantum and classical neural networks can be trained through the backpropagation of error (differentiable learning) as well as through various non-differentiable learning techniques. However, there are also fundamental differences. For example, classical neural networks derive their power from the non-linear transformation of input. In contrast, all quantum gates are linear operators and the power of quantum neural networks comes from the mapping of the input into the high-dimensional Hilbert space where classification can more easily be done.

8.1   Quantum Neural Networks

Figure 8.1 provides a schematic representation of a typical Quantum Neural Network (QNN) trained as a classifier. Let us have a look at the quantum circuit and understand how it operates. The network consists of n quantum registers, a number of one-qubit and two-qubit gates, and m measurement operators. The input is a quantum state $|\psi_k\rangle$ encoding the k-th sample from the dataset. If our dataset is classical, then every classical sample should first be encoded in the input quantum state (as explained in the previous chapter). With m measurement operators, the output is a bitstring that can encode up to 2^m integer values (class labels). In the case of a binary classifier, it is sufficient to perform measurement on a single qubit. Alternatively, it is possible to measure several qubits, postprocess the measured bitstring, and map it to a binary value.

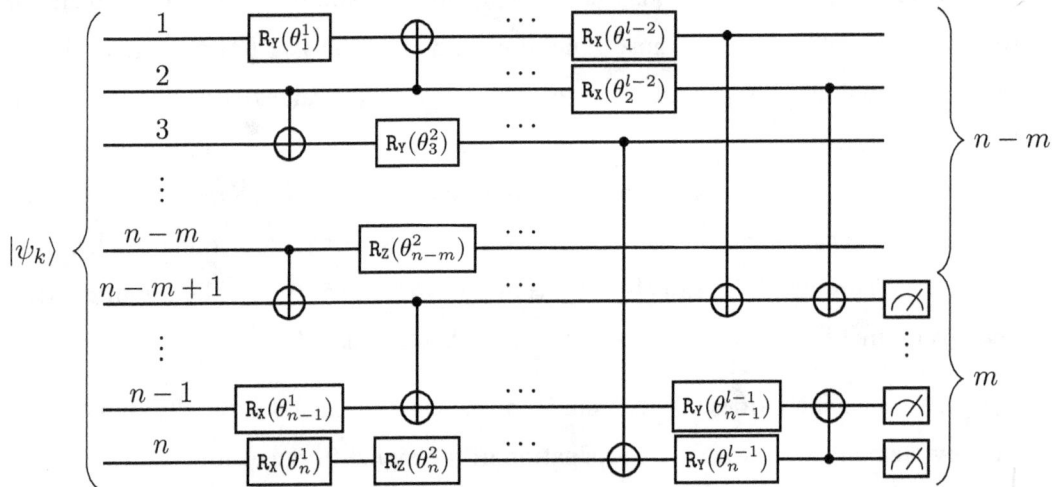

Figure 8.1: Schematic representation of a quantum neural network – parameterised quantum circuit – consisting of one-qubit and two-qubit gates and measurement operators on one or more quantum registers. The initial state $|\psi_k\rangle$ encodes the k-th sample from the dataset.

The measurement process produces a single sample from the probability distribution encoded in the quantum state. Therefore, we need to run the quantum circuit many times for the same input in order to collect sufficient statistics for each qubit on which we perform measurement.

For example, if our QNN is organised as a classifier that should be able to predict one of the four possible classes ("0", "1", "2", and "3"), then we would need to perform measurements on 2 qubits with possible outcomes $|00\rangle$ corresponding to class "0", $|01\rangle$ corresponding to class "1", $|10\rangle$ corresponding to class "2", and $|11\rangle$ corresponding to class "3". Let us assume that we have run the quantum circuit 1,000 times and observed the following results as shown in Table 8.1:

Measured bitstring	Class label	Number of observations
00	0	100
01	1	550
10	2	200
11	3	150

Table 8.1: 1,000 runs of the quantum circuit.

Then we can conclude that the most likely class label for the given input is class "1" (with a probability of 55%). At the same time, we also obtain probabilities for all other possible class values, which may be useful in some cases.

The network is organised as l layers of one-qubit and two-qubit gates. The gates can be *adjustable*, meaning that they can be controlled by adjustable parameters, such as rotation angles, or they can be *fixed*. The two-qubit gates in Figure 8.1 are fixed CX gates but, in principle, they can be adjustable controlled rotation gates. Although the network shown schematically in Figure 8.1 can have up to $n \times l$ adjustable parameters $(\theta_i^j)_{i=1,...,n;\ j=1,...,l}$, it is often the case that the two-qubit gates are fixed and we only have one-qubit rotations as available degrees of freedom in training the network.

Similar to classical neural networks, QNNs can be trained through either differentiable learning (for example, backpropagation of error with gradient descent) or non-differentiable learning (e.g., evolutionary search heuristics). Both approaches have their relative strengths and weaknesses. In theory, differentiable learning can be faster, but convergence is not guaranteed due to the well-known problem of "barren plateaus" associated with the gradients becoming vanishingly small [222] and is problem dependent. Non-differentiable learning is, as a rule, slower but avoids being trapped in local minima and works well in situations where the cost function is not smooth. Sections 8.3 and 8.4 provide detailed descriptions of the QNN training procedures.

Obviously, the strongest motivation for using quantum classifiers is their ability to process quantum data. The input quantum states that must be classified may be outputs of some other quantum circuits. As we may not be able to store the information encoded in these quantum states classically, a quantum classifier becomes an indispensable tool. However, quantum classifiers have a realistic chance to demonstrate their advantage on purely classical data too. There are several considerations that motivate our interest in trying to apply QNNs to classical datasets.

First, parameterised quantum circuits possess a larger expressive power than equivalent classical neural networks. Second, they are structurally able to efficiently fight overfitting. Finally, quantum speedup is achievable on some types of quantum hardware for specific use cases even at these very early stages of quantum computing development. Chapter 13 investigates these questions in more detail.

In this chapter, we focus on using QNNs to efficiently solve specific finance-related classification use cases and provide a comparison with a number of standard classical classifiers. While experimentally proving quantum speedup and larger expressive power of QNNs requires powerful quantum hardware, the way QNNs fight overfitting can be verified on relatively small and shallow quantum circuits with the help of quantum simulators.

QNNs are PQCs trained as ML models such as classifiers. QNNs have a natural advantage over classical neural networks when it comes to classifying quantum data. However, classical datasets can also be encoded as quantum states and processed by QNNs with their larger expressive power, their ability to efficiently fight overfitting, and, ultimately, their quantum speedup.

8.2   Universal Approximation

While empirical results have supported the power of classical neural networks, advances in *Universal Approximation Theorems* have provided strong theoretical foundations on which to build them. These essentially ensure that – for example, in the context of feedforward neural networks – any function (subject to some conditions) can be approximated as accurately as desired by a neural network with sufficient width [75, 149, 150] or depth [123, 176, 333], and precise quantitative bounds have also been derived [50, 192, 248].

Given the clear analogy between classical neural networks and parameterised quantum circuits – which we investigate more precisely in the context of classification in the rest of this chapter – it is therefore natural to wonder whether such universal approximation results carry over. Before providing some clues, one may first ask the sort of outputs a PQC provides. Since the output is an n-qubit quantum state, of the form

$$|\psi\rangle = \sum_{i=0}^{2^n-1} \alpha_i |i\rangle,$$

an obvious answer is a discrete probability distribution over the labels "i", each associated to the probability mass $|\alpha_i|^2$, for $i = 0, \ldots, 2^n - 1$. Leveraging the angle encoding scheme from Section 7.2, we can go beyond this first result. Indeed, take $x \in [0, \pi)$ and consider the rotation gate $R_X(2x) := e^{-ixX} = \cos(x)I - i\sin(x)X$, so that

$$R_X(2x)|0\rangle = \{\cos(x)I - i\sin(x)X\}|0\rangle = \cos(x)|0\rangle - i\sin(x)|1\rangle.$$

By estimating the probability of observing state "0", we can therefore construct (an estimation of) the function $f(x) \equiv \cos(x)^2$. Going yet one step further, taking $x \in [0, \pi)$ and $\theta \in [0, \pi)$, the quantum algorithm

$$\mathrm{R}_\mathrm{X}(2\theta)\mathrm{R}_\mathrm{X}(2x)\,|0\rangle = \cos(\theta + x)\,|0\rangle - \mathrm{i}\sin(\theta + x)\,|1\rangle$$

yields (an estimation of) the *parametric* function

$$f_\theta(x) \equiv \cos(\theta + x)^2, \tag{8.2.1}$$

by estimating the probability of observing state "0".

Instead of computing the value function through the probability amplitudes, one could also compute it (parametrically or not) as an expectation, with the help of Section 10.2.

Regarding universal approximation, only a few results are available so far for PQCs/QNNs, but research is growing fast. In [250, 251], the authors proved such an approximation by constructing a one-qubit quantum circuit able to arbitrarily approximate any continuous complex-valued function. Similarly, Schuld and co-authors [281] showed that data encoding can be approximated by infinitely repeating simple encoding schemes based on Pauli gates.

Both previous papers rely on Fourier series representations of the function to be approximated, a very natural route in light of (8.2.1). A more recent development was proposed by Gonon and Jacquier [117], who provide a constructive approach for a quantum universal approximation with error bounds that do not depend on the dimension of the problem. More precisely, they show that, for a large class of functions in $\mathbb{R}^d$, it is possible to construct a PQC (with parameters θ) such that the L^2 distance between the function and its approximation is bounded by some constant divided by $\sqrt{n}$, for any positive integer $n \in \mathbb{N}$, with $\lceil \log_2(4n) \rceil$ qubits.

Two important points need to be made though:

- The approximation function is evaluated as a linear combination of the probabilities of observing different states, akin to the formulation above.

- While the proposed quantum circuit is explicit, the precise values of the optimal parameters θ are not.

Regarding the first item, other formulations are most likely possible, and we believe that future research will enlighten us on potentially more optimal ones. About the second item, two routes can be followed: On the one hand, Gonon and Jacquier also provide a reservoir version of the previous result, whereby the parameters are randomly generated and only a linear regression is required on the final (classical) layer. One the other hand, finding the optimal vector θ only then boils down to classical optimisation, similarly to classical neural networks. We investigate this in the following sections in the context of quantum classifiers.

8.3   Training QNN with Gradient Descent

Since we are not only interested in building QNNs as standalone QML tools but also in comparing and contrasting them with classical neural networks, we start our review of QNN training methods with gradient descent – a ubiquitous classical ML algorithm.

8.3.1   The finite difference scheme

Training QNNs consists of specifying and executing a procedure that finds an optimal configuration of the adjustable rotation parameters θ. Assume that a QNN is specified on n quantum registers with l layers of adjustable quantum gates, where each adjustable gate is controlled by a single parameter $(\theta_i^j)_{i=1,\ldots,n;\ j=1,\ldots,l}$. In this case, $\theta \in \mathcal{M}_{n,l}$ is an $n \times l$ matrix of adjustable network parameters:

$$\theta = \begin{bmatrix} \theta_1^1 & \cdots & \theta_1^l \\ \vdots & \ddots & \vdots \\ \theta_n^1 & \cdots & \theta_n^l \end{bmatrix}. \tag{8.3.1}$$

Without loss of generality, we assume that we work with a binary classifier. The latter takes an input (a quantum state that encodes a sample from the dataset), applies a sequence of quantum gates (the parameterised quantum circuit controlled by at most $n \times l$ adjustable parameters), and performs the measurement of an observable M on the chosen quantum register. An example of an observable is the Pauli Z gate and the result of a single measurement is ± 1 for a qubit found in the $|0\rangle$ or $|1\rangle$ state, respectively. The value of the measured observable is mapped to a value of a binary variable $\{0, 1\}$. This process is repeated N times for each sample in order to collect sufficient statistics for the classification result.

The first step in finding an optimal configuration of adjustable parameters θ is to choose an appropriate cost function – an objective function that represents the total error in classifying samples from the training dataset and which can be minimised by changing the adjustable network parameters. Let $\mathbf{y} := (y_1, \ldots, y_K)$ be a vector of binary labels and $\mathbf{f}(\theta) := (f_1(\theta), \ldots, f_K(\theta))$ a vector of binary classifier predictions for the training dataset consisting of K samples. The cost function $L(\theta)$ can then be defined, for example, as the sum of squared errors across all samples in the training dataset:

$$L(\theta) := \frac{1}{2} \sum_{k=1}^{K} (y_k - f_k(\theta))^2 . \qquad (8.3.2)$$

The next step is an iterative update of the adjustable parameters in the direction that reduces the value of the cost function. That direction is given by the cost function gradient – hence the name of the method. The parameters are updated towards the direction of the steepest descent of the cost function. At step $u + 1$, we update the system to

$$_{u+1}\theta_i^j \longleftarrow {}_u\theta_i^j - \eta \frac{\partial L(\theta)}{\partial \theta_i^j}, \quad \text{for each } i = 1, \ldots, n, \ j = 1, \ldots, l,$$

where η is the learning rate, namely a hyperparameter controlling the magnitude of the update. For each $i = 1, \ldots, n, j = 1, \ldots, l$, the derivative can be calculated numerically using a finite difference scheme:

$$\frac{\partial L(\theta)}{\partial \theta_i^j} \approx \frac{L(\theta_1^1, \ldots, \theta_i^j + \Delta\theta_i^j, \ldots, \theta_n^l) - L(\theta_1^1, \ldots, \theta_i^j - \Delta\theta_i^j, \ldots, \theta_n^l)}{2\Delta\theta_i^j},$$

with an error of order $\mathcal{O}((\Delta\theta_i^j)^2)$, where $\Delta\theta_i^j$ is a small rotation angle increment. The physical characteristics of the NISQ devices put restrictions on how small this increment can be: in most cases $\Delta\theta_i^j$ should not be smaller than 0.1 radians. The rest of the training routine follows the standard classical algorithm of training neural networks through the backpropagation of error with gradient descent.

8.3.2   The analytic gradient approach

An alternative to the finite difference method, which can be unstable and ill-conditioned due to truncation and round-off errors (for parameterised quantum circuits [28] or, in fact, for classical neural networks [26]), is the analytic gradient approach. It can be a viable choice for parameterised quantum circuits with adjustable one-qubit gates and fixed multi-qubit gates. From (8.3.2), the cost function gradient with respect to the parameter θ_i^j is given by

$$\frac{\partial L(\theta)}{\partial \theta_i^j} = -\sum_{k=1}^{K} (y_k - f_k(\theta)) \frac{\partial f_k(\theta)}{\partial \theta_i^j},$$

so that the task of calculating the gradient of the cost function is reduced to the task of calculating the partial derivative of the expected value of the measurement operator for each sample quantum state that encodes the classical sample from the training dataset. Let $|\psi_k\rangle$ be the quantum state that encodes the k-th sample from the training dataset and let $U(\theta)$ denote the unitary operator that represents the sequence of QNN gates transforming the initial state $|\psi_k\rangle$. Then the expected value of the measurement operator M is given by

$$f_k(\theta) = \langle\psi_k| U^\dagger(\theta)MU(\theta) |\psi_k\rangle.$$

According to the conventions we used in constructing the QNN ansatz, the parameter θ_i^j only affects a single gate, which we will denote as $G(\theta_i^j)$. Therefore, the sequence of gates $U(\theta)$ can be represented as

$$U(\theta) = VG(\theta_i^j)W,$$

where W and V are gate sequences that precede and follow gate $G(\theta_i^j)$. Let us absorb V into the Hermitian observable $Q = V^\dagger M V$ and W into the quantum state $|\phi_k\rangle = W|\psi_k\rangle$:

$$f_k(\theta) = \langle\phi_k| G^\dagger(\theta_i^j)QG(\theta_i^j) |\phi_k\rangle .$$

Then the partial derivative of $f_k(\theta)$ with respect to parameter θ_i^j is calculated as

$$\frac{\partial f_k(\theta)}{\partial \theta_i^j} = \frac{\partial}{\partial \theta_i^j} \langle\phi_k| G^\dagger(\theta_i^j)QG(\theta_i^j) |\phi_k\rangle$$

$$= \langle\phi_k| \left(\frac{\partial G(\theta_i^j)}{\partial \theta_i^j}\right)^\dagger QG(\theta_i^j) |\phi_k\rangle + \langle\phi_k| G^\dagger(\theta_i^j)Q \left(\frac{\partial G(\theta_i^j)}{\partial \theta_i^j}\right) |\phi_k\rangle . \quad (8.3.3)$$

Let us denote

$$B := G(\theta_i^j) \quad \text{and} \quad C := \frac{\partial G(\theta_i^j)}{\partial \theta_i^j},$$

and notice that

$$\langle\phi_k| C^\dagger QB |\phi_k\rangle + \langle\phi_k| B^\dagger QC |\phi_k\rangle$$
$$= \frac{1}{2} \left(\langle\phi_k| (B+C)^\dagger Q(B+C) |\phi_k\rangle - \langle\phi_k| (B-C)^\dagger Q(B-C) |\phi_k\rangle\right) . \quad (8.3.4)$$

Therefore, if we can find a way to implement the operator $B \pm C$ as part of an overall unitary evolution then we can evaluate (8.3.3) directly.

8.3.3   The parameter shift rule for analytic gradient calculation

Following [279], we outline the parameter shift rule for gates with generators with two distinct eigenvalues – this covers all one-qubit gates. Being unitary, the gate $G(\theta_i^j)$ above can be represented as

$$G(\theta_i^j) = \exp\left(-i\theta_i^j \Gamma\right),$$

for some Hermitian operator Γ (Theorem 6). The partial derivative with respect to θ_i^j reads

$$\frac{\partial G(\theta_i^j)}{\partial \theta_i^j} = -i\Gamma \exp\left(-i\theta_i^j \Gamma\right) = -i\Gamma G(\theta_i^j). \qquad (8.3.5)$$

Substituting (8.3.5) into (8.3.3) yields

$$\frac{\partial f_k(\theta)}{\partial \theta_i^j} = \langle \phi_k' | i\Gamma Q | \phi_k' \rangle + \langle \phi_k' | Q(-i\Gamma) | \phi_k' \rangle, \qquad (8.3.6)$$

where $|\phi_k'\rangle = G(\theta_i^j) |\phi_k\rangle$. If Γ has just two distinct eigenvalues, we can shift the eigenvalues to $\pm r$, since the global phase is unobservable [279]. With I denoting the identity operator, we can rewrite (8.3.6) as

$$\frac{\partial f_k(\theta)}{\partial \theta_i^j} = r \left(\langle \phi_k' | \frac{i\Gamma}{r} Q I | \phi_k' \rangle - \langle \phi_k' | I Q \frac{i\Gamma}{r} | \phi_k' \rangle \right). \qquad (8.3.7)$$

Denoting

$$B := I \quad \text{and} \quad C := -\frac{i}{r}\Gamma,$$

and using (8.3.4) we obtain from (8.3.7):

$$\frac{\partial f_k(\theta)}{\partial \theta_i^j} = \frac{r}{2} \left[\langle \phi_k' | \left(I - \frac{i}{r}\Gamma \right)^\dagger Q \left(I - \frac{i}{r}\Gamma \right) | \phi_k' \rangle - \langle \phi_k' | \left(I + \frac{i}{r}\Gamma \right)^\dagger Q \left(I + \frac{i}{r}\Gamma \right) | \phi_k' \rangle \right].$$

A straightforward computation [279, Theorem 1] shows that if the Hermitian generator Γ of the unitary operator $G(\theta) = \exp(-i\theta\Gamma)$ has at most two unique eigenvalues $\pm r$, then

$$G\left(\mp \frac{\pi}{4r}\right) = \frac{1}{\sqrt{2}} \left(I \pm \frac{i}{r}\Gamma \right).$$

In this case, the gradient can be estimated using two additional evaluations of the quantum circuit. Either the gate $G(\pi/(4r))$ or the gate $G(-\pi/(4r))$ should be placed in the original circuit next to the gate we are differentiating. Since for unitarily generated one-parameter gates $G(a)G(b) = G(a+b)$, this is equivalent to shifting the gate parameter, and we obtain the "parameter shift rule" [279] with the shift $s = \pi/(4r)$:

$$\frac{\partial f_k(\theta)}{\partial \theta_i^j} = r \left(\langle \phi_k | \, \mathsf{G}^\dagger(\theta_i^j + s) \mathsf{QG}(\theta_i^j + s) \, |\phi_k\rangle - \langle \phi_k | \, \mathsf{G}^\dagger(\theta_i^j - s) \mathsf{QG}(\theta_i^j - s) \, |\phi_k\rangle \right).$$

If Γ is a one-qubit rotation generator given by Pauli X, Y, and Z operators, then $r = 1/2$ and $s = \pi/2$ [228,279]:

$$
\begin{aligned}
\frac{\partial f_k(\theta)}{\partial \theta_i^j} = \frac{1}{2} \Big(&\langle \phi_k | \, \mathsf{G}^\dagger \left(\theta_i^j + \frac{\pi}{2} \right) \mathsf{QG} \left(\theta_i^j + \frac{\pi}{2} \right) |\phi_k\rangle \\
&- \langle \phi_k | \, \mathsf{G}^\dagger \left(\theta_i^j - \frac{\pi}{2} \right) \mathsf{QG} \left(\theta_i^j - \frac{\pi}{2} \right) |\phi_k\rangle \Big).
\end{aligned}
\tag{8.3.8}
$$

Therefore, what we need to do in order to estimate the gradient is to execute two circuits N times to collect statistics and to calculate the expectations on the right-hand side of (8.3.8). The first circuit will have the gate parameter shifted by $\pi/2$ and the second circuit will have the gate parameter shifted by $-\pi/2$.

Although this procedure is not necessarily faster than the finite difference scheme, it can produce a more accurate estimate of the cost function gradient. The main argument here is the fact that the NISQ hardware operates with limited precision. The state-of-the-art superconducting qubits have one-qubit gate fidelity $\leq 99.9\%$ and two-qubit gate fidelity $\leq 99.7\%$ with rotation angle precision of order 0.05 radians. Therefore, the finite difference scheme cannot assume infinitesimal rotation angles $\Delta\theta$ – they should not be smaller than about 0.1 radians (and, probably, materially larger in most cases). This means that gradients obtained with the finite difference scheme have some degree of built-in uncertainty that can only be fixed with further improvements in the NISQ hardware.

QNNs can be trained with the gradient descent algorithm in full analogy with the backpropagation of error in classical neural networks. The gradients can be either calculated analytically or estimated numerically.

8.4 Training QNN with Particle Swarm Optimisation

Having specified the gradient descent scheme for training QNNs in the previous section, we now turn our attention to a non-differentiable learning method based on the powerful evolutionary search algorithm.

8.4.1 The Particle Swarm Optimisation algorithm

The Particle Swarm Optimisation (PSO) algorithm belongs to a wide class of evolutionary search heuristics where at each algorithm iteration ("generation" in the language of evolutionary algorithms), the population of solutions ("chromosomes" or "particles") is evaluated in terms of their fitness with respect to the environment. In the standard PSO formulation [258], a number of particles are placed in the solution space of some problem and each evaluates the fitness at its current location. Each particle then determines its movement through the solution space by combining some aspects of the history of its own fitness values with those of one or more members of the swarm, and then moves through the solution space with a velocity determined by the locations and processed fitness values of those other members, along with some random perturbations.

It is a standard procedure [134, 186] to follow three steps in specifying the PSO algorithm. First, we initialise the positions $x_k^i := (x_k^i(1), \ldots, x_k^i(n)) \in \mathbb{R}^n$ of each particle i at time k moving through the n-dimensional search space and taking values in some range $[x_{\min}, x_{\max}]$. Next, we initialise the velocities $v_k^i := (v_k^i(1), \ldots, v_k^i(n)) \in \mathbb{R}^n$ of each particle in the swarm. The initialisation process consists of distributing swarm particles randomly across the solution space:

$$x_0^i = x_{\min} + \omega_x \left(x_{\max} - x_{\min}\right), \quad v_0^i = \frac{x_{\min} + \omega_v \left(x_{\max} - x_{\min}\right)}{\Delta t}, \tag{8.4.1}$$

where ω_x and ω_v are uniformly distributed random variables on $[0, 1]$ and Δt is the time step between algorithm iterations.

We then update the velocities of all particles at time $k + 1$ according to the specified objective function, which depends on the particles' current positions in the solution space at time k. The value of the objective function determines which particle has the best position p_k^{global} in the current swarm and also determines the best position p^i of each particle over time, that is, in the current and all previous moves. The velocity update formula uses these two pieces of information for each particle in the swarm along with the effect of the current motion v_k^i to provide a search direction p_{k+1}^i for the next iteration. The velocity update formula includes random parameters to ensure good coverage of the solution space and to avoid entrapment in local optima. The three values that affect the new search direction are the current motion, the particle's own memory, and the swarm influence. They are incorporated via a summation approach with three weight factors: inertia w, self-confidence c_1, and swarm confidence c_2:

$$v_{k+1}^i = w v_k^i + c_1 \omega_1 \frac{\left(p^i - x_k^i\right)}{\Delta t} + c_2 \omega_2 \frac{\left(p_k^{global} - x_k^i\right)}{\Delta t},$$

where ω_1 and ω_2 are uniformly distributed random variables on $[0, 1]$.

Finally, the position of each particle is updated using its velocity vector:

$$x_{k+1}^i = x_k^i + v_{k+1}^i \Delta t.$$

These steps are repeated until either a desired convergence criterion is met or we reach the maximum number of iterations. Various reflection rules (stopping at the boundary, mirror reflection back into the allowed domain, etc.) [205] can be designed for the new position x_{k+1}^i falling outside the $[x_{\min}, x_{\max}]$ bounds and the dynamics can be normalised with $\Delta t \equiv 1$. If K is the last iteration of the algorithm, then the best solution found by the PSO is p_K^{global}. Figure 8.2 provides a schematic illustration of the particle movement through the solution space under the influence of three forces: momentum, attraction to the globally best solution found by all particles at the previous iteration, and attraction to the best solution found by the given particle across all previous iterations.

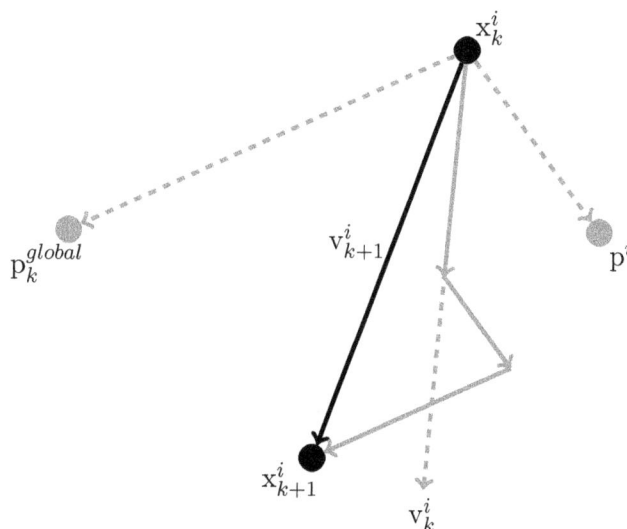

Figure 8.2: Schematic illustration of the PSO algorithm. Each particle moves through the solution space under the influence of three forces: momentum, own memory, and swarm influence.

8.4.2 PSO algorithm for training quantum neural networks

We are now ready to specify the PSO algorithm to train QNNs. We consider the most general case of an $n \times l$ matrix of adjustable parameters (rotations) θ, where n is the number of quantum registers and l is the number of network layers. The solution we look for is the matrix (8.3.1) of adjustable parameters that minimises the chosen cost function.

The cost function can be specified in many different ways depending on what particular aspects we want to encourage or penalise. Given the training dataset, we would like to find a configuration of adjustable parameters θ such that as many samples as possible are classified correctly. One possible choice of cost function, for example, may be the ratio of incorrect to correct classification decisions. However, the classification process is probabilistic in nature – we decide on the sample label after many runs of the quantum circuit, which generate sufficient statistics. Therefore, each classification decision is not just right or wrong but can be seen as "more right" or "more wrong". If the correct sample label is "1" and we get "0" 51% of the time then the classifier is slightly wrong: the chances are that similar samples would be classified correctly or only a small change to the adjustable network parameters is required to rectify the classification process. However, if we get "0", say, 90% of the time, then the classifier is "very wrong" and we need to penalise the outcome more aggressively.

One possible realisation of the cost function that takes into account the above argument is as follows. Without loss of generality, assume that we work with the binary class labels "0" and "1", and let $y := (y_1, \ldots, y_K)$ be a vector of sample labels (either "0" or "1") from the training dataset. Further, let $\mathbb{P}(\theta) := (\mathbb{P}_1(\theta), \ldots, \mathbb{P}_K(\theta))$ be a vector of QNN estimated probabilities of predicting class "1" for the given sample (i.e., the number of quantum circuit runs that returned "1" after measurement divided by the total number of quantum circuit runs). Then the cost function $L(\theta)$ is given by the following pseudo code:

```
cost_function = 0
for i in range(K):
    if y[i] == 0:
        if P[i] > 0.7:
            cost_function += 4
        if P[i] > 0.6 and P[i] <= 0.7:
            cost_function += 2
        if P[i] > 0.5 and P[i] <= 0.6:
            cost_function += 1
    else:
        if P[i] < 0.5 and P[i] >= 0.4:
            cost_function += 1
        if P[i] < 0.4 and P[i] >= 0.3:
            cost_function += 2
        if P[i] < 0.3:
            cost_function += 4
```

This cost function penalises large errors in the class probability estimate more than small errors and represents the total error across all samples in the training dataset.

We can now formulate the QNN training algorithm, which has the following inputs:

Variable	Meaning
$\mathbf{X} := (\mathbf{X}_1, \ldots, \mathbf{X}_K) \in \mathbb{R}^{M \times K}$	training dataset of features encoded as rotation angles on $[0, \pi]$
$\mathbf{y} := (y_1, \ldots, y_K) \in \{0, 1\}^K$	vector of binary labels
N_{iter}	number of iterations
N_{runs}	number of quantum circuit runs
M	number of particles (solutions)
w	momentum coefficient
c_1	particle memory coefficient
c_2	swarm influence coefficient
n	number of quantum registers
l	number of QNN layers

Table 8.2: Inputs of the QNN training algorithm

The algorithm operates on the following objects, where $m = 1, \ldots, M$ denotes the m-th particle, and $t = 0, \ldots, N_{iter}$ represents the algorithm iteration step:

- $\theta(t; m) \in \mathcal{M}_{nl}([-\pi, \pi])$: position of particle m at time t;

- $v(t; m) \in \mathcal{M}_{nl}([-\pi, \pi])$: velocity of particle m at time t;

- $\Xi(m) \in \mathcal{M}_{nl}([-\pi, \pi])$: best position found by particle m across all iterations;

- $\Phi(t) \in \mathcal{M}_{nl}([-\pi, \pi])$: the globally best position found by all particles at time t;

- $L(\theta)$: value of the cost function for the solution θ.

Algorithm 5: Particle Swarm Optimisation

Result: Optimal configuration of adjustable QNN parameters $\theta^* := \operatorname{argmin} L(\theta)$.

Initialisation and evaluation of the first set of solutions
(*we set Δt in (8.4.1) equal to* 1):

for *each particle $m = 1, \ldots, M$* **do**

    for $i = 1, \ldots, n, j = 1, \ldots, l$ **do**

        Randomly draw the rotation angle $\theta_i^j(0; m)$ from $\mathcal{U}([-\pi, \pi])$.

        Randomly draw the rotation angle $v_i^j(0; m)$ from $\mathcal{U}([-\pi, \pi])$.

    end

    Initialise the individually best solution:

        $\Xi(m) \leftarrow \theta(0; m)$

    for $k = 1, \ldots, K$ **do**

        Run the quantum circuit N_{runs} times with configuration $\theta(0; m)$ on sample

        X_k to estimate the probability $\mathbb{P}_k$ of reading out "1" on the target qubit.

    end

    Evaluate the cost function $L(\theta(0; m))$ given the probabilities $\mathbb{P} := (\mathbb{P}_1, \ldots, \mathbb{P}_K)$.

end

Order solutions from best (minimal cost function) to worst (maximal cost function).

$\Phi(0) \leftarrow$ configuration corresponding to the minimum of the cost function.

Initialise the optimal configuration:

    $\theta^* \leftarrow \Phi(0)$

Iterations:

for $t = 1, \ldots, N_{iter}$ **do**

    for $m = 1, \ldots, M$ **do**

        for $i = 1, \ldots, n, j = 1, \ldots, l$ **do**

            Generate independent random numbers $\omega_1 \sim U[0,1]$ and $\omega_2 \sim U[0,1]$.

            momentum $\leftarrow w v_i^j(t-1; m)$

            particle $\leftarrow c_1 \omega_1 [\Xi_i^j(m) - \theta_i^j(t-1; m)]$

            swarm $\leftarrow c_2 \omega_2 [\Phi_i^j(t-1) - \theta_i^j(t-1; m)]$

            $v_i^j(t; m) \leftarrow$ momentum $+$ particle $+$ swarm

            $\theta_i^j(t; m) \leftarrow \theta_i^j(t-1; m) + v_i^j(t; m)$

        end

        for $k = 1, \ldots, K$ **do**

            Run the quantum circuit N_{runs} times with configuration $\theta(t; m)$ on

            sample X_k to estimate the probability $\mathbb{P}_k$ of reading out "1" on the target

            qubit.

        end

        Evaluate the cost function $L(\theta(t; m))$ given $\mathbb{P} := (\mathbb{P}_1, \ldots, \mathbb{P}_K)$.

        if $L(\theta(t; m)) < L(\Xi(m))$ **then**

            $\Xi(m) \leftarrow \theta(t; m)$

        end

    end

    Order solutions from best (minimum value of the cost function) to worst

    (maximum value of the cost function).

    $\Phi(t) \leftarrow$ configuration corresponding to the minimum of the cost function.

    if $L(\theta^*) < L(\Phi(t))$ **then**

        $\theta^* \leftarrow \Phi(t)$

    end

end

> The non-differentiable learning based on the evolutionary search heuristic
> works well for irregular, non-convex objective functions with many
> local minima.

8.5   QNN Embedding on NISQ QPU

Ideally, parameterised quantum circuits should be constructed in a hardware-agnostic way,
only driven by the characteristics of the problem being solved. This, however, would require
the existence of large and exceptionally well-connected quantum computing systems with
very high qubit fidelity and coherence time. In other words, we would need QPUs with
capabilities that significantly exceed those of existing NISQ devices. The time for such
powerful quantum computing systems may come sooner than one may expect but we still
have to find a way of running PQCs efficiently on NISQ QPUs.

8.5.1   NISQ QPU connectivity

A typical approach to designing a PQC executable on the NISQ QPU would start with
observing two main characteristics of quantum computing systems: the graph (qubit connectivity) and the set of native gates. We can illustrate these points by looking at Rigetti's
Aspen system [72] in Figure 8.3.

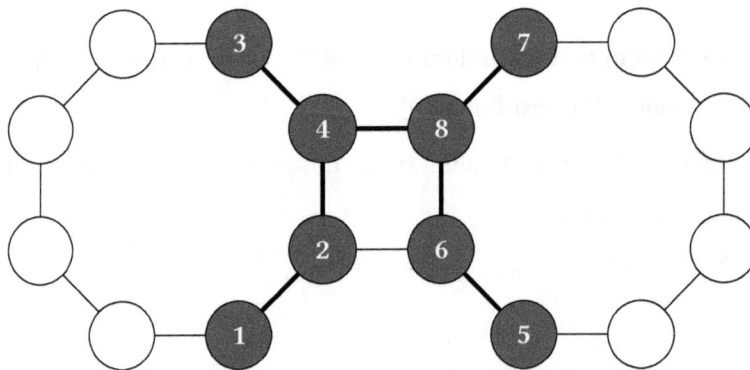

Figure 8.3: Rigetti's Aspen system.

As we can see, most qubits are only connected to their nearest neighbours on the linear grid, with only four qubits having three connections. These extra connections form a bridge between two 8-qubit islands that, otherwise, would be completely independent.

8.5.2 QNN embedding scheme

The shaded qubits in Figure 8.3 can be used to construct the 8-qubit tree network capable of processing a dataset with up to 16 continuous features (two features per quantum register) as shown in Figure 8.4. The thick lines in Figure 8.3 represent qubits connectivity used in constructing the QNN. The thin lines represent all other available qubit connections that have not been utilised in the QNN ansatz.

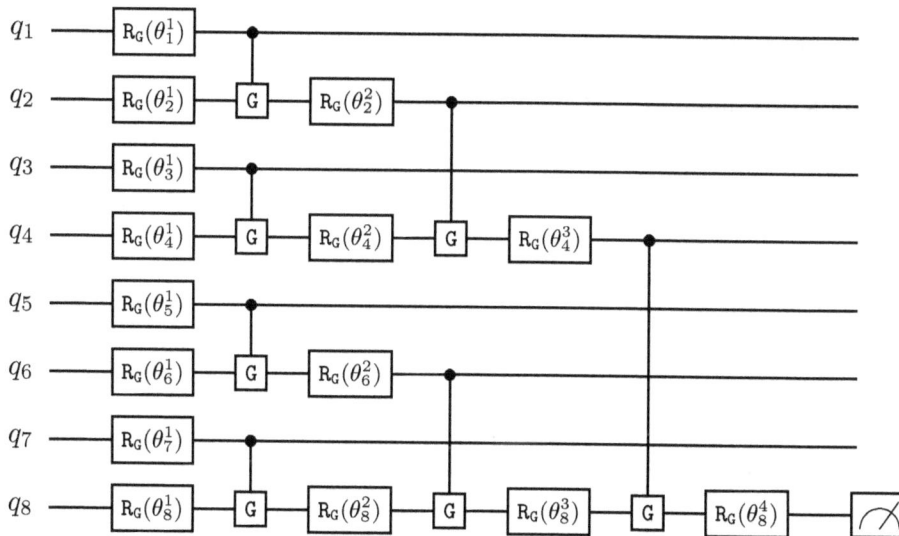

Figure 8.4: QNN for the Aspen system; the gate G is any of the {X, Y, Z} gates.

With the limited connectivity of existing QPUs, we need to fully utilise the graph structure of the quantum chips to implement the most efficient QNN embedding and extract the best possible performance.

8.6   QNN Trained as a Classifier

We now demonstrate how a binary QNN classifier can be trained on a classical credit approval dataset using the non-differentiable learning approach.

8.6.1   The ACA dataset and QNN ansatz

One of the most fundamental use cases for a binary classifier in finance is credit approval. The UCI Machine Learning Database [262, 263] holds the Australian Credit Approval (ACA) dataset consisting of 690 samples. There are 14 features (binary, integer, continuous) representing various attributes of potential borrowers and a binary class label (accept/reject credit application). The dataset is reasonably hard for classical classifiers due to the limited predictive power of the features and its relatively small size. This makes it ideal for testing and benchmarking the QNN performance.

We start with the simplest tree network that can be mapped onto Rigetti's Aspen system graph described in the previous section. Figure 8.5 shows the full quantum circuit consisting of sample encoding and sample processing modules [185]. The proposed scheme allows us to encode up to two continuous features per quantum register with the help of rotations around the x and the y axes.

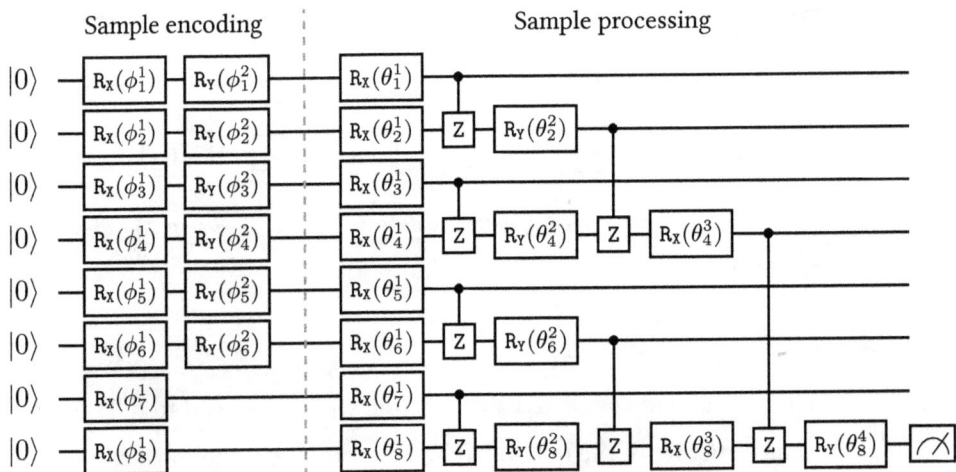

Figure 8.5: PQC for the credit approvals classifier.

The features are encoded as rotation angles $\phi \in [0, \pi]$ according to the encoding scheme described in Section 7.2. With all qubits initialised as $|0\rangle$ in the computational basis, this ensures the uniqueness of the encoded samples. The sample processing module consists of layers of adjustable one-qubit gates (rotations around the x and y axes) and fixed two-qubit gates (CZ). We split the ACA dataset 50:50 into a training and a testing dataset using the `train_test_split()` function provided by the `sklearn.model_selection` module. Our objective is to train the QNN and various classical classifiers (classical benchmarks) on the training dataset and compare their out-of-sample performance on the testing dataset. The classical classifiers have a number of hyperparameters that can be fine-tuned to optimise the classifier performance on the given dataset. In contrast, the QNN architecture (location and types of one-qubit and two-qubit gates) is fixed.

8.6.2 Training an ACA classifier with the PSO algorithm

We first verify that the QNN can be efficiently trained with the Particle Swarm Optimisation algorithm – a non-differentiable learning approach. Figure 8.6 illustrates PSO convergence for the set of PSO parameters given in Table 8.3.

Parameter	Notation	Value
Inertia coefficient	w	0.25
Self-confidence coefficient	c_1	0.25
Swarm confidence coefficient	c_2	0.25
Number of particles	M	10
Number of iterations	N_{iter}	20
Number of quantum circuit runs	N_{runs}	1000

Table 8.3: PSO parameters.

The sample algorithm run has reached the minimum of the objective function in just four iterations with only ten particles, exploring the search space using the `Qiskit` quantum simulator.

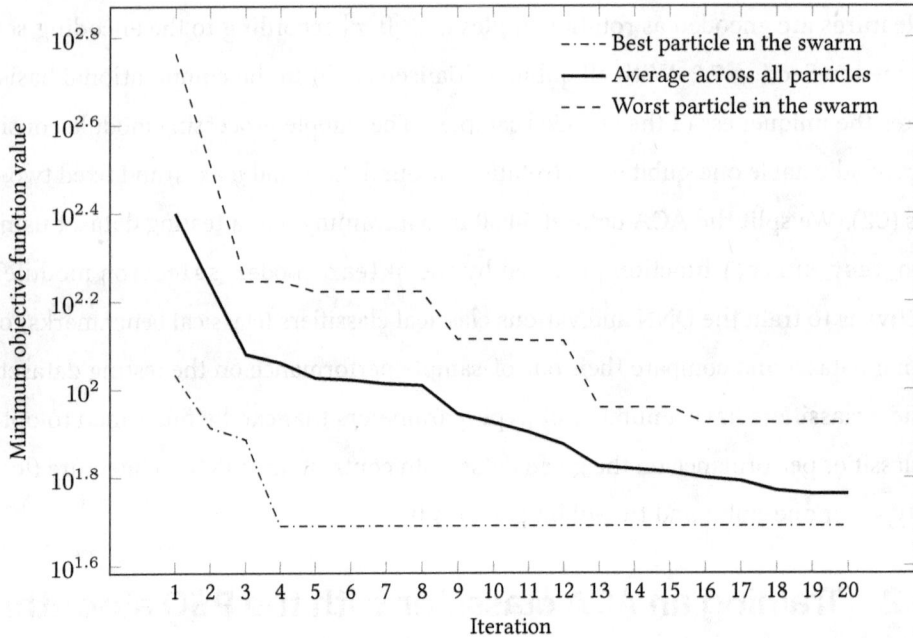

Figure 8.6: Minimum objective function values found by individual particles.

The configuration of adjustable parameters (rotations) that corresponds to the minimum of the objective function found by the PSO algorithm is given by (8.6.1).

$$
\theta = \begin{bmatrix}
0.16\pi & & & \\
-0.55\pi & 0.66\pi & & \\
-0.13\pi & & & \\
0.08\pi & 0.72\pi & 0.02\pi & \\
0.33\pi & & & \\
0.06\pi & 0.95\pi & & \\
0.48\pi & & & \\
0.19\pi & -0.91\pi & -0.83\pi & 0.59\pi
\end{bmatrix}. \tag{8.6.1}
$$

Figure 8.7 displays the in- and out-of-sample confusion matrices for the QNN classifier obtained with the Qiskit quantum simulator, assuming that Class 0 is the positive class.

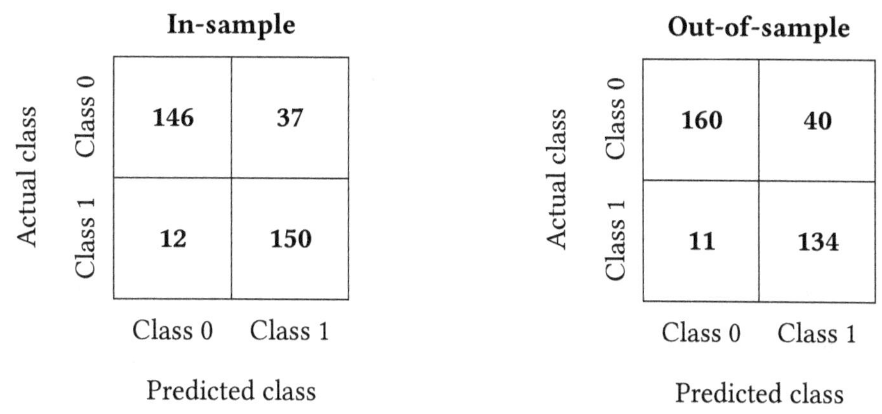

In-sample

Out-of-sample

Figure 8.7: Confusion matrix for the QNN classifier (ACA dataset).

The results are robust with an in-sample accuracy of 0.86 and an out-of-sample accuracy of 0.85. Interestingly, the in-sample and out-of-sample results are very close, indicating that the QNN provides strong regularisation. The question of quantum and classical neural networks regularisation will be tackled in Chapter 13.

8.7 Classical Benchmarks

In Chapter 4, we introduced two classical classifiers: a feedforward artificial neural network (Multi-Layer Perceptron) and a decision tree algorithm. We now expand the range of classical benchmark classifiers by adding Support Vector Machine (SVM) [70], Logistic Regression [31], and Random Forest [144]. The SVM approach based on the *kernel method* is covered in Chapter 14. Here, we briefly explain the main principles of logistic regression and random forest classifiers.

8.7.1 Logistic Regression and Random Forest

Logistic regression can be seen as a special case of a feedforward neural network with a single hidden layer consisting of an activation unit with the logistic activation function. The model operates as shown in Figure 4.3 with

$$y(s) = \left(1 + e^{-s}\right)^{-1}.$$

The standard logistic regression model is a *linear classifier* because the outcome always depends on the sum of the (weighted) inputs. Therefore, logistic regression performs well when working with a dataset where the classes are more or less linearly separable.

Random forest is an *ensemble learning* model and, as the name suggests, is based on combining the classification results of multiple decision trees. The ensemble technique used by random forest is known as *bootstrap aggregation*, or *bagging*, by choosing random subsets from the dataset. Hence, each decision tree is generated from samples drawn from the original dataset with replacement (row sampling). This step of row sampling with replacement is called the *bootstrap*. Each decision tree is trained independently. The final output for the given samples is based on *majority voting* after combining the results of all individual decision trees. This is the *aggregation* step.

8.7.2 Benchmarking against standard classical classifiers

The classical benchmarking can be done by training several popular `scikit-learn` models. Table 8.4 provides classical benchmarking results in terms of out-of-sample F_1 scores for the following (weakly) optimised `scikit-learn` classifiers:

- A feedforward neural network (MLP) classifier: `neural_network.MLPClassifier`
- A support vector machine classifier: `svm.SVC`
- An ensemble learning model: `ensemble.RandomForestClassifier`
- A logistic regression classifier: `linear_model.LogisticRegression`

The F_1 score is a harmonic average of two performance metrics, precision and recall:

$$F_1 := 2 \frac{\text{Precision} \times \text{Recall}}{\text{Precision} + \text{Recall}}.$$

Both were introduced in Chapter 4. In the context of credit approvals, optimising for recall helps with minimising the chance of approving a credit application that should be rejected. However, this comes at the cost of not approving credit applications for some high-quality borrowers. If we optimise for precision, then we improve the overall correctness of our decisions at the cost of approving some applicants with bad credits. The F_1 score is used to balance the positives and negatives in optimising precision and recall.

Classifier	Average F_1 score
Logistic Regression Classifier	0.88
Random Forest Classifier	0.87
MLP Classifier	0.86
QNN Classifier	0.85
Support Vector Classifier	0.84

Table 8.4: Out-of-sample F_1 scores for the classical and QNN classifiers trained on the ACA dataset.

The QNN classifier performance, as measured by the average F_1 score for Class 0 and Class 1, falls somewhere in the middle of the range of out-of-sample F_1 scores for the chosen classical benchmarks. This is encouraging since the QNN ansatz was fixed and we did not optimise the QNN hyperparameters – the placement and types of the two-qubit gates. The classifier performance can be further improved by deploying the standard ensemble learning techniques, as explained in the following section.

QNNs can be productively used for classification tasks on classical finance-related datasets.

8.8 Improving Performance with Ensemble Learning

The ensemble learning methods combine different weak classifiers into a strong classifier that has better generalisation capabilities than each individual standalone classifier. In Chapter 4, we saw how the principles of ensemble learning can be used in combination with the methods of quantum annealing. Here, we look at them from the QNN perspective.

8.8.1   Majority voting

The popular ensemble learning methods are majority voting (binary classification) and plurality voting (multiclass classification). Majority voting means what it says: the class label for the given sample is the one that receives more than half of the individual votes. Plurality voting chooses the class that receives the largest number of votes (the mode).

The ensemble of the individual classifiers can be built from different classification algorithms, for example, by combining neural network classifiers, support vector machines, decision trees, and so on. On the other hand, the same basic classification algorithm can be used to produce multiple classifiers by choosing different configurations of hyperparameters and different subsets of the training dataset. The random forest classifier, which combines different decision tree classifiers, illustrates the latter approach.

With these considerations in mind, we build a strong classifier from several individual QNN classifiers by changing the QNN ansatz within the restrictions imposed by the QPU qubit connectivity. In order to test the majority voting approach, we build two new QNN classifiers by adding a few more two-qubit CZ gates to the baseline parameterised quantum circuit, as shown in Figures 8.8 and 8.9.

In the case of PQC #2, we add two extra CZ gates, exploiting the "bridge" structure of the Aspen system (Figure 8.3). This improves the overall system entanglement and allows for a richer set of achievable quantum states. PQC #3 has three extra CZ gates in comparison with the baseline circuit. The new classifiers can be trained with the same algorithm (PSO) on the same training dataset but will have different optimal configurations of the adjustable parameters and will make slightly different classification decisions on the testing dataset.

With three QNN classifiers, the majority voting leads to either a unanimous or a 2:1 decision. Performance on the ACA dataset improves marginally with all three classifiers generally in full agreement with each other. There are only a handful of instances where majority voting adds value, but this improves the average out-of-sample F_1 score from 0.85 to 0.87 – on par with the random forest classifier trained on the same dataset.

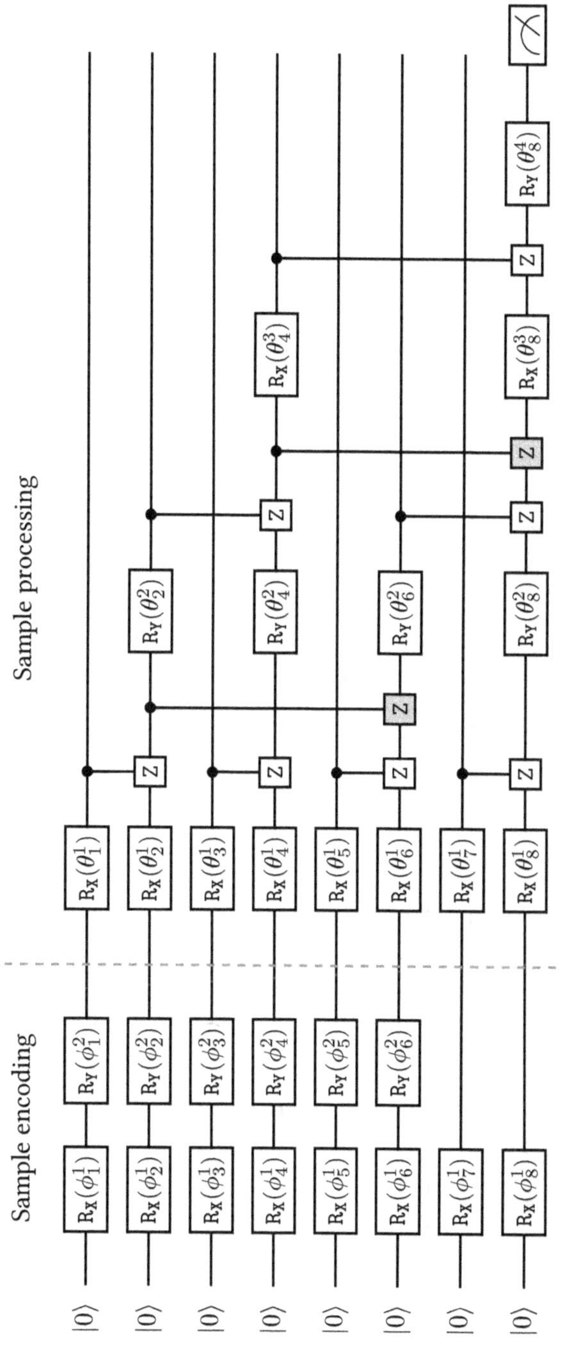

Figure 8.8: PQC #2 for the credit approvals classifier. New fixed two-qubit gates are shaded grey.

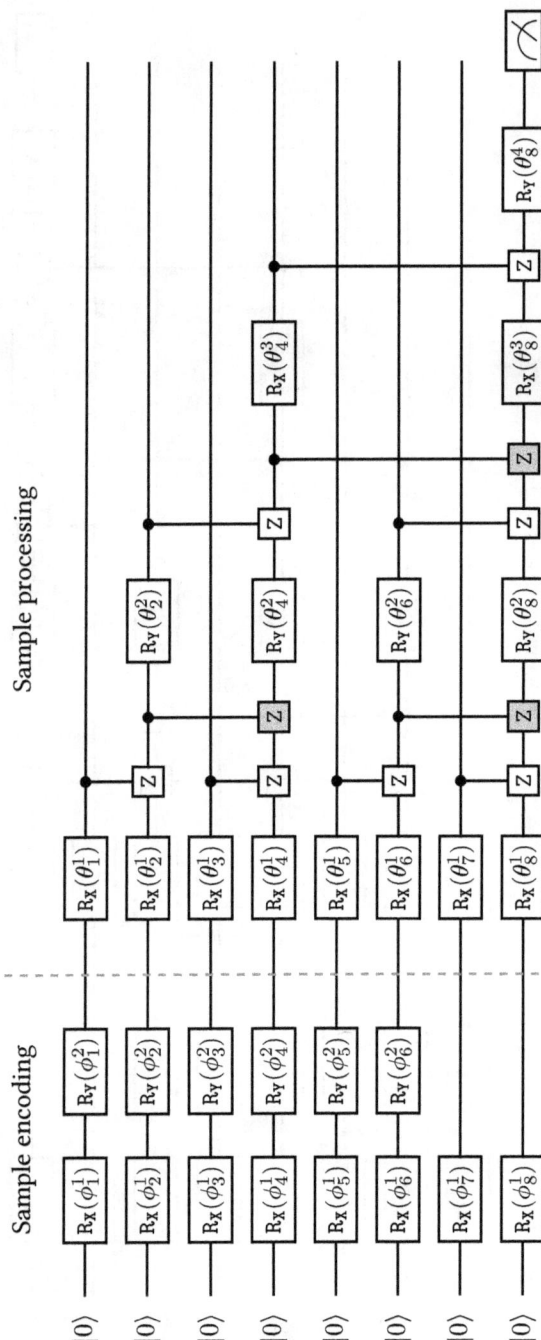

Figure 8.9: PQC #3 for the credit approvals classifier. New fixed two-qubit gates are shaded grey.

Similar results can be achieved with the original QNN classifier trained on different subsets of the training dataset. These subsets are produced by drawing the bootstrap samples – random samples with replacement – from the original training dataset. The differently trained QNN classifiers can then be combined into a single strong classifier using the majority voting approach, as described above.

8.8.2 Quantum boosting

We started by introducing the concept of ensemble learning where predictions produced by various QNNs are combined into a more robust unified prediction via a classical majority voting method. However, we can take a different approach to ensemble learning: predictions of various classical classifiers can be treated as an input into the QNN that performs their aggregation and comes up with a unified prediction. In other words, the QNN operates as a quantum booster similar to the QUBO-based QBoost model introduced in Chapter 4.

Let us come back to the classical benchmarks used in Section 8.6. There are four different machine learning models performing binary classifications. Their outputs ("0" for Class 0 and "1" for Class 1) are inputs into a 4-qubit QNN classifier. Since all quantum registers are initialised as $|0\rangle$, the outputs of individual classifiers can be encoded by either doing nothing for Class 0 output (which is equivalent to applying an identity operator I) or by applying a NOT gate X for Class 1 output.

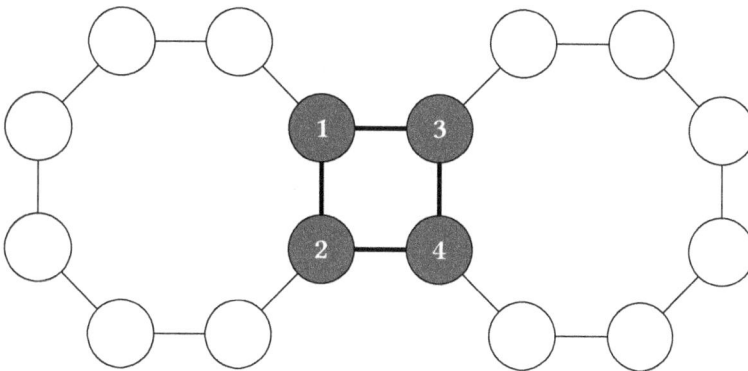

Figure 8.10: Embedding of a 4-qubit QNN onto the bridge section of Rigetti's Aspen system.

Figure 8.10 shows how a 4-qubit QNN can be efficiently embedded on the QPU and Figure 8.11 shows the corresponding parameterised quantum circuit with adjustable one-qubit gates (R_X, R_Y) and fixed two-qubit gates (CZ).

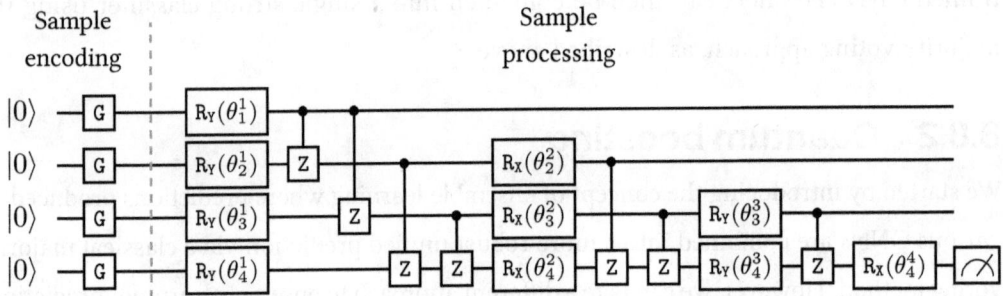

Figure 8.11: QBoost circuit. The sample encoding gate G *is either an identity gate* I *if the input is "0", or a* NOT *gate* X *if the input is "1".*

> Ensemble learning can improve QNN performance in the same way it improves the performance of the classical weak learners, with QNN performing at par with random forest in terms of F1 score on the out-of-sample dataset.

Summary

In this chapter, we introduced the concept of a quantum neural network as a parameterised quantum circuit trained as a classifier. We considered two approaches to training QNNs: differentiable (gradient descent) and non-differentiable (Particle Swarm Optimisation) methods. Gradient descent is generally faster but can face the problem of barren plateaus (vanishing gradients). The evolutionary search heuristics may be slower but can handle the presence of multiple local minima and strike the right balance between exploration and exploitation.

We also explored the embedding of QNNs on the NISQ QPUs with limited connectivity between the qubits. As an example, we considered Rigetti's Aspen system and proposed an efficient embedding scheme that mirrors the "tree structure" architecture of the QNN.

Once our QNN was fully specified and embedded into a QPU graph, we investigated its performance on a real-world dataset of credit approvals and provided comparisons with several standard classical classifiers.

Finally, we looked at several ensemble learning techniques that assist in improving QNN performance in the context of a hybrid quantum-classical protocol.

In the next chapter, we will study a powerful generative QML model – the Quantum Circuit Born Machine – which is a direct quantum counterpart of the classical Restricted Boltzmann Machine we considered in Chapter 5.

9

Quantum Circuit Born Machine

The arrival of the new computational paradigm of quantum computing and the progress achieved in developing quantum computing hardware prompted intensive research in exploring the capabilities of quantum machine learning models and, more specifically, quantum generative models that can be viewed as quantum counterparts of the classical RBMs introduced in Chapter 5. Classical generative models form one of the most important classes of unsupervised machine learning techniques, with numerous applications in finance, such as the generation of synthetic market data [48, 187], the development of systematic trading strategies [190], and data anonymisation [188], to name just a few.

Quantum generative models have all the necessary qualities needed to establish quantum advantage on NISQ devices. Probably the most well-known example of such models is the Quantum Circuit Born Machine (QCBM), which consists of several layers of adjustable and fixed gates followed by measurement operators. The input is a quantum state where all qubits are initialised as $|0\rangle$ in the computational basis. The output is a bitstring, which is a sample from the probability distribution encoded in the final state constructed by the application of adjustable and fixed gates to the initial state.

The expectation of experimental proof of the quantum advantage is motivated by the following observations. First, QCBMs have greater expressive power than classical RBMs when only a polynomial number of parameters is allowed (the number of qubits in a QCBM or the number of visible activation units in an RBM) [88]. Second, generating an independent sample from the learned distribution can be done in a single run of the quantum circuit in the case of QCBM – this compares favorably with the up to 10^3–10^4 forward and backward passes through the network in the case of RBM, which are needed to achieve a state of thermal equilibrium [187]. This points toward material quantum speedup. Third, quantum generative models can be used to load data into a quantum state, thus facilitating realisations of many promising quantum algorithms [341].

9.1   Constructing QCBM

As we saw in Chapter 8, the art of building a QML model that can be run on a NISQ computer consists of finding an optimal PQC architecture that can be embedded into the chosen QPU graph. In this section, we show how it can be done for a QCBM compatible with IBM's Melbourne and Rochester systems.

9.1.1   QCBM architecture

The QCBM is a parameterised quantum circuit where a layer of adjustable one-qubit gates is followed by a layer of fixed two-qubit gates. Such a pattern can be repeated any number of times, building a progressively deeper circuit. The input is a quantum state where all qubits are initialised as $|0\rangle$ in the computational basis. The final layer consists of measurement operators producing a bitstring sample from the learned distribution. Therefore, to specify the QCBM architecture means to specify the number of layers, the type of adjustable gates, and the type of fixed gates for each layer. Since the theory of PQC is still being developed [28], we can rely on similarities and analogies between PQCs and classical neural networks to come up with some initial guesses about the possible QCBM architecture.

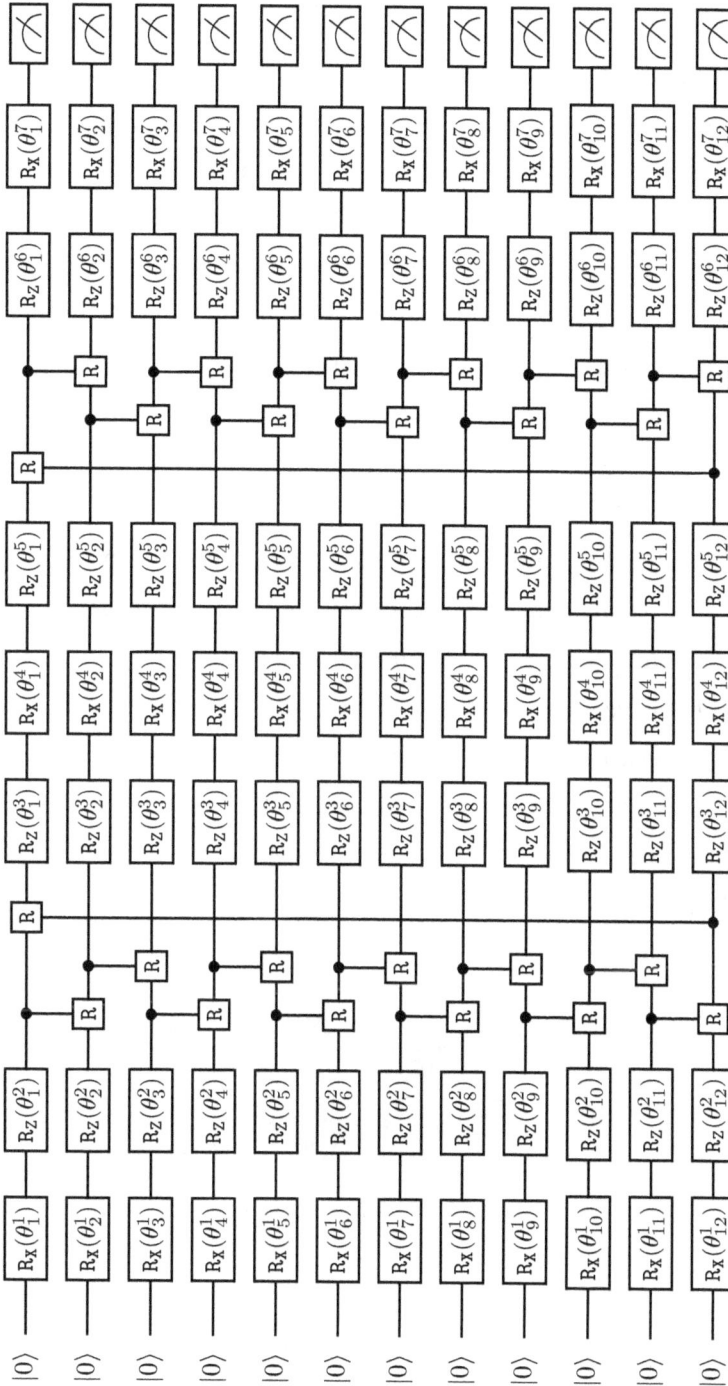

Figure 9.1: QCBM(12, 7).

Figure 9.1 displays a 12-qubit QCBM with two layers of controlled rotation gates $R = R_G(\phi)$ for $G \in \{X, Y, Z\}$ and $\phi \in [-\pi, \pi]$, where G and ϕ are fixed, and three layers of one-qubit gates $R_X(\theta)$ and $R_Z(\theta)$ with a total of seven adjustable gates per quantum register. The circuit is wide enough and deep enough to learn a complex distribution of a continuous random variable while remaining implementable on existing NISQ devices: the 12-digit binary representation of a continuous random variable provides sufficient precision and seven adjustable parameters (rotation angles) per qubit provide sufficient flexibility. At the same time, the circuit is not too deep to be compromised by the gate fidelity achievable in existing quantum hardware [46, 178].

9.1.2 QCBM embedding

The chosen QCBM architecture is compatible with the limited connectivity observed in the current generation of quantum processors. For example, the proposed circuit requires sequential qubit connectivity where qubit n is directly connected with qubits $n - 1$ and $n + 1$ but does not have to be directly connected with other qubits. This architecture can, for example, be supported by IBM's Melbourne system [223], as can be seen in Figure 9.2, where the 12 shaded qubits correspond to the 12 quantum registers in Figure 9.1. The thick lines represent connections used in the QCBM ansatz while the thin lines represent all other available qubit connections.

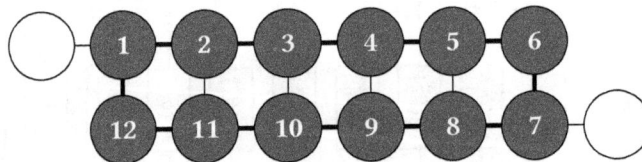

Figure 9.2: IBM's Melbourne system.

The 53-qubit Rochester device [223] in Figure 9.3 can also be used to implement this QCBM architecture. Here, we have several choices for embedding the QCBM circuit (12 linearly connected qubits forming a closed loop); shaded qubits show one such possibility.

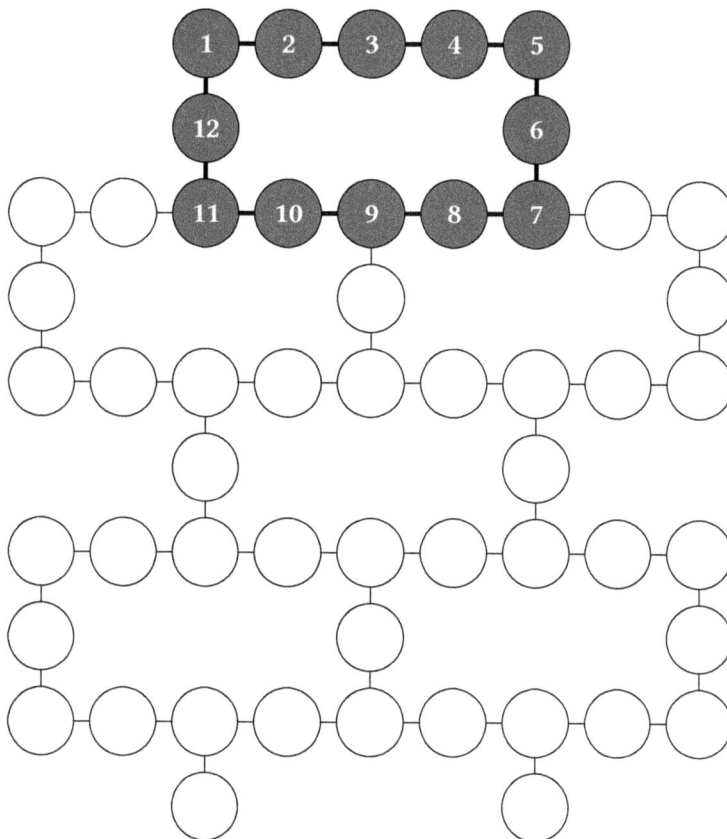

Figure 9.3: IBM's Rochester system.

IBM systems, such as Melbourne and Rochester, are based on superconducting qubits. The choice of the underlying technology means that there is a set of native gates – the quantum gates derived directly from the types of interactions that occur in the given technical realisation of the quantum chip.

In the case of IBM devices, the cross-resonance gate generates a ZX interaction that leads to a CNOT gate. When it comes to single-qubit gates, we note that R_Z is a diagonal gate given by (6.3.3) and can be implemented virtually in hardware via frame change (at zero error and duration). Therefore, it is sufficient to have just an X drive to rotate the qubit on the Bloch sphere (one can move a qubit between two arbitrary points on the Bloch sphere with the help of just two gates, R_X and R_Z).

This means that we can introduce the concept of a *hardware-efficient* architecture not only in terms of connectivity but also in terms of the choice of one-qubit and two-qubit gates. Taking into account the CNOT and CPHASE gate decomposition shown in Figures 6.19 and 6.20, the hardware-efficient QCBM architecture for the Melbourne and Rochester systems would consist of a combination of R_X and R_Z adjustable single-qubit gates and CNOT and CPHASE fixed two-qubit gates [29, 165].

> The QCBM is a PQC trained as a generative ML model. A QCBM operating on N quantum registers transforms the initial quantum state $|0\rangle^{\otimes N}$ into the quantum state encoding the learned probability distribution.

9.2 Differentiable Learning of QCBM

The output of a QCBM circuit is a bitstring that represents a sample from the probability distribution encoded in the quantum state. The circuit itself is, essentially, a mechanism of transforming an initial state $|0\rangle^{\otimes n}$ into a final state from which a sample is generated by means of measuring the qubits in the computational basis.

Different configurations of one-qubit and multi-qubit gates encode different probability distributions – the training of a QCBM consists of finding an optimal circuit configuration (ansatz) and an optimal set of adjustable parameters that minimise the distance between the probability distribution encoded in the final quantum state (before measurement, or "before sampling") and the probability distribution of the training dataset.

Following the structure we adopted in Chapter 8, we start with the differentiable learning approach, before moving to the non-differentiable learning method based on a different kind of evolutionary search heuristic – the genetic algorithm.

9.2.1 Sample encoding

In the most general case, a training dataset consists of samples containing continuous, integer, and categorical features. However, a QCBM operates on binary variables. Therefore, we need to design a method to convert continuous features into binary ones and a method for converting generated binary QCBM output (sampling) into continuous variables. The integer and binary features can be treated as special cases of continuous features and categorical features can be first converted into binary features through one-hot encoding. Such a method can be realised as a two-step routine (see Algorithm 6):

  1: Conversion of a continuous variable into the corresponding integer variable.

  2: Conversion of the integer variable into the corresponding binary variable.

Given the generated binary output, the same routine can be used in reverse mode to produce continuous samples (see Algorithm 7):

  1: Conversion of the generated binary QCBM output into integer samples.

  2: Conversion of integer samples into the corresponding continuous samples.

Algorithm 6: Continuous to integer to binary transformation (training phase)

Result: Conversion of continuous variables into M-digit binary features.

Input: $\left(X_{\text{real}}^{(n)}(l)\right)_{l=1,\ldots,N_{\text{samples}};\, n=1,\ldots,N_{\text{variables}}}$ – continuous data sample.

for $n = 1, \ldots, N_{variables}$ **do**

$\quad\left|\; X_{\min}^{(n)} \leftarrow \min_{l=1,\ldots,N_{\text{samples}}}\left(X_{\text{real}}^{(n)}(l)\right) - \varepsilon_{\min}^{(n)}, \text{ for } \varepsilon_{\min}^{(n)} \geq 0 \right.$

$\quad\left|\; X_{\max}^{(n)} \leftarrow \max_{l=1,\ldots,N_{\text{samples}}}\left(X_{\text{real}}^{(n)}(l)\right) + \varepsilon_{\max}^{(n)}, \text{ for } \varepsilon_{\max}^{(n)} \geq 0 \right.$

$\quad\left|\;\right.$ **for** $l = 1, \ldots, N_{samples}$ **do**

$$X_{\text{integer}}^{(n)}(l) \leftarrow \text{int}\left((2^M - 1)\, \frac{X_{\text{real}}^{(n)}(l) - X_{\min}^{(n)}}{X_{\max}^{(n)} - X_{\min}^{(n)}} \right)$$

$$X_{\text{binary}}^{(n)}(l) \leftarrow \text{bin}\left(X_{\text{integer}}^{(n)}(l) \right)$$

$\quad\left|\;\right.$ **end**

end

Each data sample is represented by an M-digit binary number with every digit becoming a separate feature. The total number of features is $M \times N_{\text{variables}}$.

Algorithm 7: Binary to integer to continuous transformation (sampling phase)

Result: Conversion of the generated M-digit binary sample into continuous sample.

Input: $\left(\widehat{X}_{[m]}^{(n)} \right)_{m=0,\dots,M-1;\ n=1,\dots,N_{\text{variables}}}$ – generated M-digit binary sample.

for $n = 1, \dots, N_{variables}$ **do**

$$\widehat{X}_{\text{integer}}^{(n)} := \sum_{m=0}^{M-1} 2^m \widehat{X}_{[M-1-m]}^{(n)}$$

$$\widehat{X}_{\text{real}}^{(n)} \leftarrow X_{\min}^{(n)} + \frac{1}{2^M - 1} \widehat{X}_{\text{integer}}^{(n)} \left(X_{\max}^{(n)} - X_{\min}^{(n)} \right)$$

end

Algorithms 6 and 7 describe the transformations of continuous variables into M-digit binary variables and then back into continuous variables [187]. It is important to note the role of the parameters $\varepsilon_{\min}$ and $\varepsilon_{\max}$. They are non-negative and expand the interval on which the variables are defined. In the case where $\varepsilon_{\min} = \varepsilon_{\max} = 0$, this interval is determined by the minimum and maximum values of the variable as observed in the training dataset. By allowing $\varepsilon_{\min}$ and $\varepsilon_{\max}$ to take positive values, we expand the interval of possible values the variable can take. This allows the model to generate a wider range of possible scenarios: with some (small) probability the generated values can fall outside the interval given by the samples from the training dataset.

The precision of the binary representation is feature specific. More important features can have more granular representation. The right choice of precision is important for NISQ devices that operate with a limited number of quantum registers. For example, the QCBM ansatz shown in Figure 9.1 can be used to encode two continuous variables with 6-digit binary precision each. Alternatively, the more important variable can be encoded with, e.g., 8-digit binary precision and the less important one with only 4-digit binary precision.

Figure 9.4 illustrates how the readout from 12 quantum registers can be translated into a sample consisting of two continuous variables: the value of the first one is encoded as a 7-digit binary number and the value of the second one is encoded as a 5-digit binary number. In this example, we assume that both variables take values in the interval $[-1, 1]$.

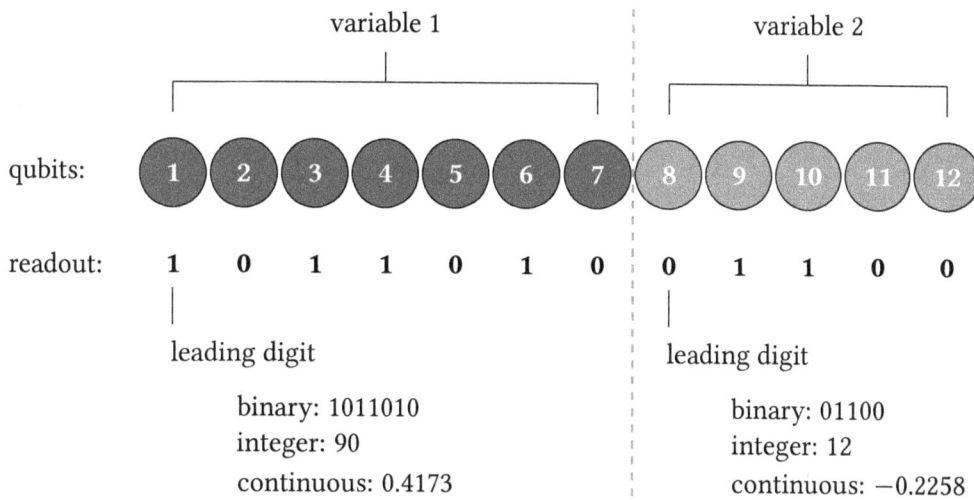

Figure 9.4: Sample QCBM readout and data transformation for two continuous variables taking values in the interval $[-1, 1]$ and where we set $\varepsilon_{\min} = \varepsilon_{\max} = 0$.

9.2.2 Choosing the right cost function

The differentiable learning of QCBM follows the same principles as that of training the quantum neural networks outlined in Chapter 8: minimisation of the cost function with the gradient descent method. The main difference is the form of the cost function. In the case of a QNN-based classifier, the cost function represents the classification error, while the cost function for QCBM represents the distance between two probability distributions: the distribution of samples in the training dataset and the distribution of samples in the generated dataset.

Let θ denote the set of adjustable QCBM parameters, $p_\theta(\cdot)$ the QCBM distribution, and $\pi(\cdot)$ the data distribution. Then we can define the cost function $L(\theta)$ as

$$L(\theta) := \sum_x |p_\theta(x) - \pi(x)|,$$

where the sum goes over all samples x in the dataset. This cost function is a strong metric but may not be the easiest to deal with [73]. An efficient alternative choice of the cost function is the *maximum mean discrepancy* [204]:

$$L(\theta) := \underset{x \sim p_\theta, y \sim p_\theta}{\mathbb{E}}[K(x, y)] - 2 \underset{x \sim p_\theta, y \sim \pi}{\mathbb{E}}[K(x, y)] + \underset{x \sim \pi, y \sim \pi}{\mathbb{E}}[K(x, y)],$$

where $K(\cdot, \cdot)$ is a *kernel function*, i.e., a measure of similarity between points in the sample space. A popular choice of kernel function is the Gaussian mixture:

$$K(x, y) = \frac{1}{c} \sum_{i=1}^{c} \exp\left(-\frac{\|x - y\|^2}{2\sigma_i^2}\right),$$

for some $c \in \mathbb{N}$ and where $(\sigma_i)_{i=1,\ldots,c}$ are the bandwidth parameters of each Gaussian kernel and $\|\cdot\|$ is the L_2 norm.

We can also explore the possibility of using *quantum kernels*. Quantum kernels can provide an advantage over classical methods for kernels that are difficult to compute on a classical device. For example, we can consider a non-variational quantum kernel method [253], which uses a quantum circuit $U(x)$ to map real data into a quantum state $|\phi\rangle$ via a *feature map*:

$$|\phi(x)\rangle = U(x) |0\rangle^{\otimes n}. \tag{9.2.1}$$

The kernel function is then defined as the squared inner product

$$K(x, y) = |\langle \phi(x) | \phi(y) \rangle|^2.$$

This quantum kernel is evaluated on a quantum computer and is hard to compute on a classical one [137]. We investigate the question of expressive power of various models in Chapter 13 and provide a detailed analysis of the quantum kernel approach in Chapter 14. Taking into account the mapping (9.2.1) and denoting $|0\rangle = |0\rangle^{\otimes n}$, the kernel becomes

$$K(x, y) = |\langle 0 | U^\dagger(x) U(y) |0\rangle|^2,$$

which is the probability of measuring the all-zero outcome. It can be calculated by measuring, in the computational basis, the state that results from running the circuit given by $U(y)$, followed by that of $U^\dagger(x)$.

9.3 Non-Differentiable Learning of QCBM

The hardware-efficient ansatz we proposed for the QCBM architecture, while simple and intuitive, may be vulnerable to barren plateaus, or regions of exponentially vanishing gradient magnitudes that make training untenable [54, 147, 324]. This provides a strong motivation for exploring a non-differentiable learning alternative such as a Genetic Algorithm (GA).

9.3.1 The principles of Genetic Algorithm

The GA is a powerful evolutionary search heuristic [229] that was introduced in Chapter 3. It performs a multi-directional search by maintaining a population of proposed solutions (chromosomes) for a given problem. Each solution is represented in a fixed alphabet with an established meaning (genes). The population undergoes a simulated evolution, with relatively good solutions producing offspring, which subsequently replace the worse ones, and the quality of a solution is estimated with some objective function (environment). GAs have found applications is such diverse fields as quantitative finance (for portfolio optimisation problems [186]) and experiments with adiabatic quantum computing (as a classical benchmark [320]).

The simulation cycle is performed in three basic steps. During the selection step, a new population is formed by stochastic sampling (with replacement). Then, some of the members of the newly selected populations recombine. Finally, all new individuals are re-evaluated. The mating process (recombination) is based on the application of two operators: mutation and crossover. Mutation introduces random variability into the population, and crossover exchanges random pieces of two chromosomes in the hope of propagating partial solutions.

The training of the QCBM specified in Figure 9.1 consists of finding an optimal configuration of the rotation angles $(\theta_i^j)_{i=1,...,12; \, j=1,...,7}$ that would minimise a chosen cost function given a particular choice of the fixed two-qubit gates. Since we only deal with 84 adjustable parameters (rather than tens of thousands), we do not need to implement the crossover mechanism and can rely on parameter mutations to achieve GA convergence to the minimum of the cost function. This significantly simplifies the algorithm.

9.3.2   Training QCBM with a Genetic Algorithm

Algorithm 8 outlines the proposed approach. However, before we provide a formal description of the algorithm, we have to specify the main individual components.

- **Solution.** The solution is a 12×7 matrix of rotation angles:

$$\theta = \begin{bmatrix} \theta_1^1 & \cdots & \theta_1^7 \\ \vdots & \ddots & \vdots \\ \theta_{12}^1 & \cdots & \theta_{12}^7 \end{bmatrix}.$$

In GA language, the matrix θ plays the role of a chromosome and its components θ_i^j play the roles of individual genes.

- **Mutation.** The genes can mutate from generation to generation. The mutation rate can be either constant or time dependent. For example, the mutation rate can start at some large value and then decrease exponentially such that it halves after each κ generations. In Algorithm 8, we adopt the following mutation dynamics:
  - A rotation angle (gene) can mutate to any of the allowed discrete values with equal probability.
  - Mutation is controlled by a single global parameter $\alpha \in (0, 1]$, which can be either constant or exponentially decreasing at some fixed rate $\beta \geq 0$.
  - Mutations happen independently for each column in θ.
  - For each column in θ, at each generation, a single rotation angle mutation happens with probability α. All rotation angles are equally likely to mutate. After that, one more mutation can happen with probability $\alpha/2$. Again, all rotation angles are equally likely to mutate. This ensures that we can have scenarios where two rotation angles within the same column can mutate simultaneously.

- **Search Space.** The rotation angles θ_i^j are defined in $[-\pi, \pi]$, which we split into 2^m equal subintervals, so that the possible values for θ_i^j are $(-\pi + n\pi/2^{m-1})_{n=0,\ldots,2^m-1}$. A rotation angle can mutate into any of these values. The search space can quickly become enormous even for the relatively small values of m. For example, for $m = 7$ we have 128 possible values for each rotation angle making the total number of possible configurations $\sim 10^{177}$. The GA can only explore a tiny fraction of the search space. But due to the GA's ability to propagate best solutions and to avoid being trapped in local minima, the algorithm can achieve reasonably fast convergence to the solution in the vicinity of the global minimum. For a detailed analysis of the rate of convergence of genetic algorithms, we refer the interested reader to [138, 288].

- **Cost Function.** A cost function is a measure of how far the distribution of generated samples is from the distribution of original samples provided by the training dataset. Let $u := (u_1, \ldots, u_K)$ be a sample from the training dataset and $v(\theta) := (v_1(\theta), \ldots, v_K(\theta))$ a sample from the QCBM-generated dataset that corresponds to a particular configuration of rotation angles θ. Let us order these samples from the smallest to the largest with any suitable sort($\cdot$) function:

$$\bar{u} = \text{sort}(u), \quad \bar{v}(\theta) = \text{sort}(v(\theta)). \tag{9.3.1}$$

The cost function $L(\cdot)$ can then be defined as

$$L(\theta) := \sum_{k=1}^{K} (\bar{u}_k - \bar{v}_k(\theta))^2. \tag{9.3.2}$$

The sort($\cdot$) function in (9.3.1) can be, e.g., quicksort [145] or mergesort [180], which belong to the class of divide-and-conquer algorithms. Alternatively, it can be, e.g., heapsort [328] – a comparison-based sorting algorithm.

Algorithm 8: Genetic Algorithm

Result: Optimal configuration of the set of QCBM parameters θ^* minimising the
    cost function.

Input:

- $u \in \mathbb{R}^K$: vector of sample training dataset;
- L: number of iterations (generations);
- M: number of best solutions in the given generation, chosen for further mutation;
- N: number of solutions in each generation ($N = DM, D \in \mathbb{N}$);
- α, β: mutation parameters;
- m: search space parameter.

    The possible values of rotation angles are $\left(-\pi + \dfrac{\nu\pi}{2^{m-1}}\right)_{\nu=0,\ldots,2^m-1}$.

Initialise and evaluate the first generation of solutions:

for $n = 1, \ldots, N$ **do**
    Generate a configuration $\theta(0; n)$ by randomly drawing each rotation
    angle $\theta_i^j(0; n)$ from the uniform distribution on the set of possible values of
    rotation angles given by m.

    for $k = 1, \ldots, K$ **do**
        Run the quantum circuit with configuration $\theta(0; n)$ and generate new
        sample $v_k(\theta(0; n))$.
    end

    Evaluate the cost function $L(\theta(0; n))$.
end

Order solutions from best (minimum of the cost function) to worst (maximum of the
cost function).

$\theta^* \leftarrow$ configuration corresponding to the minimum of the cost function.

Iterations:

for $l = 1, \ldots, L$ **do**

$\qquad \alpha \leftarrow \alpha e^{-\beta}$

$\qquad$ Select M best solutions from generation $l - 1$ and generate new solutions

$\qquad$ $(\theta(l; n))_{n=1,\ldots,N}$ by mutating the rotation angles using the updated mutation

$\qquad$ rate α. Each of the M best solutions is used to produce D new solutions.

$\qquad$ **for** $n = 1, \ldots, N$ **do**

$\qquad\qquad$ **for** $k = 1, \ldots, K$ **do**

$\qquad\qquad\qquad$ Run the quantum circuit with $\theta(l; n)$ and generate new

$\qquad\qquad\qquad$ sample $v_k(\theta(l; n))$.

$\qquad\qquad$ **end**

$\qquad\qquad$ Evaluate the cost function $L(\theta(l; n))$.

$\qquad$ **end**

$\qquad$ Order the solutions from best (minimum of the cost function) to worst

$\qquad$ (maximum of the cost function).

$\qquad$ $\theta^*(l) \leftarrow$ configuration corresponding to the minimum of the cost function (l-th

$\qquad$ generation).

$\qquad$ **if** $L(\theta^*(l)) < L(\theta^*)$ **then**

$\qquad\qquad$ $\theta^* \leftarrow \theta^*(l)$

$\qquad$ **end**

end

Having described the training algorithm, we now specify the classical benchmark before comparing the results obtained by the quantum and the classical generative models on the sample datasets.

9.4 Classical Benchmark

There is a deep connection between the QCBM and its classical counterpart – the Restricted Boltzmann Machine [60]. The RBM, introduced and discussed in Chapter 5 in the context of quantum annealing, is a generative model inspired by statistical physics, where the probability of a particular data sample, v, is given by the Boltzmann distribution:

$$\mathbb{P}(v) = \frac{1}{Z}e^{-E(v)}. \tag{9.4.1}$$

Here, $E(v)$ is the (positive) *energy* of the data sample (data samples with lower energy have higher probabilities) and Z is the partition function, namely the normalisation factor of the probability density:

$$Z = \sum_v e^{-E(v)}.$$

Alternatively, we can use the inherent probabilistic nature of quantum mechanics that allows us to model the probability distribution using a quantum state $|\psi\rangle$:

$$\mathbb{P}(v) = \langle\psi|\mathcal{P}_v^\dagger\mathcal{P}_v|\psi\rangle, \tag{9.4.2}$$

where $\mathcal{P}_v$ is the measurement operator introduced in Section 1.2.3 and, since the quantum state $|\psi\rangle$ is a unit vector, we have

$$\langle\psi|\psi\rangle = 1.$$

We realise this approach in the QCBM, where generative modelling of probability density is translated into learning a quantum state. The sole purpose of the QCBM's parameterised circuit is to create the quantum state $|\psi\rangle$ that encodes the desired probability distribution starting from the initial state $|0\rangle^{\otimes n}$, with sampling performed by applying the measurement operators.

Therefore, providing a classical benchmark for QCBM consists in finding a suitable RBM configuration that will allow us to compare two methods generating the probability distribution $\mathbb{P}(v)$: one given by (9.4.1) for RBM and another one given by (9.4.2) for QCBM [184].

Figure 9.5 shows an RBM with 12 stochastic binary visible activation units and 7 stochastic binary hidden activation units, where $(a_i)_{i=1,\ldots,12}$, $(b_j)_{j=1,\ldots,7}$, and $(w_{ij})_{i=1,\ldots,12;\ j=1,\ldots,7}$ denote, respectively, the biases for the visible and hidden layers and the network weights.

This network architecture makes the RBM equivalent to the QCBM as described in Section 9.1 in the sense that both generative models have the same number of adjustable parameters (the number of RBM weights is equal to the number of adjustable rotation angles in a QCBM) and the number of visible activation units is equal to the number of quantum registers. The latter ensures that both generative models can learn the empirical distribution of a continuous random variable with the same precision (12-digit binary representation).

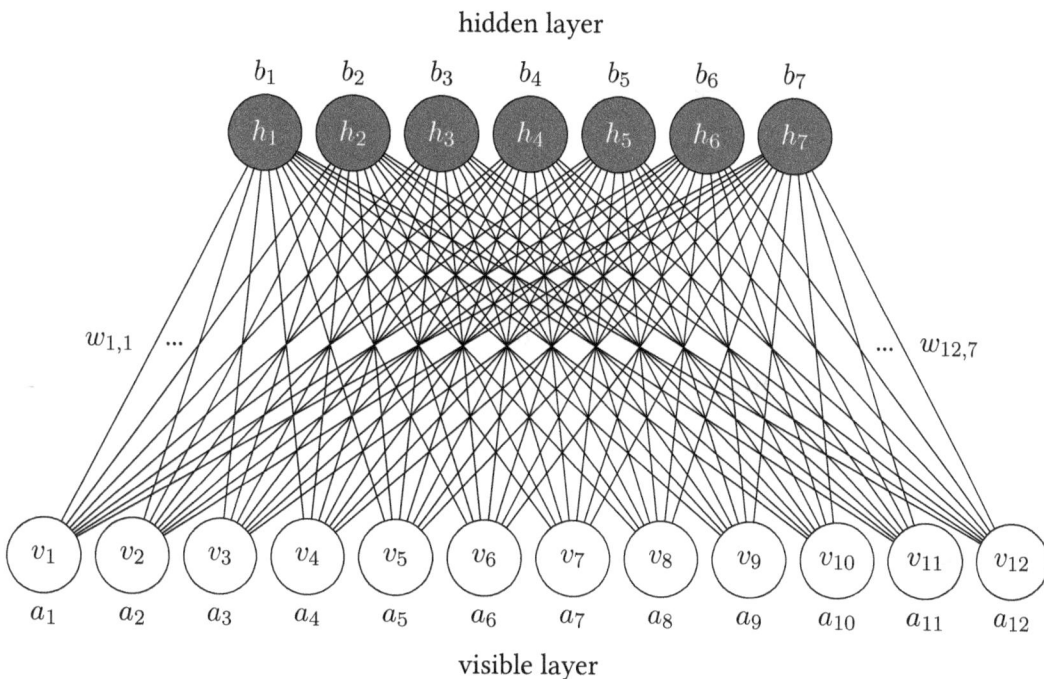

Figure 9.5: RBM(12, 7).

QCBM performance should be compared against the performance of its classical counterpart, the RBM. Both models operate on the binary representation of the dataset and have a comparable number of adjustable parameters.

9.5 QCBM as a Market Generator

The most obvious financial application of a QCBM is as a market generator. An efficient generation of realistic market scenarios, for example, sampling from the joint distribution of risk factors, is one of the most important and challenging problems in quantitative finance today. We thus need to investigate how well QCBM can execute this task, and compare it to classical benchmarks.

9.5.1 Non-parametric modelling of market risk factors

Historically, the problem of producing reliable synthetic market scenarios was solved through sampling from some easy-to-calibrate parametric models, such as the multivariate Normal distribution of risk factor log-returns (equities) or a Gaussian copula combining the multivariate Normal dependence structure with heavy-tailed univariate marginal distributions of individual risk factors (credit). However, there are well-known issues with this approach that often outweigh the benefits provided by simplicity and transparency [232].

A parametric model is often a poor approximation of reality. To be useful, it has to be relatively simple: one should be able to describe the key features of the risk factor distribution with a handful of parameters achieving the best possible fit to either the empirical distribution derived from historical data or from prices of traded instruments observed in the market at the time of model calibration. Making the parametric model too complex would lead to overfitting and poor generalisation.

It is even more difficult to model a realistic dependence structure. A typical parametric approach used in most Monte Carlo risk engines starts with modelling the dynamics of various risk factors independently, and then imposes a dependence structure by correlating the corresponding stochastic drivers. These are, almost invariably, Brownian motions, and the linear correlations between them are supposed to be sufficient to construct the joint distribution of risk factors.

An alternative approach is to use non-parametric modelling, where the joint and marginal distributions of risk factors are learned directly from the available datasets. Classically, we can realise this approach with the help of an RBM – the classical benchmark of choice described in the previous section and successfully applied to a number of financial use cases [187, 188]. Another possibility is to use the Generative Adversarial Network (GAN) framework, where the distribution learned from the dataset by a generative neural network is tested by a discriminative neural network trying to judge whether samples are coming from the true distribution (data) or from the reconstructed distribution (generated samples) [119].

Chapter 13 explores the question of the larger expressive power of QCBM in comparison with classical neural networks (RBM). However, the first step should be an experimental verification of their performance characteristics. With this in mind, we would like to test the ability of both QCBM and RBM to learn relatively complex probability distributions and then efficiently sample from them.

9.5.2 Sampling from the learned probability distributions

We are going to test the performance of QCBM and RBM on two datasets:

Dataset A. A heavy-tailed distribution of daily S&P 500 index returns observed between 5 January 2009 and 22 February 2011 (UCI Machine Learning Repository [9, 10]). The dataset consists of 536 samples.

Dataset B. A specially constructed distribution of a continuous random variable with a highly spiky probability density function (pdf) modelled as a mixture of Normal distributions. The dataset consists of 5,000 generated samples from a mixture of four Normal distributions with the following means, standard deviations, and weights:

Mean	Standard deviation	Weight
−3	0.3	0.1
−1	0.3	0.2
1	0.3	0.3
3	0.3	0.4

Table 9.1: Parameters of the mixture of standard Normal distributions.

In both cases, we convert the continuous samples into the corresponding 12-digit binary representation as per Algorithm 6. Once the networks are trained (QCBM with Algorithm 8 and RBM with Algorithm 2), we generate new samples: 536 new samples for Dataset A and 5,000 new samples for Dataset B. This allows us to visualise the quality of the generated samples (once they are converted into the corresponding continuous representation as per Algorithm 7) by producing the empirical pdf and the QQ-plots as shown in Figures 9.6 and 9.7, which display sample simulation results for the fully trained models. We can see that both QCBM(12, 7) and RBM(12, 7) can successfully learn complex empirical distributions (heavy-tailed in the case of Dataset A and light-tailed with spiky pdf in the case of Dataset B). We have chosen CX for the fixed gates in QCBM and used the Qiskit quantum simulator to simulate the quantum parts of the training and sampling algorithms.

The following sets of hyperparameters were used to train the models:

- **Genetic Algorithm for training QCBM (Algorithm 8)**

  $N = 1000$, $M = 25$, $m = 7$, $\alpha = 1.0$, $\beta = 0.013863$, $\kappa = 50$, $L = 200$.

  The value of β ensures that the mutation rate halves after each κ generations.

- **Contrastive Divergence algorithm for training RBM**

  (sklearn.neural_network.BernoulliRBM)

  n_components = 7 – number of hidden activation units for RBM(12, 7)

  learning_rate = 0.0005 – learning rate η in Algorithm 2

  batch_size = 10 – size of the training minibatches S in Algorithm 2

  n_iter = 40000 – number of iterations

Although a visual inspection of the pdf and QQ-plots in Figures 9.6 and 9.7 suggests that both the QCBM and the RBM are doing a good job in generating high-quality samples from the learned empirical distributions encoded in model parameters, we would like to have a more objective measure of the model performance. This is especially important since we deal with generative models and very little can be concluded from a single model run.

Running the quantum circuit multiple times for a particular configuration of model parameters (e.g., an optimal set of rotation angles found with the help of the GA) results in the distribution of objective function values. This gives us an idea of what metrics can be used to measure the performance of the QCBM and RBM [184]. The cost function (9.3.2) we used for training the QCBM can be calculated on the samples generated by the RBM. In other words, we can compare the performance of the QCBM and RBM by comparing the distributions of the cost function values calculated for the samples generated by these models.

Table 9.2 shows the means and standard deviations of the cost functions calculated for 100 runs of QCBM(12, 7) and RBM(12, 7). Each run generated 5,000 samples from the learned empirical distribution (the models were trained on Dataset B, which consists of 5,000 samples from the mixture of four Normal distributions).

Model	Mean	Standard deviation
QCBM(12, 7)	30.5	23.6
RBM(12, 7)	39.6	30.8

Table 9.2: Cost function statistics for the models trained on Dataset B.

It is clear from Table 9.2 that QCBM(12, 7) with a weakly optimised set of hyperparameters performs better than RBM(12, 7) trained with equally weakly optimised hyperparameters (a small learning rate combined with a large number of iterations and the small size of the minibatches [142]). Although this cannot be seen as proper evidence of quantum advantage, this nevertheless opens the door to promising further research.

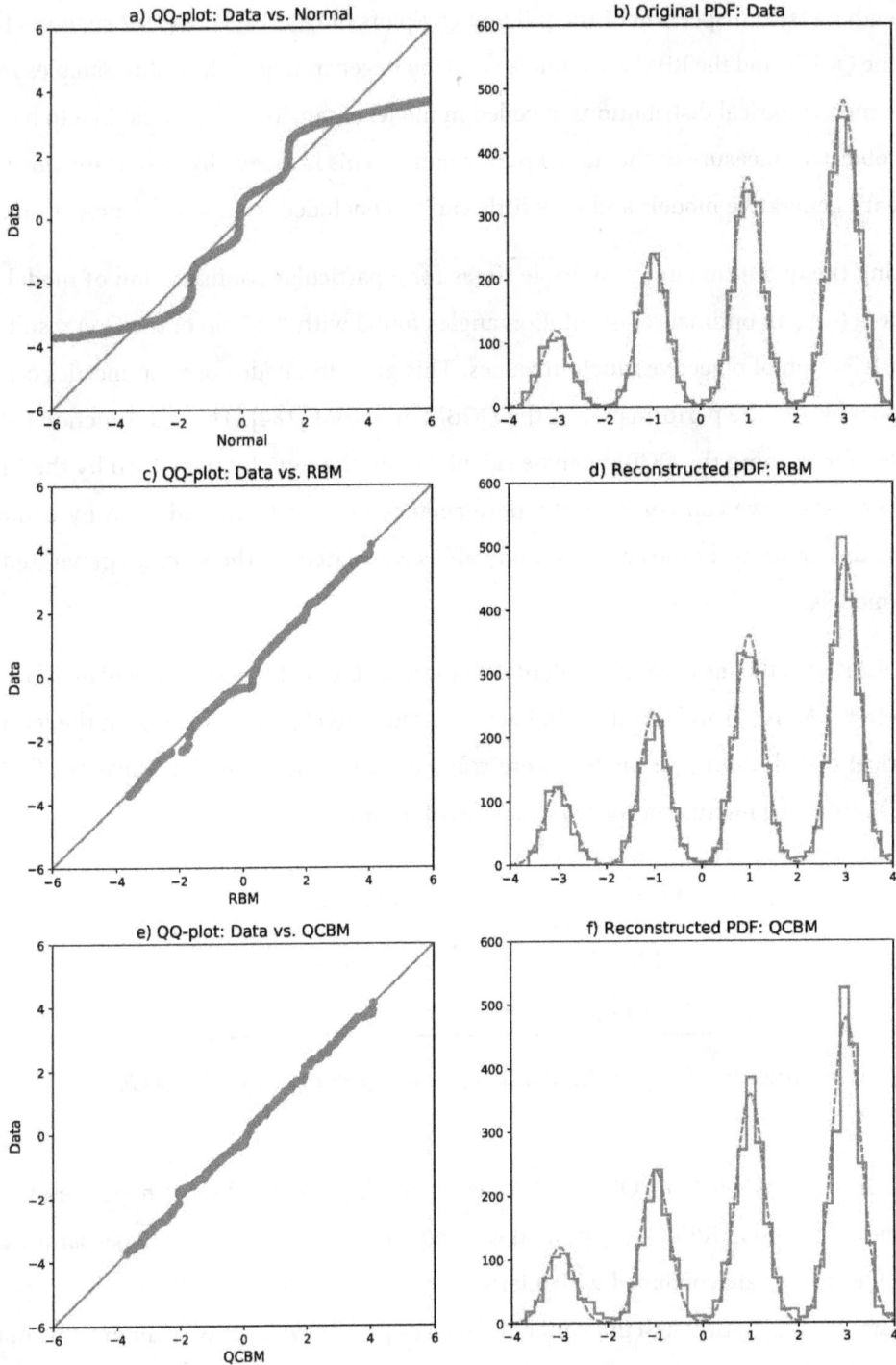

Figure 9.6: Mixture of Normal distributions.

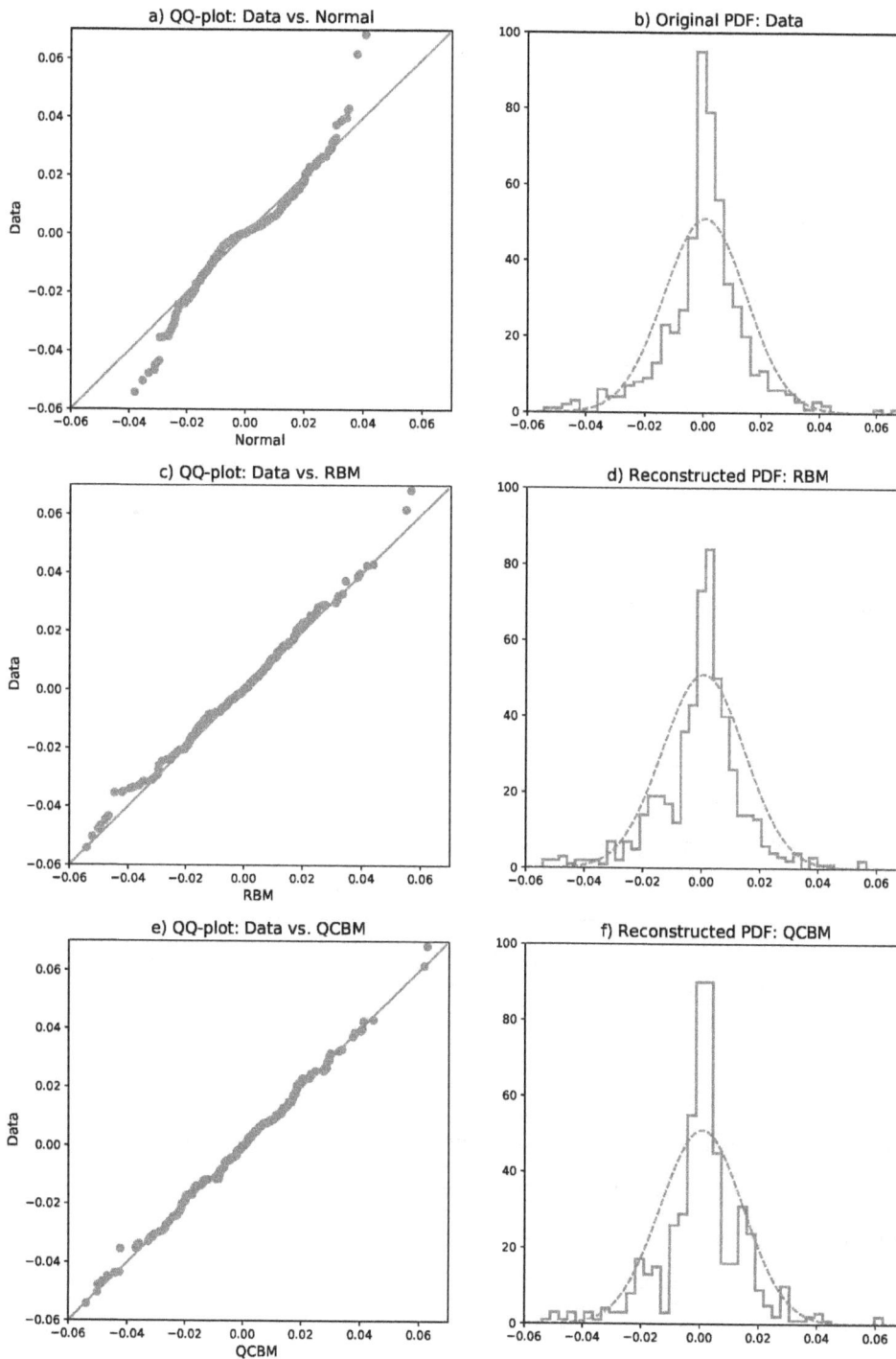

Figure 9.7: Distribution of S&P 500 index returns.

Let us now turn our attention to Dataset A. The dataset consists of just 536 samples and, as we can see in Figure 9.7, the empirical pdf displays pronounced heavy tails, which are also clearly seen in the QQ-plot against the Normal distribution. The relatively small number of samples means that we have to deal with a substantial amount of noise. Therefore, we need to use a robust statistical test to compare the QCBM and RBM. Since we are working with a univariate distribution, we can estimate the quality of generated samples with the Kolmogorov-Smirnov (KS) test [254].

Table 9.3 provides the p-values and KS statistics for the RBM- and QCBM-generated samples as well as a Normal distribution fitted to the original dataset (by matching the first two moments). The p-value represents the probability of obtaining test results supporting the null hypothesis of the two datasets coming from the same distribution. In the context of our numerical experiments, the larger the p-value, the more likely the generated samples were drawn from the correct distribution.

Distribution	p-value	KS statistic
Normal	0.004 ± 0.009	0.121 ± 0.017
RBM generated samples	0.46 ± 0.23	0.055 ± 0.011
QCBM generated samples	0.46 ± 0.11	0.053 ± 0.005

Table 9.3: p-value and KS statistic for Normal, RBM-, and QCBM-generated samples in the format: mean $\pm$ standard deviation. Number of Normal, RBM-, and QCBM-generated datasets: 20. Number of samples in each generated dataset: 536 (equal to the number of samples in the original dataset).

The KS statistic takes the largest absolute difference between the two distribution functions across all values of the random variable. The larger the KS statistic, the less likely the generated samples were drawn from the correct distribution. The KS statistic can be compared with the critical values calculated for the given confidence level and number of samples. For example, the critical value corresponding to the 95th percentile confidence level and 536 samples in both datasets is 0.0587. If the KS statistic is larger, then, with 95%

certainty, we can reject the null hypothesis that 536 generated samples were drawn from the right distribution.

The first observation is that we can definitely reject the null hypothesis that the daily S&P 500 index returns are Normally distributed. The corresponding p-value is much smaller than 1, and the KS statistic is twice the critical value. More importantly, the QCBM performs at par with the RBM in terms of both the p-value and the KS statistic: we therefore cannot reject the null hypothesis that QCBM- and RBM-generated samples were drawn from the same distribution as the original dataset.

9.5.3 Training algorithm convergence and hyperparameter optimisation

Next, we would like to explore the GA behaviour for various model configurations. In particular, it is interesting to investigate the algorithm convergence for different types of fixed gates, not just CX, and for different choices of the mutation rate. The charts in Figure 9.8 confirm our intuition about CX being the best choice of fixed gate given the configuration of one-qubit gates (Figure 9.1) and the exponentially decreasing mutation rate performing better than the constant mutation rates. Here, we continue working with Dataset B.

As we can see in Figure 9.1, the fixed gates are flanked by one-qubit gates performing rotations around the z axis. Therefore, adding another rotation around the z axis by $\phi = \pi$ ($Z = R_z(\pi)$) may not offer the same flexibility as rotation around the x axis by $\phi = \pi$ ($X = R_x(\pi)$). Controlled rotations around the z axis by an angle $\phi < \pi$ are likely to perform even worse. This is exactly what we see in Figure 9.8 (left chart) for three different types of fixed gates: CX, CZ, and $CR_z(\pi/4)$.

Our intuition about the optimal choice of mutation rate suggests that it should be productive to start the algorithm with a really large mutation rate in order to explore the search space as broadly as possible (the "exploration" phase). Then, as the algorithm finds progressively better solutions, it should be useful to reduce the mutation rate in order to perform a more detailed search in the vicinity of the best solutions found so far (the "exploitation" phase).

As the algorithm converges, we may want to perform more and more refined searches by only mutating one or two parameters. Figure 9.8 (right chart) shows that this is indeed the case. Here, the maximum value of the mutation rate is $\alpha = 1.0$ and the minimum value is $\alpha = 0.0625$ – the value reached after $L = 200$ algorithm iterations when the algorithm is run with the initial value of mutation rate $\alpha = 1.0$ and exponential decay factor $\beta = 0.013863$.

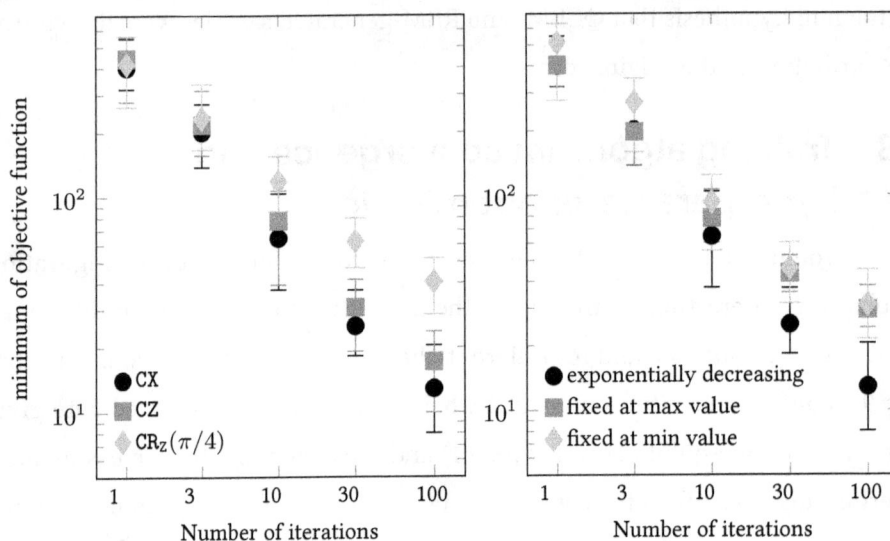

Figure 9.8: **Left:** *GA convergence as a function of fixed gate type.* **Right:** *GA convergence as a function of mutation rate for* CX *fixed gates. Dots indicate mean values and error bars indicate the 10th and the 90th percentiles. GA parameters:* $N = 1000$, $M = 25$, $m = 7$, 20 *GA runs.*

Finally, we need to investigate the convergence of the algorithm as a function of the rotation angle discretisation scheme. In principle, an arbitrary rotation poses a problem as it must be approximated by a sequence of discrete gates because only discrete sets of gates can be implemented fault-tolerantly [195]. Since a GA operates on a discrete set of rotation angles, we face a trade-off between higher accuracy achieved through a finer discretisation scheme and implementation efficiency in the case of a less granular set of rotation angles. Additionally, all rotation gates can be executed with finite precision and the discretisation scheme should take this into account. Hence, in order to facilitate the efficient implementation of the rotation gates $R_X(\theta)$ and $R_Z(\theta)$, the GA operates on the rotation angles θ that take discrete values $(-\pi + \nu\pi/2^{m-1})_{\nu=0,\dots,2^m-1}$, thus splitting the $[-\pi, \pi]$ interval into 2^m equal subintervals.

Therefore, we must answer the question of GA convergence for various values of m. Figure 9.9 shows the minimum values of the objective function (9.3.2) as a function of the number of algorithm iterations for three different values of m:

- $m = 3$, rotation angle step $\Delta\theta = \pi/4$;
- $m = 5$, rotation angle step $\Delta\theta = \pi/16$;
- $m = 7$, rotation angle step $\Delta\theta = \pi/64$.

We can see that the GA performance improves only marginally for $m > 5$. This is good news suggesting that it might be sufficient to operate with rotation angle step $\Delta\theta = \pi/16$ to achieve the desired precision in learning the target distribution.

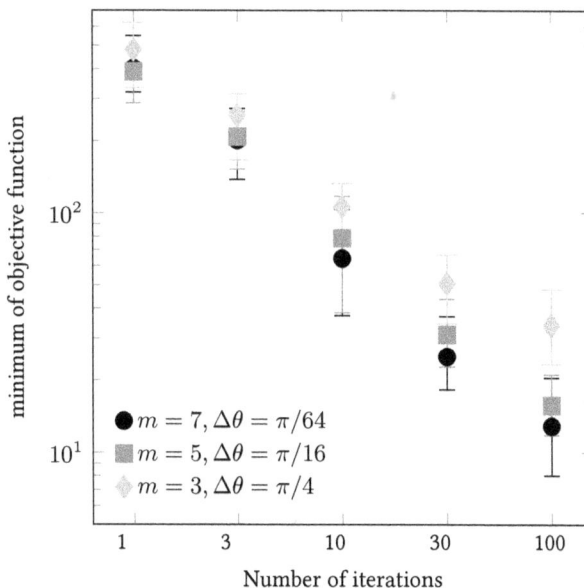

Figure 9.9: GA convergence as a function of the rotation angle discretisation scheme for CX fixed gates and exponentially decreasing mutation rate. Dots indicate mean values and error bars indicate the 10th and the 90th percentiles. GA parameters: $N = 1000$, $M = 25$, $\alpha = 1.0$, $\beta = 0.013863$, 20 GA runs.

The non-differentiable learning of a QCBM with a GA is a viable approach to the training of PQCs. A QCBM trained with a GA performs at least as well as an equivalent classical neural network (an RBM). The performance of the QCBM and its classical counterpart were tested on two different datasets (heavy-tail distributed samples derived from the financial time series and light-tail distributed samples drawn from the specially constructed distribution with a spiky pdf) and in both cases the QCBM demonstrated its ability to learn the empirical distribution and generate new synthetic samples that have the same statistical properties as the original ones, as can be seen in the pdf and QQ-plots.

Analysing the GA convergence for different sets of hyperparameters, we observe that the best results were achieved with CX fixed gates and an exponentially decreasing mutation rate (starting from the maximum value of the mutation rate and setting the decay rate at a reasonably small value). More importantly, we see that more granular rotation angle discretisation schemes provide progressively less incremental value beyond some point. This means that for many practical purposes it is sufficient to implement rotations with the step $\Delta\theta = \pi/16$ in order to encode target distribution with the desired accuracy for deep enough QCBM architectures (at least two layers of fixed two-qubit gates). Since qubit rotations on NISQ devices can be implemented with finite precision, this ensures that QCBMs can be used productively for many real-world use cases.

> The QCBM is a viable choice for building market generators. It performs at least as well as its classical counterpart, the RBM, and demonstrates potential for achieving quantum advantage on near-term quantum processors.

Summary

In this chapter, we learned how to construct and train a generative QML model – Quantum Circuit Born Machine. We started with the general concept of a PQC as a generative model, where the readout operation produces a sample from the probability distribution encoded in the PQC parameters.

Next, we introduced the concept of a hardware-efficient PQC ansatz. Additionally, to build a model that is compatible with QPU connectivity and can easily be embedded into a QPU graph, we tried to use adjustable (one-qubit) and fixed (two-qubit) gates from the set of the native quantum gates for the given system.

Then, we studied differentiable and non-differentiable learning algorithms and experimented with a QCBM trained using a GA. Comparison with the classical benchmark (an RBM) demonstrated a realistic possibility of quantum advantage for generative quantum machine learning models.

Finally, we explored the question of training algorithm convergence for various sets of model parameters.

In the next chapter, we will study another important and exceptionally promising QML model – the Variational Quantum Eigensolver.

10

Variational Quantum Eigensolver

Parameterised quantum circuits can find many possible applications outside the quantum machine learning use cases considered in the previous chapters. They can be used to solve problems as diverse as portfolio optimisation [182] and protein folding [271]. However, one aspect remains the same regardless of the specifics of the particular algorithm: the construction of a quantum state with desired characteristics through an optimal PQC configuration (ansatz) and an optimal set of adjustable PQC parameters. This, in turn, is done through the minimisation of some cost function – it can be a classification error in the case of a QNN-based classifier or a distance between two distributions in the case of QCBM.

The Variational Quantum Eigensolver (VQE) is a PQC-based algorithm that aims to find the smallest eigenvalue (the lowest energy) of a problem Hamiltonian. As we know from Chapter 3, the objective functions of many NP-hard combinatorial optimisation problems can be encoded in the Hamiltonians of quantum systems – thus, finding the ground state of the Hamiltonian gives us the minimum of the objective function. The VQE was originally proposed in [252] and quickly became one of the most popular tools for experimenting with the wide range of optimisation problems solvable on NISQ devices [165, 230]. The variational part of the algorithm refers to the systematic search for the best possible approximation of the ground state by trying various PQC ansätze and configurations of adjustable PQC parameters – the variational approach.

10.1   The Variational Approach

Let us start with recollecting the details of training discriminative (QNN) and generative (QCBM) models. In both cases, our task was to find an optimal configuration of PQC parameters (e.g., rotation angles of adjustable one-qubit gates) such that the resulting quantum state had the desired properties: we could either sample from the encoded probability distribution (generative model) or obtain a class label for the given sample (discriminative model). The process of finding an optimal configuration of PQC parameters is called *learning* when we are dealing with QML use cases. This learning can be done either in a differentiable or in a non-differentiable way, but it always consists of the minimisation of some cost function through varying the adjustable circuit parameters.

What if the cost function we want to minimise is encoded in the problem Hamiltonian and the task is formulated as finding its ground state? In Chapter 3, we saw how this problem can be solved on a quantum annealer. But it is also possible to find the lowest energy state of a quantum system (or at least a good approximation) using a gate model quantum computer within the PQC framework. The characteristic equation for the Hamiltonian $\mathcal{H}$ reads

$$\mathcal{H} \left| \psi_i \right\rangle = E_i \left| \psi_i \right\rangle,$$

where $\left| \psi_i \right\rangle$ is an eigenstate associated with the eigenvalue E_i. The objective is to find the smallest eigenvalue E_0 (the lowest energy) of $\mathcal{H}$ corresponding to the ground state (the lowest energy state) $\left| \psi_0 \right\rangle$. This would be a straightforward task if the latter was known since the eigenvalue (energy) of $\mathcal{H}$ is simply the expectation of $\mathcal{H}$:

$$\langle \psi_i | \, \mathcal{H} \, | \psi_i \rangle = \langle \psi_i | \, E_i \, | \psi_i \rangle = E_i \langle \psi_i | \psi_i \rangle = E_i.$$

We will explain below how this expectation is calculated on a quantum computer. However, in most cases, the ground state is not known. In fact, the task is to find the ground state that encodes the solution to the optimisation problem by searching for the state that minimises the expectation value of $\mathcal{H}$. What we can do is construct a progressively better approximation of the ground state, yielding a tighter and tighter upper bound for the ground state energy E_0.

The variational approach is motivated by the spectral theorem introduced in Chapter 1, which allows us to expand the Hermitian Hamiltonian $\mathcal{H}$ as

$$\mathcal{H} = \sum_i E_i \, | \psi_i \rangle \, \langle \psi_i | \,. \tag{10.1.1}$$

Assume that we constructed a state $| \psi \rangle$, which is an approximation of the true ground state $| \psi_0 \rangle$. As we know from Chapter 1, the expectation value of $\mathcal{H}$ in state $| \psi \rangle$ is $\langle \psi | \, \mathcal{H} \, | \psi \rangle$. Substituting $\mathcal{H}$ given by (10.1.1) into this, we obtain, by linearity,

$$\begin{aligned}
\langle \psi | \, \mathcal{H} \, | \psi \rangle &= \langle \psi | \left(\sum_i E_i \, | \psi_i \rangle \, \langle \psi_i | \right) | \psi \rangle \\
&= \sum_i E_i \, \langle \psi | \psi_i \rangle \, \langle \psi_i | \psi \rangle = \sum_i E_i \, | \langle \psi, \psi_i \rangle |^2 \,.
\end{aligned} \tag{10.1.2}$$

The expression (10.1.2) shows that the expectation of $\mathcal{H}$ on any state $| \psi \rangle$ can be expressed as a linear combination of the eigenvalues of $\mathcal{H}$ with all weights greater than or equal to zero, since $| \langle \psi, \psi_i \rangle |^2 \geq 0$ for each i. Therefore, we obtain

$$\langle \psi | \, \mathcal{H} \, | \psi \rangle \geq E_0,$$

since E_0 is the smallest eigenvalue of $\mathcal{H}$ and all coefficients (weights) in the linear combination (10.1.2) are non-negative.

The role of PQC is to produce the candidate state $|\psi\rangle$. The *variational* part of the algorithm consists of iterative improvements of the candidate state (iterative updates of the adjustable parameters). This is something that can be done as a classical part of the hybrid quantum-classical protocol. The quantum part of the algorithm consists of running the PQC and then measuring $\mathcal{H}$ on the constructed quantum state in order to obtain the expectation value of $\mathcal{H}$.

> The variational approach allows us to solve hard optimisation problems encoded in the Hamiltonian on the digital gate model quantum computer – an alternative to adiabatic quantum computing since not all optimisation problems can be efficiently formulated in a QUBO format.

10.2 Calculating Expectations on a Quantum Computer

The key element of the VQE algorithm is the calculation of the expectation value. We now show how this can be performed on a quantum computer. We start with the one-qubit case and then generalise the proposed approach to the two-qubit and multi-qubit cases.

10.2.1 The one-qubit case

Consider the simplest case of a one-qubit system. Since any 2×2 unitary and Hermitian matrix can always be decomposed into a sum of the Pauli matrices X, Y, Z, and an identity matrix I (more on this in Section 10.2.3), we can represent any one-qubit Hamiltonian as

$$\mathcal{H} = a\text{X} + b\text{Y} + c\text{Z} + d\text{I}, \tag{10.2.1}$$

where a, b, c, and d are some real coefficients. For a given state $|\psi\rangle$, the expectation value of the Hamiltonian (10.2.1) is given by

$$\langle \mathcal{H} \rangle \equiv \langle \psi| \, \mathcal{H} \, |\psi\rangle = a \, \langle \psi| \, \text{X} \, |\psi\rangle + b \, \langle \psi| \, \text{Y} \, |\psi\rangle + c \, \langle \psi| \, \text{Z} \, |\psi\rangle + d \, \langle \psi| \, \text{I} \, |\psi\rangle . \tag{10.2.2}$$

The expectation value of $\mathcal{H}$ is computed by adding the expectation values of all its terms, which means that we can compute the expectation values of the Pauli terms independently and then sum them up to obtain $\langle \mathcal{H} \rangle$. We can do it by first constructing the state $|\psi\rangle$ with the help of a PQC and then by performing measurement in the computational basis. The cycle of constructing the state and performing measurement should be repeated a sufficient number of times in order to obtain accurate statistics. Let us go through the terms of the Hamiltonian (10.2.1) one by one to see how it can be done.

We start with the last term, which is the identity operator $\mathbf{I}$ multiplied by the coefficient d. This is a trivial case and we do not even need to run a quantum circuit to compute its expectation value, since the expectation value of $\mathbf{I}$ is 1:

$$\langle \psi | \, \mathbf{I} \, | \psi \rangle = \langle \psi | \psi \rangle = 1,$$

so that this term will contribute d to $\langle \mathcal{H} \rangle$.

We move now to the next term, $c\mathbf{Z}$. The measurement is performed in the computational basis, which is the z basis. In this basis, $|\psi\rangle$ can be represented as a superposition of the basis states $|0\rangle$ and $|1\rangle$ as

$$|\psi\rangle = \alpha_z |0\rangle + \beta_z |1\rangle,$$

with $\alpha_z, \beta_z \in \mathbb{C}$. The expectation $\langle \psi | \, \mathbf{Z} \, | \psi \rangle$ is then calculated as

$$
\begin{aligned}
\langle \psi | \, \mathbf{Z} \, | \psi \rangle &= |\alpha_z|^2 \, \langle 0 | \, \mathbf{Z} \, | 0 \rangle + \alpha_z^* \beta_z \, \langle 0 | \, \mathbf{Z} \, | 1 \rangle + \alpha_z \beta_z^* \, \langle 1 | \, \mathbf{Z} \, | 0 \rangle + |\beta_z|^2 \, \langle 1 | \, \mathbf{Z} \, | 1 \rangle \\
&= |\alpha_z|^2 \, \langle 0 | 0 \rangle - \alpha_z^* \beta_z \, \langle 0 | 1 \rangle + \alpha_z \beta_z^* \, \langle 1 | 0 \rangle - |\beta_z|^2 \, \langle 1 | 1 \rangle \\
&= |\alpha_z|^2 - |\beta_z|^2,
\end{aligned}
$$

using the definition of the $\mathbf{Z}$ gate (Chapter 6) and orthogonality of the basis states.

By definition, $|\alpha_z|^2$ and $|\beta_z|^2$ are the probabilities that after the z basis measurement, the quantum state $|\psi\rangle$ will become $|0\rangle$ or $|1\rangle$, respectively. In order to find that value, we should run the quantum circuit (to construct state $|\psi\rangle$) and then perform measurement N

times. The probability of finding a qubit in state $|0\rangle$ is then estimated as n_0/N, where n_0 is the number of state $|0\rangle$ measurements. Similarly, the probability of finding a qubit in state $|1\rangle$ can be estimated as n_1/N, where n_1 is the number of state $|1\rangle$ measurements.

Therefore, the contribution of the Z term to $\langle \mathcal{H} \rangle$ is given by

$$c \langle \psi | \, Z \, | \psi \rangle = c \frac{n_0 - n_1}{N}.$$

Now, we can move to the first two terms on the right side of (10.2.2). Recall that $|0\rangle$ and $|1\rangle$ are the eigenstates of Z with corresponding eigenvalues $+1$ and -1, namely,

$$Z \, |0\rangle = |0\rangle \quad \text{and} \quad Z \, |1\rangle = - \, |1\rangle. \tag{10.2.3}$$

Furthermore, the eigenstates of X are

$$|+\rangle = \frac{|0\rangle + |1\rangle}{\sqrt{2}} \quad \text{and} \quad |-\rangle = \frac{|0\rangle - |1\rangle}{\sqrt{2}},$$

and the eigenstates of Y are

$$|R\rangle = \frac{|0\rangle + i \, |1\rangle}{\sqrt{2}} \quad \text{and} \quad |L\rangle = \frac{|0\rangle - i \, |1\rangle}{\sqrt{2}}.$$

Their corresponding eigenvalues also are $+1$ and -1, so that

$$X \, |+\rangle = |+\rangle, \quad X \, |-\rangle = - \, |-\rangle, \quad Y \, |R\rangle = |R\rangle, \quad Y \, |L\rangle = - \, |L\rangle. \tag{10.2.4}$$

Therefore, the quantum state $|\psi\rangle$ can also be decomposed into the superposition of the basis states $\{ |R\rangle , |L\rangle \}$ (y basis) and $\{ |+\rangle , |-\rangle \}$ (x basis):

$$|\psi\rangle = \alpha_x \, |+\rangle + \beta_x \, |-\rangle = \alpha_y \, |R\rangle + \beta_y \, |L\rangle.$$

If we can perform measurement in the x basis and the y basis, the expectations $\langle \psi | \, X \, | \psi \rangle$ and $\langle \psi | \, Y \, | \psi \rangle$ can be calculated in exactly the same way as the expectation $\langle \psi | \, Z \, | \psi \rangle$, namely,

$$a \, \langle \psi | \, X \, | \psi \rangle = a \frac{n_+ - n_-}{N}, \quad b \, \langle \psi | \, Y \, | \psi \rangle = b \frac{n_R - n_L}{N}.$$

Here, n_+ and n_- are the numbers of measurements in the x basis that correspond, respectively, to the $|+\rangle$ and $|-\rangle$ outcomes, and n_R and n_L are the numbers of measurements in the y basis that correspond, respectively, to $|R\rangle$ and $|L\rangle$ outcomes.

However, it may be the case that we can only perform measurement in the z basis. In this case, we need to apply some additional gates to $|\psi\rangle$ before the measurement, such that the probability of measuring $|0\rangle$ in the z basis is the same as the probability of measuring $|+\rangle$ in the x basis if we are calculating $\langle \psi | \, X \, | \psi \rangle$, or the probability of measuring $|0\rangle$ in the z basis is the same as the probability of measuring $|R\rangle$ in the y basis if we are calculating $\langle \psi | \, Y \, | \psi \rangle$. Denoting these gates H and G, we have

$$\text{H} \, |\psi\rangle = \text{H} \, (\alpha_x \, |+\rangle + \beta_x \, |-\rangle) = \alpha_x \, |0\rangle + \beta_x \, |1\rangle \, ,$$

with $\text{H} \, |+\rangle = |0\rangle$ and $\text{H} \, |-\rangle = |1\rangle$ and

$$\text{G} \, |\psi\rangle = \text{G} \, (\alpha_y \, |R\rangle + \beta_y \, |L\rangle) = \alpha_y \, |0\rangle + \beta_y \, |1\rangle \, ,$$

with $\text{G} \, |R\rangle = |0\rangle$ and $\text{G} \, |L\rangle = |1\rangle$.

The operators H (which is simply the Hadamard operator) and G admit the following matrix representations:

$$\text{H} = \frac{1}{\sqrt{2}} \begin{bmatrix} 1 & 1 \\ 1 & -1 \end{bmatrix} \quad \text{and} \quad \text{G} = \frac{1}{\sqrt{2}} \begin{bmatrix} 1 & -i \\ 1 & i \end{bmatrix} = \text{HS}^\dagger.$$

10.2.2   The two-qubit case

What if the problem Hamiltonian has terms involving more than a single qubit? Consider a Hamiltonian with terms consisting of tensor products of Pauli matrices such as $X \otimes Y$, $Y \otimes Z$, and so on. The general approach remains the same: the expectation value of the Hamiltonian consists of the sum of expectation values of all its terms. Thus, we need to know how to calculate the expectation value of the product of Pauli matrices. Without loss of generality, consider the $X \otimes Y$ term – as the very same logic applies to all other Pauli tensor products.

Recall that $X \otimes Y$ is the tensor product of the two Pauli operators X and Y, each acting on their own qubits, not a sequential application of gates X and Y to the same qubit. Indeed, given for two unitary operators U_1 and U_2, the tensor product $U_1 \otimes U_2$ acts on the state of a two-qubit system as

$$(U_1 \otimes U_2) \, |\psi_1\rangle \otimes |\psi_2\rangle = \left(U_1 \, |\psi_1\rangle \right) \otimes \left(U_2 \, |\psi_2\rangle \right). \tag{10.2.5}$$

We immediately see from (10.2.5) that the tensor product of eigenvectors of X and Y is an eigenvector of $X \otimes Y$. Indeed, if $U \, |\psi_U\rangle = E_U \, |\psi_U\rangle$ with $E_U \in \mathbb{C}$, for $U \in \{X, Y\}$, then

$$(X \otimes Y) \, |\psi_X\rangle \otimes |\psi_Y\rangle = X \, |\psi_X\rangle \otimes Y \, |\psi_Y\rangle = E_X \, |\psi_X\rangle \otimes E_Y \, |\psi_Y\rangle = E_X E_Y \, |\psi_X\rangle \otimes |\psi_Y\rangle \, .$$

We also remember that all eigenvectors of Pauli operators have eigenvalues equal to either $+1$ or -1 (as detailed in (10.2.3) and (10.2.4)). Now, for the $X \otimes Y$ Pauli term, the eigenvectors with eigenvalue $+1$ are

$$|+\rangle \otimes |R\rangle = |+R\rangle \quad \text{and} \quad |-\rangle \otimes |L\rangle = |-L\rangle \, ,$$

and the eigenvectors with eigenvalue -1 are

$$|+\rangle \otimes |L\rangle = |+L\rangle \quad \text{and} \quad |-\rangle \otimes |R\rangle = |-R\rangle \, ,$$

which follow directly from the computations

$$(X \otimes Y) \ket{+R} = (X \otimes Y) \ket{+} \otimes \ket{R} = X \ket{+} \otimes Y \ket{R} = \ket{+} \otimes \ket{R} = \ket{+R},$$

$$(X \otimes Y) \ket{+L} = (X \otimes Y) \ket{+} \otimes \ket{L} = X \ket{+} \otimes Y \ket{L} = \ket{+} \otimes (-\ket{L}) = -\ket{+L},$$

$$(X \otimes Y) \ket{-R} = (X \otimes Y) \ket{-} \otimes \ket{R} = X \ket{-} \otimes Y \ket{R} = (-\ket{-}) \otimes \ket{R} = -\ket{-R},$$

$$(X \otimes Y) \ket{-L} = (X \otimes Y) \ket{-} \otimes \ket{L} = X \ket{-} \otimes Y \ket{L} = (-\ket{-}) \otimes (-\ket{L}) = \ket{-L}.$$

Let us write down the representation of a quantum state of the two-qubit system $\ket{\psi} = \ket{\psi_1} \otimes \ket{\psi_2}$ in the basis of $X \otimes Y$ eigenvectors:

$$\begin{aligned}
\ket{\psi} &= \ket{\psi_1} \otimes \ket{\psi_2} \\
&= (\alpha_x \ket{+} + \beta_x \ket{-}) \otimes (\alpha_y \ket{R} + \beta_y \ket{L}) \\
&= \alpha_x \alpha_y \ket{+R} + \alpha_x \beta_y \ket{+L} + \beta_x \alpha_y \ket{-R} + \beta_x \beta_y \ket{-L},
\end{aligned}$$

with $(\alpha_x, \beta_x, \alpha_y, \beta_y) \in \mathbb{C}^4$.

We want to apply an operator allowing us to perform measurements in the z basis such that the probability amplitudes of the corresponding states remain the same. It is easy to see that this operator is a tensor product of the H and G gates:

$$\begin{aligned}
(H \otimes G) \ket{\psi} &= (H \otimes G) (\alpha_x \alpha_y \ket{+R} + \alpha_x \beta_y \ket{+L} + \beta_x \alpha_y \ket{-R} + \beta_x \beta_y \ket{-L}) \\
&= \alpha_x \alpha_y (H \otimes G) \ket{+} \otimes \ket{R} + \alpha_x \beta_y (H \otimes G) \ket{+} \otimes \ket{L} \\
&\quad + \beta_x \alpha_y (H \otimes G) \ket{-} \otimes \ket{R} + \beta_x \beta_y (H \otimes G) \ket{-} \otimes \ket{L} \\
&= \alpha_x \alpha_y (H \ket{+} \otimes G \ket{R}) + \alpha_x \beta_y (H \ket{+} \otimes G \ket{L}) \\
&\quad + \beta_x \alpha_y (H \ket{-} \otimes G \ket{R}) + \beta_x \beta_y (H \ket{-} \otimes G \ket{L}) \\
&= \alpha_x \alpha_y (\ket{0} \otimes \ket{0}) + \alpha_x \beta_y (\ket{0} \otimes \ket{1}) + \beta_x \alpha_y (\ket{1} \otimes \ket{0}) + \beta_x \beta_y (\ket{1} \otimes \ket{1}) \\
&= \alpha_x \alpha_y \ket{00} + \alpha_x \beta_y \ket{01} + \beta_x \alpha_y \ket{10} + \beta_x \beta_y \ket{11}.
\end{aligned}$$

The eigenvalues of $Z \otimes Z$ corresponding to the eigenstates $\{|00\rangle, |01\rangle, |10\rangle, |11\rangle\}$ are the same as the eigenvalues of $X \otimes Y$ corresponding to the eigenstates $\{|+R\rangle, |+L\rangle, |-R\rangle, |-L\rangle\}$:

$$(Z \otimes Z) |00\rangle = (Z \otimes Z) |0\rangle \otimes |0\rangle = Z |0\rangle \otimes Z |0\rangle = |0\rangle \otimes |0\rangle = |00\rangle,$$

$$(Z \otimes Z) |01\rangle = (Z \otimes Z) |0\rangle \otimes |1\rangle = Z |0\rangle \otimes Z |1\rangle = |0\rangle \otimes (-|1\rangle) = -|01\rangle,$$

$$(Z \otimes Z) |10\rangle = (Z \otimes Z) |1\rangle \otimes |0\rangle = Z |1\rangle \otimes Z |0\rangle = (-|1\rangle) \otimes |0\rangle = -|10\rangle,$$

$$(Z \otimes Z) |11\rangle = (Z \otimes Z) |1\rangle \otimes |1\rangle = Z |1\rangle \otimes Z |1\rangle = (-|1\rangle) \otimes (-|1\rangle) = |11\rangle.$$

Therefore, the expectation $\langle\phi| Z \otimes Z |\phi\rangle$, with

$$|\phi\rangle = \alpha_x \alpha_y |00\rangle + \alpha_x \beta_y |01\rangle + \beta_x \alpha_y |10\rangle + \beta_x \beta_y |11\rangle,$$

in the z basis is given by

$$|\alpha_x \alpha_y|^2 - |\alpha_x \beta_y|^2 - |\beta_x \alpha_y|^2 + |\beta_x \beta_y|^2.$$

The values of the probabilities $|\alpha_x \alpha_y|^2$, $|\alpha_x \beta_y|^2$, $|\beta_x \alpha_y|^2$, and $|\beta_x \beta_y|^2$ can be found using quantum computers in exactly the same way we found probabilities in the one-qubit case. By counting the numbers n_{ij} of outcomes $|ij\rangle$ (for $i, j \in \{0, 1\}$, with $\sum_{i,j\in\{0,1\}} n_{ij} = N$), the expectation value of $X \otimes Y$ is given by

$$\langle X \otimes Y \rangle = \frac{n_{00} - n_{01} - n_{10} + n_{11}}{N}.$$

10.2.3   The multi-qubit case

It is straightforward to scale this approach to more complex Pauli products and larger Hamiltonians since any Hamiltonian may be written as

$$\mathcal{H} = \sum_{i\alpha} h_\alpha^i \sigma_\alpha^i + \sum_{ij\alpha\beta} h_{\alpha\beta}^{ij} \sigma_\alpha^i \sigma_\beta^j + \cdots$$

for real h, where the superscripts $i, j, \ldots$ identify the subsystem (qubit) on which the operator acts, and the subscripts $\alpha, \beta, \ldots$ identify the Pauli operator. For example, $i = 1$, $\alpha = x$, and $\sigma_x^1 = $ X acting on qubit 1. No assumption about the dimension or structure of the Hermitian Hamiltonian is needed for this expansion to be valid [252].

We have already used the linearity of quantum observables that allows us to calculate the expectation of the Hamiltonian as a sum of expectations of the individual terms:

$$\langle \mathcal{H} \rangle = \sum_{i\alpha} h_\alpha^i \langle \sigma_\alpha^i \rangle + \sum_{ij\alpha\beta} h_{\alpha\beta}^{ij} \langle \sigma_\alpha^i \sigma_\beta^j \rangle + \ldots$$

As long as we consider Hamiltonians that can be written as a polynomial number of terms with respect to the system size, the evaluation of $\langle \mathcal{H} \rangle$ is reduced to the sum of a polynomial number of expectation values of simple Pauli operators for some quantum state $|\psi\rangle$, multiplied by some real constants. As we have seen, a quantum computer can efficiently evaluate the expectation value of a tensor product of an arbitrary number of simple Pauli operators [242].

> Quantum computers can be used to efficiently calculate expectation values of Hamiltonians consisting of tensor products of Pauli operators. Any Hamiltonian may be represented as a sum of tensor products of Pauli operators (X, Y, Z, and I gates).

10.3 Constructing the PQC

The question of how to construct a high-quality candidate state used to calculate expectations is of fundamental importance. Unless we have some prior knowledge about the ground state and where to search for it in the Hilbert space of the n-qubit system, the first task would be to generate a range of candidate states that will cover the whole Hilbert space without being heavily concentrated in any one region. Let us see how this can be done for the single-qubit and multi-qubit systems.

10.3.1   One-qubit ansatz

We return to the Bloch sphere that visualises the possible states of a one-qubit system. Figure 10.1 shows how the qubit state can change from its initial state $|0\rangle$ to the intermediate state $|\psi_i\rangle$ and then to the final state $|\psi_f\rangle$ through a rotation around the y axis followed by a rotation around the z axis.

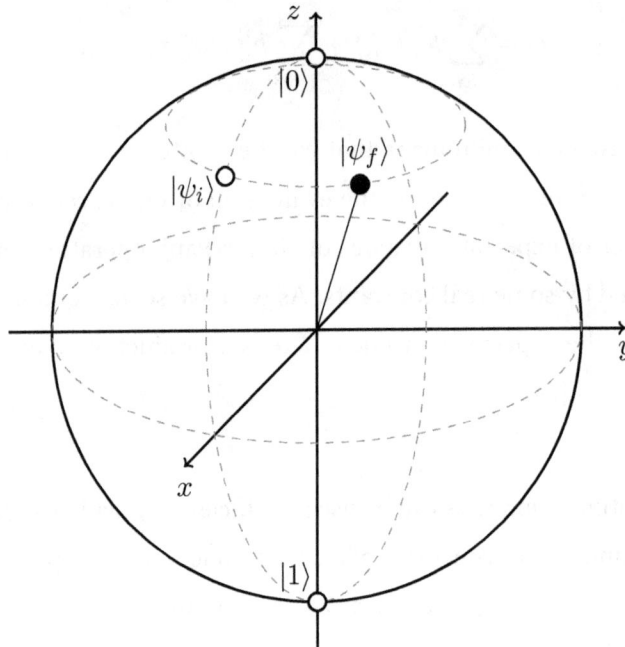

Figure 10.1: Bloch sphere: visualisation of one-qubit rotations.

It is possible to reach any point on the Bloch sphere starting from $|0\rangle$ with just two rotations around any two orthogonal axes. The corresponding circuit is shown in Figure 10.2.

Figure 10.2: PQC for a one-qubit system.

The PQC shown in Figure 10.2 is everything we need in the one-qubit case if we only have the Z and I terms in the problem Hamiltonian. If we want to calculate the expectation value of the X term, we have to add an H gate to the circuit, as shown in Figure 10.3.

$$|0\rangle \quad - \boxed{R_Y(\theta_1)} - \boxed{R_Z(\theta_2)} - \boxed{H} -$$

Figure 10.3: PQC with H *gate to calculate* $\langle X \rangle$.

Similarly, if we want to calculate the expectation value of the Y term, we have to add a G gate to the circuit, as shown in Figure 10.4.

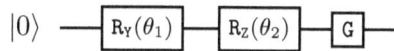

$$|0\rangle \quad - \boxed{R_Y(\theta_1)} - \boxed{R_Z(\theta_2)} - \boxed{G} -$$

Figure 10.4: PQC with G *gate to calculate* $\langle Y \rangle$.

10.3.2 Multi-qubit ansatz

We now look at the multi-qubit case and assume that the optimisation problem is encoded in the two-qubit Hamiltonian

$$\mathcal{H} = a X \otimes Y + b Y \otimes Z + c Z \otimes X, \tag{10.3.1}$$

for some coefficients $a, b, c \in \mathbb{R}$. As we know, the expectation value of the Hamiltonian (10.3.1) is given by the sum of expectation values of individual terms:

$$\langle \mathcal{H} \rangle \equiv \langle \psi | \, \mathcal{H} \, | \psi \rangle = a \, \langle \psi | X \otimes Y \, | \psi \rangle + b \, \langle \psi | Y \otimes Z \, | \psi \rangle + c \, \langle \psi | Z \otimes X \, | \psi \rangle. \tag{10.3.2}$$

We need to calculate all these expectations for the same quantum state $|\psi\rangle$. To do so, we need to construct a quantum circuit with sufficiently flexible adjustable gates to support a wide range of possible candidate states. Since the problem Hamiltonian operates on two-qubit states, the PQC that constructs the candidate states may look like the one shown in Figure 10.5:

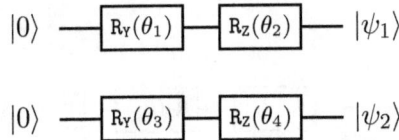

Figure 10.5: PQC for the construction of candidate states.

In the PQC shown in Figure 10.5, the parameters (rotation angles) θ_1 and θ_2 uniquely specify state $|\psi_1\rangle$ of the first qubit, and the parameters θ_3 and θ_4 uniquely specify state $|\psi_2\rangle$ of the second qubit. The full quantum circuit for the calculation of the expectation value $\langle X \otimes Y \rangle$ is shown in Figure 10.6, where the gates H (first quantum register) and G (second quantum register) form the change of basis layer before the measurement in the computational basis. Figures 10.7 and 10.8 display the PQCs for the calculation of $\langle Y \otimes Z \rangle$ and $\langle Z \otimes X \rangle$.

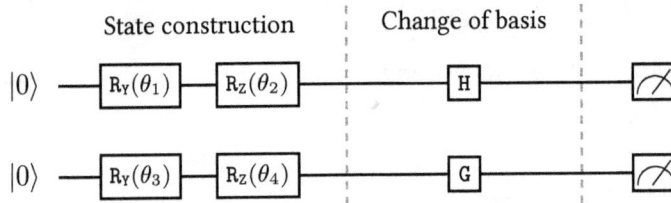

Figure 10.6: PQC for the calculation of the $\langle X \otimes Y \rangle$ term.

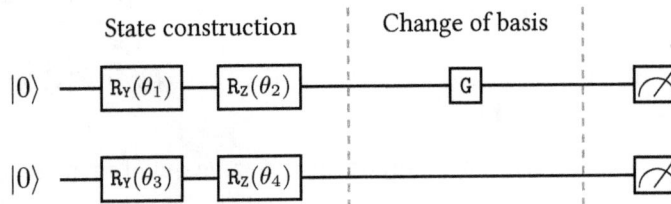

Figure 10.7: PQC for the calculation of the $\langle Y \otimes Z \rangle$ term.

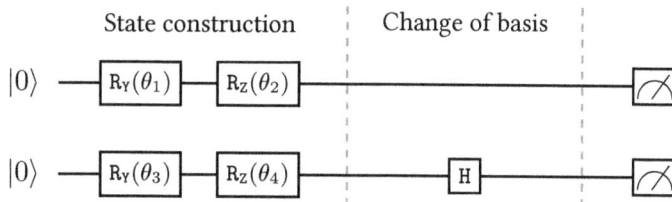

State construction | Change of basis

Figure 10.8: PQC for the calculation of the $\langle Z \otimes X \rangle$ term.

Note that the state construction circuits shown in Figures 10.6-10.8 consist of one-qubit gates and do not contain two-qubit gates, which would create entanglement. Adding two-qubit gates, such as CNOT and CPHASE, would help explore a wider range of possible quantum states and should be done as a matter of course, as we know from the previous chapters. However, our task here is to illustrate the general principle and compare the results obtained by running the PQC with the calculations done "by hand", as we shall see in the next section.

10.4   Running the PQC

We now run some numerical experiments for the optimisation problem encoded in the Hamiltonian (10.3.1) and compare the obtained results with direct calculations to better understand the mechanics of the algorithm and to build intuition.

10.4.1   Experimenting with the two-qubit ansatz

In line with the variational approach and taking into account the PQC architecture specified in Figure 10.5, we need to build the candidate states controlled by the four adjustable parameters θ_1, θ_2, θ_3, and θ_4. In Chapters 8 and 9, we considered two possible approaches to train the PQC: differentiable and non-differentiable learning. Both methods have their strong and weak points and can be used to find an optimal set of adjustable parameters for the PQC of arbitrary size. However, the PQC shown in Figure 10.5 is only two layers deep and two quantum registers wide – it is perfectly feasible in this case to apply the brute force method. The brute force method consists of discretising the range of possible values of the rotation angles with the elementary step (increment) kept reasonably small. The parameters θ_1 and θ_3 are rotation angles around the y axis and are defined on the interval $[0, \pi]$, which we discretise as

$$\left\{ \frac{(2k+1)\pi}{2m} \right\}_{k=0,\ldots,m-1} , \tag{10.4.1}$$

for some integer m. Similarly, θ_2 and θ_4 are rotation angles around the z axis defined on $[0, 2\pi]$, with range

$$\left\{ \frac{(2k+1)\pi}{2m} \right\}_{k=0,\ldots,2m-1} . \tag{10.4.2}$$

In both cases, the increment is π/m. Taking, for example, $m = 8$ is a sensible compromise between speed and accuracy. What is more, once the optimal configuration of adjustable parameters is found, we can run an additional, more granular search in the vicinity of the candidate optimal configuration in order to further improve it.

The algorithm looks as follows. For the given configuration of R_Y and R_Z rotation angles drawn from the sets (10.4.1) and (10.4.2), we execute 100,000 runs of PQCs specified in Figures 10.6, 10.7, and 10.8 on the Qiskit quantum simulator. This gives us the expectation values of $X \otimes Y$, $Y \otimes Z$, and $Z \otimes X$ for the quantum state $|\psi\rangle = |\psi_1\rangle \otimes |\psi_2\rangle$, where

$$|\psi_1\rangle = \cos\left(\frac{\theta_1}{2}\right)|0\rangle + e^{i\theta_2}\sin\left(\frac{\theta_1}{2}\right)|1\rangle$$

and

$$|\psi_2\rangle = \cos\left(\frac{\theta_3}{2}\right)|0\rangle + e^{i\theta_4}\sin\left(\frac{\theta_3}{2}\right)|1\rangle .$$

We select the state $|\psi^*\rangle$ with the smallest value of $\langle \mathcal{H} \rangle$ given by (10.3.2), and denote the corresponding rotation angles $\theta_1^*, \ldots, \theta_4^*$. After that, we perform a more refined search in the neighbourhood of $|\psi^*\rangle$. The new set values of the rotation angles are now

$$\left\{ \theta_i^* + \frac{(k-4)\pi}{4m} \right\}_{k=0,\ldots,m} , \quad i = 1, \ldots, 4,$$

with increment $\pi/(4m)$. Again, we select the quantum state $|\psi'\rangle$ with the smallest value of $\langle \mathcal{H} \rangle$ and denote the corresponding rotation angles $\theta'_1, \ldots, \theta'_4$. With $m = 8$, $a = 4$, $b = 3$, and $c = 2$, we obtain the results in Table 10.1:

1st search:	$\theta^*_1 = 1.7671$	$\theta^*_2 = 3.0434$	$\theta^*_3 = 1.7671$	$\theta^*_4 = 1.4726$	$\langle \mathcal{H} \rangle_{\min} = -3.93$
2nd search:	$\theta'_1 = 1.5708$	$\theta'_2 = 3.1416$	$\theta'_3 = 1.5708$	$\theta'_4 = 1.5708$	$\langle \mathcal{H} \rangle_{\min} = -4.00$

Table 10.1: Optimal configurations of adjustable PQC parameters that minimise the expectation value of the Hamiltonian.

In Table 10.1, the values of $\langle \mathcal{H} \rangle$ are in the units of coefficients a, b, and c, and the values of rotation angles are in radians. Note that $3.1416 = \pi$ and $1.5708 = \pi/2$. Therefore, the optimal configuration of rotation angles that minimises $\langle \mathcal{H} \rangle$ is

$$\theta'_1 = \theta'_3 = \theta'_4 = \frac{\pi}{2} \quad \text{and} \quad \theta'_2 = \pi.$$

The corresponding quantum states are

$$|\psi_1\rangle = \frac{1}{\sqrt{2}}|0\rangle - \frac{1}{\sqrt{2}}|1\rangle, \quad |\psi_2\rangle = \frac{1}{\sqrt{2}}|0\rangle + \frac{i}{\sqrt{2}}|1\rangle,$$

$$|\psi\rangle = |\psi_1\rangle \otimes |\psi_2\rangle = \frac{1}{2}|00\rangle + \frac{i}{2}|01\rangle - \frac{1}{2}|10\rangle - \frac{i}{2}|11\rangle.$$

10.4.2 Analysis of the obtained results

Do the obtained results make sense? Since the problem size is small and the circuit is not too deep, we can verify the results by direct manual calculations. First, we visualise the states $|\psi_1\rangle$ and $|\psi_2\rangle$. Figure 10.9 shows their positions on the Bloch sphere. State $|\psi_1\rangle$ is the black dot at the intersection of the x axis and the equator. We get to $|\psi_1\rangle$ from state $|0\rangle$ by performing a $\pi/2$ radian rotation around the y axis and then a π radian rotation around the z axis. State $|\psi_2\rangle$ is the gray dot at the intersection of the y axis and the equator, which is reached from $|0\rangle$ by performing a $\pi/2$ radian rotation around the y axis and then a $\pi/2$ radian rotation around the z axis.

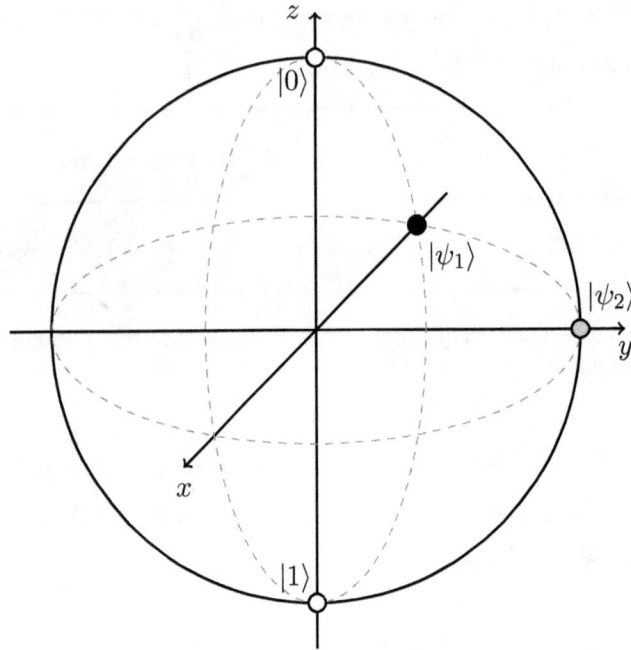

Figure 10.9: Visualisation of states $|\psi_1\rangle$ (black dot) and $|\psi_2\rangle$ (grey dot).

Consider the expectation $\langle X \otimes Y \rangle$. The operator X acts on $|\psi_1\rangle$, which is one of its eigenstates in the x basis:

$$|\psi_1\rangle = \frac{1}{\sqrt{2}}|0\rangle - \frac{1}{\sqrt{2}}|1\rangle = |-\rangle \,,$$

and Y acts on $|\psi_2\rangle$, which is also one of its eigenstates in the y basis:

$$|\psi_2\rangle = \frac{1}{\sqrt{2}}|0\rangle + \frac{i}{\sqrt{2}}|1\rangle = |R\rangle \,.$$

We then perform measurements in the computational basis (z basis). Before measurement, the Hadamard operator H will transform $|\psi_1\rangle = |-\rangle$ into $|1\rangle$, which is a basis state in the z basis. Similarly, the operator G will transform $|\psi_2\rangle = |R\rangle$ into $|0\rangle$, which is also a basis state in the z basis.

Thus, measurement in the z basis will give us the state $|1\rangle \otimes |0\rangle = |10\rangle$ with probability 1. If we perform N measurements, we will obtain the state $|10\rangle$ N times and the corresponding expectation value $\langle X \otimes Y \rangle$ will be

$$\langle X \otimes Y \rangle = \frac{n_{00} - n_{01} - n_{10} + n_{11}}{N} = \frac{0 - 0 - N + 0}{N} = -1.$$

The value of the coefficient a in front of the $X \otimes Y$ term was set equal to 4. Therefore, the contribution of the first term to the expectation value of the Hamiltonian (10.3.1) is -4.

The expectations $\langle Y \otimes Z \rangle$ and $\langle Z \otimes X \rangle$ are equally straightforward to calculate. Let us start with $\langle Y \otimes Z \rangle$. The operator Y acts on the first qubit in the state $|\psi_1\rangle$. Since we measure the resulting state in the z basis, we need to apply G before measurement. The operator G transforms $|\psi_1\rangle$ into the state

$$\frac{1+i}{2} |0\rangle + \frac{1-i}{2} |1\rangle .$$

Measurement in the z basis will result in $|0\rangle$ and $|1\rangle$ with equal probability $1/2$.

The operator Z acts on the second qubit in state $|\psi_2\rangle$. Since we measure the operator Z in the z basis, we do not need to apply any gates. Measuring $|\psi_2\rangle$ in the z basis will also result in obtaining $|0\rangle$ and $|1\rangle$ with equal probability $1/2$. Therefore, we are equally likely to measure all four basis states ($|00\rangle$, $|01\rangle$, $|10\rangle$, and $|11\rangle$) with probability $1/4$, and, as N tends to infinity, the expectation value of $\langle Y \otimes Z \rangle$ in our experiment should converge to

$$\langle Y \otimes Z \rangle = \frac{n_{00} - n_{01} - n_{10} + n_{11}}{N} = \frac{\frac{1}{4}N - \frac{1}{4}N - \frac{1}{4}N + \frac{1}{4}N}{N} = 0.$$

We are left now with $\langle Z \otimes X \rangle$. The operator Z acts on $|\psi_1\rangle$. The measurement is done in the z basis, so no transformation is needed. The measurement will result in obtaining the states $|0\rangle$ and $|1\rangle$ with equal probability $1/2$. The operator X acts on the second qubit in the state $|\psi_2\rangle$. The measurement is performed in the z basis, so we need to apply the Hadamard gate H before measurement. Applying the H gate to $|\psi_2\rangle$ transforms it into

$$\frac{1+i}{2} |0\rangle + \frac{1-i}{2} |1\rangle ,$$

with measurement in the z basis producing outcomes $|0\rangle$ and $|1\rangle$ with probability $1/2$. Thus, we find ourselves in the same situation as with $\langle Y \otimes Z \rangle$: all basis states are equally likely. As N tends to infinity, the expectation value of $\langle Z \otimes X \rangle$ should also converge to

$$\langle Z \otimes X \rangle = \frac{n_{00} - n_{01} - n_{10} + n_{11}}{N} = \frac{\frac{1}{4}N - \frac{1}{4}N - \frac{1}{4}N + \frac{1}{4}N}{N} = 0.$$

This is exactly what we observed in our numerical experiment. The total contribution of all three terms of the Hamiltonian $\mathcal{H}$ given by (10.3.1) with $a = 4$, $b = 3$, and $c = 2$ is equal to -4. The $X \otimes Y$ term has the largest coefficient; therefore, it makes sense that the ground state of $\mathcal{H}$ is the one that minimises the expectation value $\langle X \otimes Y \rangle$ (with expectation values $\langle Y \otimes Z \rangle$ and $\langle Z \otimes X \rangle$ being zero).

> PQC can be used to construct the candidate states for the VQE algorithm. The selection and improvement of the candidate states are performed classically. This makes VQE a perfect example of a hybrid quantum-classical algorithm.

10.5   Discrete Portfolio Optimisation with VQE

In Chapter 3, we investigated quantum annealing for NP-hard discrete portfolio optimisation problems. The same type of QUBO problems can be solved on gate model quantum computers with the help of a hybrid VQE algorithm. The QUBO formulation of the discrete portfolio optimisation problem consists of minimising the cost function (3.3.2):

$$L(\mathsf{q}) = \sum_{i=1}^{N} a_i q_i + \sum_{i=1}^{N} \sum_{j=i+1}^{N} b_{ij} q_i q_j,$$

where $\mathsf{q} := (q_1, \ldots, q_N)$ is a vector of binary decision variables indicating which (equally weighted) assets are selected (from the universe of N investable assets): $q_i = 1$ means that asset i is selected and $q_i = 0$ means that asset i is not selected. The task is to find a configuration of q that minimises $L(\mathsf{q})$.

For each $i, j = 1, \ldots N$, the coefficients a_i, a_j, and b_{ij} reflect, respectively, the individual and joint attractiveness of assets i and j. For example, assets with larger expected returns and lower volatilities would be assigned large negative values of a. Similarly, pairs of assets with low positive or even negative correlation would be assigned negative values of b to reward diversification. Assets with lower expected returns, higher volatility, and strongly positively correlated with other assets would be penalised with positive values of a and b.

Quantum annealers solve QUBO problems in their Ising model formulation where binary decision variables $q := (q_1, \ldots, q_N)$ are translated into spin variables $s := (s_1, \ldots, s_N)$ taking values $\{+1, -1\}$ through the transformation $s_i = 2q_i - 1$. We analyse the simplest case of an investable universe consisting of just two assets. In this case, the QUBO cost function reads

$$L(q) = a_1 q_1 + a_2 q_2 + b_{12} q_1 q_2, \tag{10.5.1}$$

or, in the Ising model formulation,

$$L(s) = g_1 s_1 + g_2 s_2 + J_{12} s_1 s_2 + \text{const}, \tag{10.5.2}$$

where

$$g_1 = \frac{1}{2} a_1 + \frac{1}{4} b_{12}, \quad g_2 = \frac{1}{2} a_2 + \frac{1}{4} b_{12}, \quad J_{12} = \frac{1}{4} b_{12}. \tag{10.5.3}$$

The constant term in (10.5.2) does not depend on the decision variables s_1 and s_2 and can hence be ignored. The cost function we want to minimise thus becomes

$$L(s) = g_1 s_1 + g_2 s_2 + J_{12} s_1 s_2. \tag{10.5.4}$$

As we know from Chapter 2, the transition from the classical to the quantum mechanical description of the system consists of replacing variables corresponding to physical observables by their respective operators. In the case of the QUBO problem in its Ising model formulation, it means replacing classical spin variables with the corresponding Pauli operators σ_x, σ_y, and σ_z, which are represented in the quantum circuit by, respectively, the quantum gates X, Y, and Z.

As we remember from Chapter 3, the Ising cost function (10.5.4) corresponds to the following *final* Hamiltonian that encodes the same optimisation problem:

$$\mathcal{H}_F = g_1 \sigma_z^1 + g_2 \sigma_z^2 + J_{12} \sigma_z^1 \sigma_z^2. \tag{10.5.5}$$

Here, the classical spin variables s_1 and s_2 are replaced by the σ_z operators, and σ_z^1 is the Z gate acting on qubit 1 while σ_z^2 is the Z gate acting on qubit 2.

Note that the mapping between the binary QUBO decision variables q, classical spin variables s, and the eigenstates of Z is as follows:

$$q = 0 \rightarrow s = -1 \rightarrow |1\rangle \quad \text{since} \quad Z|1\rangle = -|1\rangle,$$

and

$$q = 1 \rightarrow s = +1 \rightarrow |0\rangle \quad \text{since} \quad Z|0\rangle = |0\rangle.$$

We now solve the QUBO problem (10.5.1) with

$$a_1 = -2, \quad a_2 = 3, \quad b_{12} = -2,$$

using both the classical method (in this case, a simple exhaustive search given that the solution space consists of just four possible solutions) and the VQE approach.

The classical exhaustive search results are straightforward to obtain via direct calculations and are summarised in Table 10.2 with the optimal solution $q^* = (1, 0)$: asset 1 is selected, while asset 2 is not.

q_1	q_2	$L(q)$
0	0	0
0	1	3
1	0	-2
1	1	-1

Table 10.2: Classical exhaustive search results.

The VQE calculations are as follows. We first rewrite the Hamiltonian (10.5.5) in the quantum gate form:

$$\mathcal{H}_F = g_1 Z^1 + g_2 Z^2 + J_{12} Z^1 \otimes Z^2,$$

where Z^1 is the Z gate acting on qubit 1, Z^2 is the Z gate acting on qubit 2, and $Z^1 \otimes Z^2$ is the tensor product of the Z gates acting on qubits 1 and 2, respectively.

The next step is to calculate the expectation values $\langle Z \rangle$ and $\langle Z \otimes Z \rangle$. As we know, Z is a PHASE gate that flips the phase of a qubit:

$$Z\,|0\rangle = |0\rangle, \quad Z\,|1\rangle = -\,|1\rangle.$$

Therefore, we have

$$
\begin{aligned}
\langle 0|\, Z^1\, |0\rangle &= +1, & \langle 1|\, Z^1\, |1\rangle &= -1, \\
\langle 0|\, Z^2\, |0\rangle &= +1, & \langle 1|\, Z^2\, |1\rangle &= -1, \\
\langle 00|\, Z^1 \otimes Z^2\, |00\rangle &= +1, & \langle 01|\, Z^1 \otimes Z^2\, |01\rangle &= -1, \\
\langle 10|\, Z^1 \otimes Z^2\, |10\rangle &= -1, & \langle 11|\, Z^1 \otimes Z^2\, |11\rangle &= +1.
\end{aligned}
$$

Here we calculated the expectation values of operators Z and Z⊗Z analytically but we would obtain exactly the same values if we were to calculate them using a quantum computer, as described in Section 10.2. Now we need to calculate the values of the coefficients g_1, g_2, and J_{12} using the transformation (10.5.3):

$$g_1 = -1.5, \quad g_2 = 1, \quad J_{12} = -0.5.$$

The expectation values of $\mathcal{H}_F$ in states $|00\rangle$, $|01\rangle$, $|10\rangle$, and $|11\rangle$ can then be calculated as

$$
\begin{aligned}
|00\rangle : \quad \langle \mathcal{H} \rangle &= g_1 \cdot (+1) + g_2 \cdot (+1) + J_{12} \cdot (+1) = -1, \\
|01\rangle : \quad \langle \mathcal{H} \rangle &= g_1 \cdot (+1) + g_2 \cdot (-1) + J_{12} \cdot (-1) = -2, \\
|10\rangle : \quad \langle \mathcal{H} \rangle &= g_1 \cdot (-1) + g_2 \cdot (+1) + J_{12} \cdot (-1) = 3, \\
|11\rangle : \quad \langle \mathcal{H} \rangle &= g_1 \cdot (-1) + g_2 \cdot (-1) + J_{12} \cdot (+1) = 0.
\end{aligned}
$$

The best solution found using VQE is therefore $|01\rangle = |0\rangle \otimes |1\rangle$, that is, asset 1 is selected while asset 2 is not, which is the same as the best solution found by exhaustive search.

A VQE is a viable tool for solving finance-related NP-hard optimisation problems.

Summary

In this chapter, we introduced the Variational Quantum Eigensolver algorithm, a powerful QML model based on the variational approach that allows us to solve hard optimisation problems. We learned how to calculate expectation values using a quantum computer and how to construct a PQC, creating the candidate quantum states.

We also experimented with running the VQE model on a sample problem encoded in a two-qubit Hamiltonian and analysed and verified the results by performing manual calculations. Finally, we demonstrated the applicability of the VQE to finance-related optimisation problems, in particular, to a stylised example of a discrete portfolio optimisation.

In the next chapter, we will introduce the Quantum Approximate Optimisation Algorithm, another example of a hybrid quantum-classical approach to solving hard optimisation problems.

11

Quantum Approximate Optimisation Algorithm

As the name suggests, the Quantum Approximate Optimisation Algorithm (QAOA) is an optimisation algorithm. It is motivated by and draws upon two optimisation algorithms considered in previous chapters: AQC and VQE. From AQC it borrows the concept of solving an optimisation problem through encoding the corresponding objective function in the problem Hamiltonian and then evolving the system in such a way that the ground state of the final Hamiltonian provides the solution we are after (in a bitstring format). From VQE it borrows the variational principle applied to the parameterised quantum circuit. Roughly speaking, QAOA is a gate-model version of an optimisation solver that otherwise could have been tackled with an analog AQC approach. We can also look at QAOA as a special case of VQE with the constraints on the form of the Hamiltonian.

QAOA was introduced in the pioneering work by Farhi, Goldstone, and Gutmann [96] in 2014, and its potential for establishing quantum supremacy was investigated by Farhi and Harrow in [99]. QAOA and its sister algorithm that generalises it, the Quantum Alternating

Operator Ansatz (with the same acronym!), have been tested on a number of financial use cases. Here we can mention the work by Hodson, Ruck, Ong, Garvin, and Dulma [146] on portfolio rebalancing experiments and the one by Barkoutsos, Nannicini, Robert, Tavernelli, and Woerner [25] using Conditional Value-at-Risk (CVaR) as the QAOA objective function. The algorithm has significant potential and promises to become a standard tool in the arsenal of quantum computing methods aimed at financial applications.

11.1   Time Evolution

Consider again the description of the dynamics of quantum mechanical systems, briefly covered in Chapter 1 (as one of the postulates of quantum mechanics) and Chapter 2 (where we introduced the principles of Adiabatic Quantum Computing). These dynamics are governed by the Schrödinger equation (1.2.1):

$$i\hbar \frac{d\,|\psi(t)\rangle}{dt} = \mathcal{H}\,|\psi(t)\rangle\,,$$

with some initial condition $|\psi(0)\rangle$, where $|\psi(t)\rangle$ is the quantum state at time t and $\mathcal{H}$ is the time-independent Hamiltonian. Its solution is given by (1.2.2), namely

$$|\psi(t)\rangle = \mathcal{U}(0,t)\,|\psi(0)\rangle\,,$$

where the operator $\mathcal{U}(0,t)$ is obtained from the Hamiltonian $\mathcal{H}$ by (1.2.3):

$$\mathcal{U}(0,t) = \exp\left(-\frac{i\mathcal{H}t}{\hbar}\right).$$

We work with units where $\hbar$ is set to 1, so that the system dynamics reads

$$|\psi(t)\rangle = e^{-i\mathcal{H}t}\,|\psi(0)\rangle\,. \tag{11.1.1}$$

If the initial state of the system $|\psi(0)\rangle$ is known, then the state of the system at time t is also known and is determined by the action of the Hamiltonian $\mathcal{H}$ over the period of time t.

However, the solution (11.1.1) assumes that the system Hamiltonian is time-independent.

At the same time, AQC works with time-dependent Hamiltonians of the form (2.2.1):

$$\mathcal{H}(t) = \left(1 - \frac{t}{T}\right)\mathcal{H}_0 + \frac{t}{T}\mathcal{H}_F,$$

for some *initial* Hamiltonian $\mathcal{H}_0$ and some *final* or *problem* (encoding the optimisation problem) Hamiltonian $\mathcal{H}_F$. How do we reconcile this mismatch? The answer is that we can *approximate* [307] the time-dependent Hamiltonian $\mathcal{H}(t)$ that transforms the state over $[0, T]$ by a sequence of time-independent Hamiltonians:

$$\mathcal{H}_1, \ \mathcal{H}_2, \ \ldots, \ \mathcal{H}_m,$$

transforming the state over the corresponding shorter time intervals:

$$[t_0 = 0, t_1], \ [t_1, t_2], \ \ldots, \ [t_{m-1}, t_m = T].$$

A good analogy is the approximation of a continuous function (e.g., $\sin(\cdot)$) by a piecewise linear function as shown in Figure 11.1. The more granular the time intervals $[t_{i-1}, t_i]$, the better the approximation.

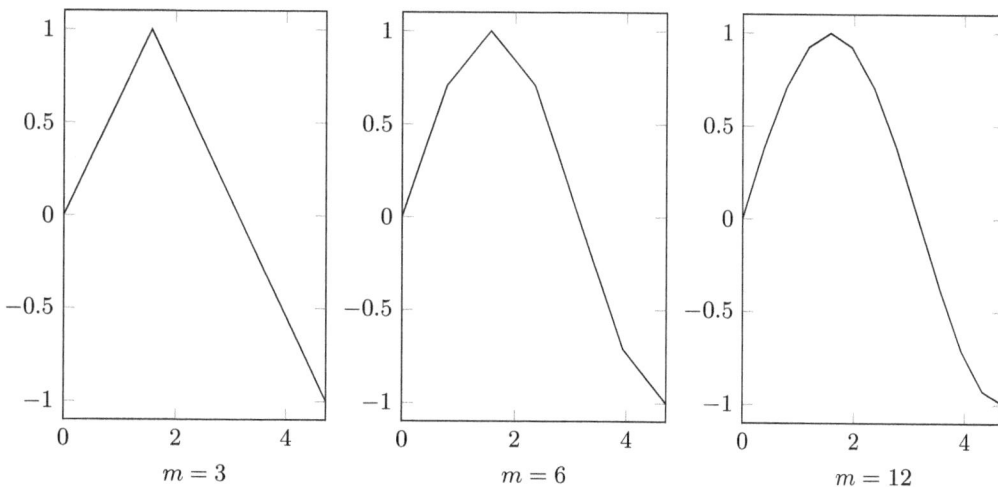

Figure 11.1: Piecewise linear approximation of $t \mapsto \sin(t)$.

Similarly, we can approximate the operator $\mathcal{U}(0, T)$ as

$$\mathcal{U}(0, T) \approx \mathcal{U}(t_{m-1}, t_m)\mathcal{U}(t_{m-2}, t_{m-1}) \cdots \mathcal{U}(t_2, t_1)\mathcal{U}(t_0, t_1).$$

Again, larger values of n give a better approximation.

> The evolution of a quantum mechanical system over a long time interval can be modelled as a sequence of time-independent Hamiltonians transforming the system state over the corresponding shorter time intervals.

11.2   The Suzuki-Trotter Expansion

A particularly useful approximation of $\mathcal{U}(0, T)$ can be obtained using the Suzuki-Trotter expansion [303]. If $\mathcal{A}_1, \mathcal{A}_2, \ldots, \mathcal{A}_k$ are operators that do not necessarily commute, then

$$\exp\left(\mathcal{A}_1 + \mathcal{A}_2 + \ldots + \mathcal{A}_k\right) = \lim_{m \to \infty} \left[\exp\left(\frac{\mathcal{A}_1}{m}\right) \exp\left(\frac{\mathcal{A}_2}{m}\right) \cdots \exp\left(\frac{\mathcal{A}_k}{m}\right)\right]^m.$$

Recall that two operators $\mathcal{A}$ and $\mathcal{B}$ are said to commute if $\mathcal{A}\mathcal{B} = \mathcal{B}\mathcal{A}$. Many operators introduced in previous chapters do not commute, for example, rotations around different axes do not, and the end result (the end quantum state) depends on how rotations are ordered.

As mentioned in Chapter 1, the expectation values of Hermitian operators are real and correspond to physical observables (e.g., the expectation of a Hermitian Hamiltonian is the physically observable energy). If operators commute, we can *measure* them in an arbitrary order and obtain the same answer. There is no uncertainty in the values of the corresponding physical observables.

The Suzuki-Trotter expansion, however, does not require operators to commute to remain valid. This has important implications for QAOA as we shall see below. If $\mathcal{U}(0, T)$ has the form $\exp([\mathcal{A} + \mathcal{B}]T)$, then we can use the Suzuki-Trotter expansion to obtain

$$\exp([\mathcal{A} + \mathcal{B}]T) = \lim_{m\to\infty} \left[\exp\left(\frac{\mathcal{A}T}{m}\right) \exp\left(\frac{\mathcal{B}T}{m}\right) \right]^m,$$

namely, the time evolution of $[\mathcal{A} + \mathcal{B}]T$ can be approximated by applying alternatively $\mathcal{A}$ and $\mathcal{B}$ over time intervals of length T/m.

In Chapter 3, we introduced detailed specifications of the AQC, where the Hamiltonians $\mathcal{H}_0$ and $\mathcal{H}_F$ have the general form

$$\mathcal{H}_0 = \sum_{i=1}^{n} \sigma_x^i \quad \text{and} \quad \mathcal{H}_F = \sum_{i=1}^{n} a_i \sigma_z^i + \sum_{i=1}^{n} \sum_{j=i+1}^{n} b_{ij} \sigma_z^i \sigma_z^j,$$

for some coefficients $(a_i)_{i=1,\dots,n}$ and $(b_{ij})_{i,j=1,\dots,n}$. We also refer the interested reader to [330] for a precise connection between QAOA and AQC, in particular in the case of a finite number of steps m.

The initial Hamiltonian $\mathcal{H}_0$ is the operator $\mathcal{A}$ – called the *mixing* Hamiltonian – and the final Hamiltonian $\mathcal{H}_F$ is the operator $\mathcal{B}$ – called the *phase* Hamiltonian. Recall from Chapter 10 that the Pauli operators σ_x, σ_y, and σ_z are represented by the quantum gates X, Y, and Z, and the superscript in σ_x^i refers to the qubit on which it acts.

The initial state is set in the equal superposition state of all possible solutions [129]:

$$|\psi(0)\rangle = \frac{1}{\sqrt{2^n}} \Big(|0\dots00\rangle + |0\dots01\rangle + \dots + |1\dots11\rangle \Big) = \frac{1}{\sqrt{2^n}} \sum_{i=0}^{2^n-1} |i\rangle,$$

which is the ground state of $\mathcal{A}$ and can be easily constructed from $|0\rangle^{\otimes n}$ by applying the Hadamard gate H to every qubit, i.e.,

$$|\psi(0)\rangle = \text{H}^{\otimes n} |0\rangle^{\otimes n}.$$

Remark: For a one-qubit system, $\mathcal{A}$ is the X gate and the initial state is

$$|\psi(0)\rangle = \frac{1}{\sqrt{2}}(|0\rangle + |1\rangle) = |+\rangle .$$

As we know from Chapter 10, state $|+\rangle$ is the eigenstate of X with an eigenvalue equal to 1, namely $X|+\rangle = |+\rangle$.

11.3   The Algorithm Specification

Now everything is in place for the formulation of the QAOA procedure [129].

Algorithm 9: Quantum Approximate Optimisation Algorithm

Input: $\mathcal{A}$ and $\mathcal{B}$.

1: A parameterised quantum state $|\psi(\beta, \gamma)\rangle$, $\beta := (\beta_1, \ldots, \beta_m)$, $\gamma := (\gamma_1, \ldots, \gamma_m)$, is created by alternately applying the operators $\mathcal{A}$ and $\mathcal{B}$ for m rounds, where the duration in round i $(i = 1, \ldots, m)$ is specified by the parameters β_i and γ_i respectively:

$$|\psi(\beta, \gamma)\rangle = e^{-i\beta_m \mathcal{A}} e^{-i\gamma_m \mathcal{B}} \cdots e^{-i\beta_2 \mathcal{A}} e^{-i\gamma_2 \mathcal{B}} e^{-i\beta_1 \mathcal{A}} e^{-i\gamma_1 \mathcal{B}} \left(H^{\otimes n} |0\rangle^{\otimes n} \right) .$$

2: A computational basis (z basis) measurement is performed on the obtained state, which returns a candidate solution. Repeating the above state preparation and measurement, the expected value of the cost function f over the returned solution samples is given by

$$\langle f \rangle = \langle \psi(\beta, \gamma)| \, \mathcal{B} \, |\psi(\beta, \gamma)\rangle ,$$

which can be statistically estimated from the samples produced (as explained in Chapter 10).

3: The above steps may then be repeated with the updated sets of time parameters β and γ – the variational part of the algorithm – within the classical optimisation loop that aims to minimise the expectation of the cost function $\langle f \rangle$.

Result: The algorithm returns the best found solution.

It is important to apply operators $\exp(-i\beta\mathcal{A})$ and $\exp(-i\gamma\mathcal{B})$ alternately to ensure that we are not trapped in a local minimum. It is also important that $\mathcal{A}$ and $\mathcal{B}$ do not commute [307]. Indeed, by applying only $\exp(-i\gamma\mathcal{B})$, we are facing ending in an eigenstate of the phase Hamiltonian. If this happens, we will be trapped there: any further application of a linear operator to its eigenvector may change its length but not its direction. The same applies to alternating between two commuting operators: if $\mathcal{A}$ and $\mathcal{B}$ commute, then we can come up with a set of basis states that are eigenstates of both $\mathcal{A}$ and $\mathcal{B}$, and once we get into one of these eigenstates, we remain trapped in it. However, since σ_x and σ_z do not commute, there is always a chance to escape from the local minimum.

The ongoing exploration of QAOA potential started with the foundational paper by Farhi, Goldstone, and Gutmann [96] where it outperformed classical algorithms on the Max-Cut problem on connected 3-regular graphs. A regular graph is one where each vertex has the same number of neighbours. In the case of a 3-regular graph (also known as a *cubic* graph), each vertex is connected with three other vertices. We consider the Max-Cut problem in the next section in its most general formulation. It was a decisive result that prompted active development of classical algorithms and, eventually, one with asymptotically much better performance was constructed by Barak *et al* [24]. This, in turn, triggered further investigations: the performance comparison between QAOA and the best classical algorithms was studied by Hastings in [136], and Bravyi, Kliesch, Koenig and Tang [44] established that the locality and the symmetry of QAOA severely limit its performance. To overcome these limitations, they proposed a non-local version that significantly outperforms standard QAOA for a frustrated Ising model on random 3-regular graphs.

11.4 The Max-Cut Problem

The Max-Cut problem is one of the special cases of the graph partitioning problem introduced in Chapter 3. The objective is to divide the vertices of the graph into two groups such that either the maximum possible number of edges going between the two groups are "cut" (if all edges have the same weight) or the total weight of these edges is maximised (if they have different weights).

The problem of maximisation of the total weight (or number of edges being cut) can be formulated as the minimisation of a cost function, which is the sum of the costs of all individual edges. Each individual cost c_{ij}, associated with the edge connecting vertices i and j, is given by

$$c_{ij} = \frac{1}{2}w_{ij}(1 - s_i s_j), \qquad (11.4.1)$$

where s_i and s_j are classical spin variables taking values $\{-1, +1\}$ and w_{ij} is the weight associated with the edge connecting vertices i and j. The two groups of vertices are those where the spin variables take the same values (either -1 or $+1$). We can see from (11.4.1) that when s_i and s_j have the same sign, the cost c_{ij} is zero; however, when s_i and s_j have opposite signs, the cost c_{ij} is equal to the weight w_{ij}.

The cost function for the whole graph then has the form

$$L(s) = \sum_{\{ij\} \in G} \frac{1}{2}w_{ij}(1 - s_i s_j), \qquad (11.4.2)$$

where $s := (s_1, \ldots, s_n)$ is the set of decision variables associated with the n-node graph G and the sum goes over all pairs of nodes connected by the graph edges.

There are many possible applications of the Max-Cut problem in finance, for example, *client clustering* or *client segmentation*, where the task can be formulated by creating a graph containing a node for each client and an edge between each pair of clients. The weight of an edge connecting any two clients is determined by the relative closeness of clients' characteristics: the closer the clients, the smaller the weight of the edge that connects them. The clusters that are formed by finding maximum weight cuts have the property that clients in one cluster are more dissimilar from clients in other clusters.

However, the flagship application of Max-Cut in finance is portfolio optimisation. Dees, Stanković, Constantinides, and Mandi [80] have shown that the graph-theoretic portfolio partitioning technique can help devise robust and tractable asset allocation schemes by virtue of a rigorous graph framework for considering smaller, computationally feasible, and economically meaningful clusters of assets, based on graph cuts. Barkoutsos, Nannicini,

Robert, Tavernelli, and Woerner [25] improved variational quantum optimisation using a Conditional Value-at-Risk technique – ubiquitous in financial risk management. A portfolio optimisation QAOA use case addressed by solving the maximum independent set problem on a quantum simulator was presented by Suchara [300]. It is also necessary to mention portfolio rebalancing experiments using the Quantum Alternating Operator Ansatz conducted by Hodson, Ruck, Ong, Garvin, and Dulma [146].

11.4.1 QAOA gates

The mixing Hamiltonian $\mathcal{A}$ and the phase Hamiltonian $\mathcal{B}$ corresponding to the cost function (11.4.2) read

$$\mathcal{A} = \sum_{i=1}^{n} \sigma_x^i \quad \text{and} \quad \mathcal{B} = \sum_{\{ij\} \in G} \frac{1}{2} w_{ij} \left(1 - \sigma_z^i \sigma_z^j\right),$$

where the spin variables s are replaced by the corresponding Pauli operators σ. We therefore need to find the quantum gate representation of the operators

$$\exp\left(-i\beta\sigma_x^i\right) \quad \text{and} \quad \exp\left(-\frac{1}{2}i\gamma\sigma_z^i\sigma_z^j\right). \tag{11.4.3}$$

To do so, we require the following (see also Lemma 1, albeit with a slightly different proof):

Theorem 9. *With $\mathcal{I}$ denoting the identity operator, the following holds for any unitary Hermitian operator $\mathcal{H}$ and any $\theta \in \mathbb{R}$:*

$$R_\theta(\mathcal{H}) \equiv \exp\left(-\frac{1}{2}i\theta\mathcal{H}\right) = \cos\left(\frac{\theta}{2}\right)\mathcal{I} - i\sin\left(\frac{\theta}{2}\right)\mathcal{H}. \tag{11.4.4}$$

Proof. Since $\mathcal{H}$ is a unitary Hermitian operator, its eigenvalues are $+1$ and -1 (Chapter 1). Let $\mathcal{P}_\pm$ be projectors onto the eigenspace of eigenvalues ± 1 respectively, so that

$$\mathcal{I} = \mathcal{P}_+ + \mathcal{P}_-, \quad \mathcal{H} = \mathcal{P}_+ - \mathcal{P}_-. \tag{11.4.5}$$

When a function (in this case R_θ) is applied to a matrix (in this case $\mathcal{H}$) it is applied to each of the eigenvalues:

$$
\begin{aligned}
R_\theta(\mathcal{H}) &= \exp\left(-\frac{1}{2}\mathrm{i}\theta(+1)\right)\mathcal{P}_+ + \exp\left(-\frac{1}{2}\mathrm{i}\theta(-1)\right)\mathcal{P}_- \\
&= \exp\left(-\frac{1}{2}\mathrm{i}\theta\right)\mathcal{P}_+ + \exp\left(\frac{1}{2}\mathrm{i}\theta\right)\mathcal{P}_-.
\end{aligned}
\tag{11.4.6}
$$

From (11.4.5) we have

$$
\mathcal{P}_+ = \frac{1}{2}(\mathcal{I}+\mathcal{H}) \quad \text{and} \quad \mathcal{P}_- = \frac{1}{2}(\mathcal{I}-\mathcal{H}).
\tag{11.4.7}
$$

Substituting (11.4.7) into (11.4.6) yields

$$
\begin{aligned}
R_\theta(\mathcal{H}) &= \frac{1}{2}\exp\left(-\frac{1}{2}\mathrm{i}\theta\right)(\mathcal{I}+\mathcal{H}) + \frac{1}{2}\exp\left(\frac{1}{2}\mathrm{i}\theta\right)(\mathcal{I}-\mathcal{H}) \\
&= \frac{1}{2}\left[\exp\left(-\frac{1}{2}\mathrm{i}\theta\right) + \exp\left(\frac{1}{2}\mathrm{i}\theta\right)\right]\mathcal{I} + \frac{1}{2}\left[\exp\left(-\frac{1}{2}\mathrm{i}\theta\right) - \exp\left(\frac{1}{2}\mathrm{i}\theta\right)\right]\mathcal{H} \\
&= \cos\left(\frac{\theta}{2}\right)\mathcal{I} - \mathrm{i}\sin\left(\frac{\theta}{2}\right)\mathcal{H}.
\end{aligned}
$$

$\square$

We can use (11.4.4) to write down expressions for the operators (11.4.3) in matrix form. We start with the first operator:

$$
\begin{aligned}
\exp(-\mathrm{i}\beta\sigma_x) &= \cos(\beta)\mathbf{I} - \mathrm{i}\sin(\beta)\mathbf{X} \\
&= \cos(\beta)\begin{bmatrix} 1 & 0 \\ 0 & 1 \end{bmatrix} - \mathrm{i}\sin(\beta)\begin{bmatrix} 0 & 1 \\ 1 & 0 \end{bmatrix} \\
&= \begin{bmatrix} \cos(\beta) & -\mathrm{i}\sin(\beta) \\ -\mathrm{i}\sin(\beta) & \cos(\beta) \end{bmatrix}.
\end{aligned}
$$

Since the operator $R_X(\theta)$ has the following matrix representation

$$R_X(\theta) = \begin{bmatrix} \cos\left(\frac{\theta}{2}\right) & -i\sin\left(\frac{\theta}{2}\right) \\ -i\sin\left(\frac{\theta}{2}\right) & \cos\left(\frac{\theta}{2}\right) \end{bmatrix},$$

(see, for example, Chapter 6 or Theorem 9), we have

$$\exp(-i\beta\sigma_x) = R_X(2\beta),$$

so that operator $\exp(-i\beta\sigma_x^i)$ should be represented in the circuit by the gate $R_X(2\beta)$ placed on the quantum register i.

We can proceed now with the gate representation of operator $\exp\left(-\frac{1}{2}i\gamma\sigma_z\sigma_z\right)$, where $\sigma_z\sigma_z$ represents the tensor product of two σ_z operators acting on two different qubits:

$$\exp\left(-\frac{1}{2}i\gamma\sigma_z\sigma_z\right) = \cos\left(\frac{\gamma}{2}\right) I \otimes I - i\sin\left(\frac{\gamma}{2}\right) Z \otimes Z$$

$$= \cos\left(\frac{\gamma}{2}\right) \begin{bmatrix} 1 & 0 & 0 & 0 \\ 0 & 1 & 0 & 0 \\ 0 & 0 & 1 & 0 \\ 0 & 0 & 0 & 1 \end{bmatrix} - i\sin\left(\frac{\gamma}{2}\right) \begin{bmatrix} 1 & 0 & 0 & 0 \\ 0 & -1 & 0 & 0 \\ 0 & 0 & -1 & 0 \\ 0 & 0 & 0 & 1 \end{bmatrix}$$

$$= \begin{bmatrix} \cos\left(\frac{\gamma}{2}\right) - i\sin\left(\frac{\gamma}{2}\right) & 0 & 0 & 0 \\ 0 & \cos\left(\frac{\gamma}{2}\right) + i\sin\left(\frac{\gamma}{2}\right) & 0 & 0 \\ 0 & 0 & \cos\left(\frac{\gamma}{2}\right) + i\sin\left(\frac{\gamma}{2}\right) & 0 \\ 0 & 0 & 0 & \cos\left(\frac{\gamma}{2}\right) - i\sin\left(\frac{\gamma}{2}\right) \end{bmatrix}$$

$$= \begin{bmatrix} e^{-i\gamma/2} & 0 & 0 & 0 \\ 0 & e^{i\gamma/2} & 0 & 0 \\ 0 & 0 & e^{i\gamma/2} & 0 \\ 0 & 0 & 0 & e^{-i\gamma/2} \end{bmatrix}.$$

Where we used the fact that

$$\exp\left(-\frac{1}{2}i\gamma\right) = \cos\left(\frac{\gamma}{2}\right) - i\sin\left(\frac{\gamma}{2}\right)$$

and

$$\exp\left(\frac{1}{2}i\gamma\right) = \cos\left(\frac{\gamma}{2}\right) + i\sin\left(\frac{\gamma}{2}\right).$$

The following lemma provides a quantum circuit for the operator $\exp\left(-\frac{1}{2}i\gamma\sigma_z\sigma_z\right)$:

Lemma 8. *The operator* $\exp\left(-\frac{1}{2}i\gamma\sigma_z^i\sigma_z^j\right)$ *can be represented by the following circuit:*

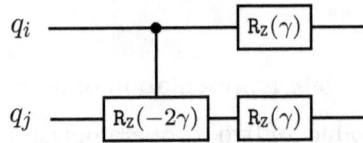

Proof. From (6.3.3), we can write

$$R_Z(\gamma) = \begin{bmatrix} e^{-i\gamma/2} & 0 \\ 0 & e^{i\gamma/2} \end{bmatrix} = e^{-i\gamma/2}\begin{bmatrix} 1 & 0 \\ 0 & e^{i\gamma} \end{bmatrix}.$$

The multiplier $\exp\left(-\frac{1}{2}i\gamma\right)$ is an unobservable global phase and can be ignored. Now,

$$R_Z(\gamma) \otimes R_Z(\gamma) = \begin{bmatrix} 1 & 0 \\ 0 & e^{i\gamma} \end{bmatrix} \otimes \begin{bmatrix} 1 & 0 \\ 0 & e^{i\gamma} \end{bmatrix} = \begin{bmatrix} 1 & 0 & 0 & 0 \\ 0 & e^{i\gamma} & 0 & 0 \\ 0 & 0 & e^{i\gamma} & 0 \\ 0 & 0 & 0 & e^{2i\gamma} \end{bmatrix}.$$

Finally, the matrix representation of the $CR_z(-2\gamma)$ gate is

$$CR_Z(-2\gamma) = \begin{bmatrix} 1 & 0 & 0 & 0 \\ 0 & 1 & 0 & 0 \\ 0 & 0 & 1 & 0 \\ 0 & 0 & 0 & e^{-2i\gamma} \end{bmatrix},$$

and the matrix representation of the whole circuit shown in the lemma is

$$
\begin{bmatrix}
1 & 0 & 0 & 0 \\
0 & e^{i\gamma} & 0 & 0 \\
0 & 0 & e^{i\gamma} & 0 \\
0 & 0 & 0 & e^{2i\gamma}
\end{bmatrix}
\begin{bmatrix}
1 & 0 & 0 & 0 \\
0 & 1 & 0 & 0 \\
0 & 0 & 1 & 0 \\
0 & 0 & 0 & e^{-2i\gamma}
\end{bmatrix}
= e^{i\gamma/2}
\begin{bmatrix}
e^{-i\gamma/2} & 0 & 0 & 0 \\
0 & e^{i\gamma/2} & 0 & 0 \\
0 & 0 & e^{i\gamma/2} & 0 \\
0 & 0 & 0 & e^{-i\gamma/2}
\end{bmatrix}.
$$

Again, the global phase can be ignored and we arrive at the same matrix expression we obtained for $\exp\left(-\frac{1}{2}i\gamma\sigma_z\sigma_z\right)$.

□

11.4.2 QAOA circuit

As pointed out in Chapters 8 and 9, existing quantum processors often have limited qubit connectivity, so that we need to use a particular embedding scheme in order to map the Max-Cut graph onto the QPU graph. The simplest case is a one-to-one mapping of the graph nodes and the connectivity edges. Figure 11.2 displays one such graph (implemented in Rigetti's Aspen system) consisting of eight nodes (embedded in qubits $1, \ldots, 8$) and eight edges. Nodes 1, 3, 5, and 7 each have one connection and nodes 2, 4, 6, and 8 each have three connections.

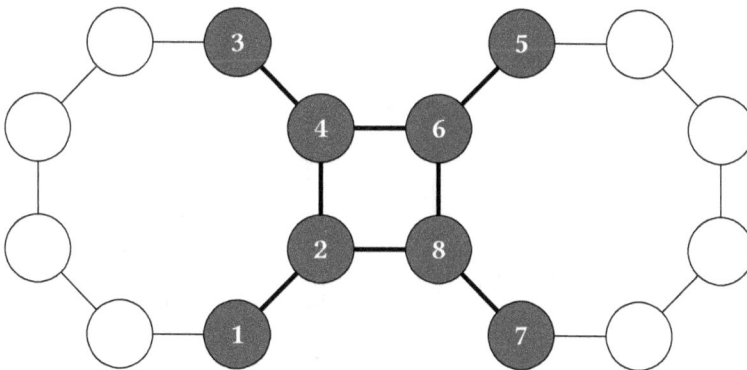

Figure 11.2: Embedding of the Max-Cut optimisation problem on Rigetti's Aspen system.

The corresponding QAOA circuit is shown in Figure 11.3.

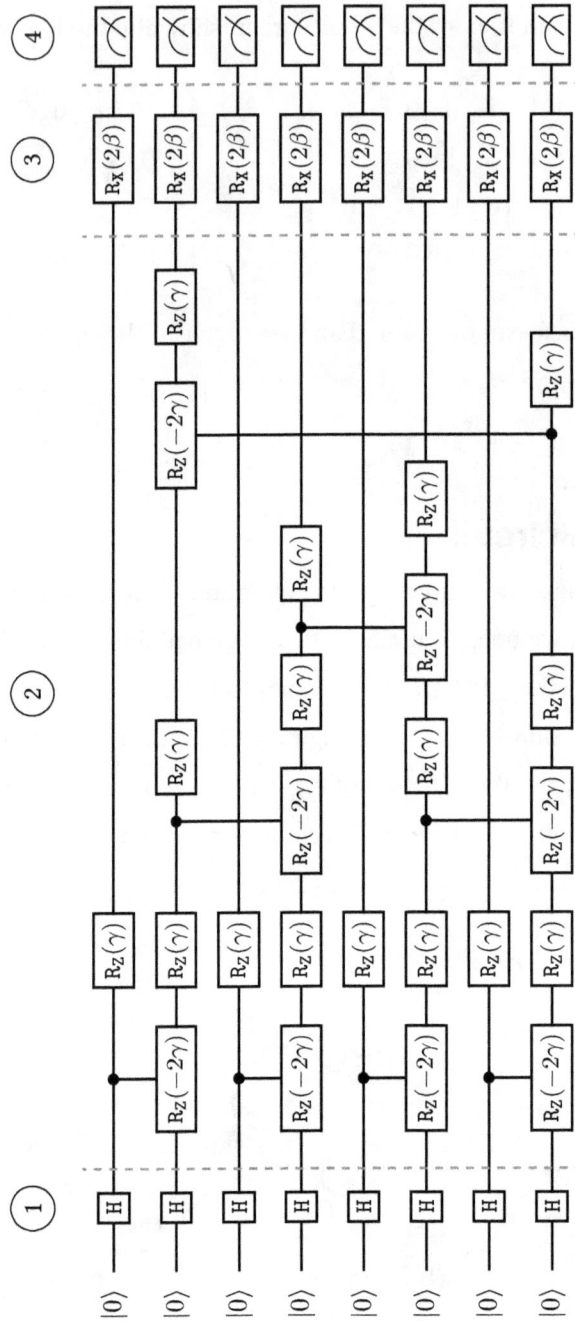

Figure 11.3: QAOA circuit for the Max-Cut problem.

The circuit consists of four distinct layers. The first one is a layer of Hadamard gates creating an equal superposition of states $|0\rangle$ and $|1\rangle$ – transformation of the basis states $|0\rangle$ into the basis states $|+\rangle$. The second layer represents the action of the phase Hamiltonian controlled by the adjustable parameter γ. The third layer represents the action of the mixing Hamiltonian controlled by the adjustable parameter β. The final layer consists of measurement operators. The second and third layers can be applied multiple times with different values of parameters γ and β.

The optimal solution found by the QAOA using the `Qiskit` quantum simulator for the case of equal weights ($w_{ij} = 1$ for all $\{i, j\} \in G$) is shown in Figure 11.4. The optimal solution reads as the bitstring `10011001`, and is represented by the dashed curve that separates nodes into two equal subsets and cuts across all edges of the graph.

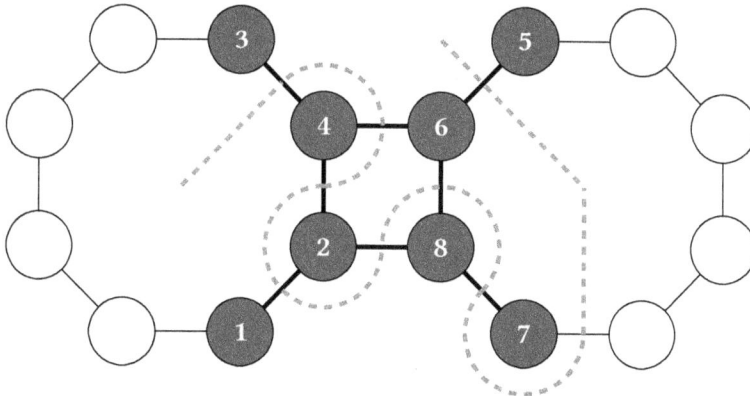

Figure 11.4: Visualisation of the Max-Cut problem solution.

Here, the graph nodes (qubits) $\{1, 4, 5, 8\}$ belong to group "1" and the graph nodes $\{2, 3, 6, 7\}$ belong to group "0".

In the case of unequal weights, the circuit layout remains the same but the adjustable gate parameters reflect the relative magnitude of the weights assigned to different edges. For example, if the weight assigned to the connection between nodes 1 and 2 is increased from 1 to 5 and all other weights remain equal to 1, then the phase Hamiltonian term that corresponds to the connection between nodes 1 and 2 changes from $\exp\left(-\frac{1}{2}i\gamma\sigma_z\sigma_z\right)$ to

$\exp\left(-\frac{5}{2}i\gamma\sigma_z\sigma_z\right)$, and the corresponding segment of the quantum circuit changes to the one shown in Figure 11.5, with the rest of the circuit remaining the same.

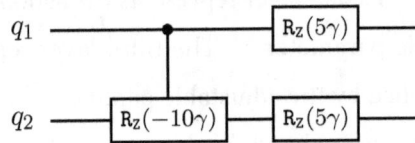

Figure 11.5: Weight w_{12} increases from 1 to 5.

QAOA can successfully solve NP-hard combinatorial optimisation problems in their QUBO formulations.

Summary

In this chapter, we studied a quantum optimisation algorithm inspired by the principles of adiabatic evolution of quantum systems. In this regard, QAOA can be seen as a quantum gate model counterpart of the AQC approach to solve classically hard optimisation problems.

We discussed the Suzuki-Trotter expansion, which provides an approximation of the time evolution of quantum mechanical systems and which lies at the heart of the algorithm. We illustrated QAOA implementation (gates and circuit) on the Max-Cut problem, which has many applications in finance.

In the next chapter, we will introduce two quantum algorithms based on the quantum feature maps that are at the cutting edge of QML: the quantum kernel method and the quantum two-sample test.

12

Quantum Kernels and Quantum Two-Sample Test

In this chapter, we discuss the possibility of achieving quantum advantage on one of the most fundamental problems of quantitative finance – classification of the probability distributions. The problem can be formulated as follows. Let us assume that we have two sets of samples (either ordered or not), in the most general case of unequal size, drawn from the unknown multivariate probability distributions. Can we say with the desired degree of confidence whether these samples were drawn from the same probability distribution or not? This problem has many direct applications to the practical use cases, especially on the buy side, such as time series analysis, detection of structural breaks, and monitoring of alpha decay, to name just a few. The problem of comparing multivariate probability distributions given by two datasets is known as a two-sample test. We already mentioned one such test in Chapter 9: the Maximum Mean Discrepancy method widely used in training generative models. In this chapter, we provide a detailed description of this method as a classical benchmark for the quantum two-sample test.

However, before specifying the quantum two-sample test that is aimed at comparing two datasets, we need to first address the point of comparing two individual samples from the datasets. In other words, we need to specify the measure of closeness of two individual samples before we build the framework for estimating the closeness of two datasets. We do it with the help of kernel methods, starting with the popular classical kernel and then describing its quantum counterpart based on the concept of parameterised quantum circuits.

12.1 Classical Kernel Method

A *kernel method* is the key element of a powerful classical supervised learning algorithm: Support Vector Machine (SVM). Unlike a feedforward neural network based classifier whose objective is to minimise the classification error, the SVM's objective is to maximise the margin, defined as the distance between a separating hyperplane (decision boundary separating samples belonging to different classes) and the training samples that are closest to this hyperplane [264]. The samples that are closest to the separating hyperplane are called *support vectors*, thus giving its name to the algorithm.

The maximisation of the margins lowers the generalisation error and helps fight overfitting. This is a very important property but finding the separating hyperplane is not an easy task for non-linearly separable data. Fortunately, the kernel method allows us to overcome this difficulty, by creating non-linear combinations of the original features and projecting them onto a higher-dimensional space where the data samples become linearly separable.

Whereas an SVM with linearly separable data operates on the inner product $\langle \mathrm{x}^i, \mathrm{x}^j \rangle$ of the training samples, the generalised version dealing with non-linearly separable data operates on the kernel function

$$k(\mathrm{x}^i, \mathrm{x}^j) := \phi(\mathrm{x}^i)^\top \phi(\mathrm{x}^j), \tag{12.1.1}$$

where $\phi : \mathbb{R}^N \to \mathbb{R}^M$, with $M \gg N$, is the feature map that projects the N-dimensional feature $\mathrm{x} := (x_1, \ldots, x_N)$ onto the M-dimensional feature space. The inner product (12.1.1) would be computationally expensive to calculate directly but the kernel function is computationally inexpensive – this is known as the *kernel trick*. The kernel function can be seen as a similarity function operating on a pair of samples. For example, the radial basis function (Gaussian kernel with scaling parameter σ)

$$k(\mathrm{x}^i, \mathrm{x}^j) = \exp\left(-\frac{\|\mathrm{x}^i - \mathrm{x}^j\|^2}{2\sigma^2}\right), \tag{12.1.2}$$

translates the distance between samples x^i and x^j (defined on $[0, \infty)^N$) into a similarity score (defined on the interval $[0, 1]$).

The right choice of kernel function can make the classification task much easier. However, some kernels may be hard to compute. This is where quantum computing may play an important role by providing efficient quantum circuits to compute them.

12.2 Quantum Kernel Method

Wang, Du, Luo, and Ta [323] have shown a close correspondence between classical and quantum kernels. The feature map $\phi(\cdot)$ coincides with the preparation of a quantum state via a parameterised quantum circuit $\mathcal{U}(\cdot)$, which maps the input data sample into a high-dimensional Hilbert space described by n qubits:

$$\phi(\mathrm{x}) \to |\psi(\mathrm{x})\rangle = \mathcal{U}(\mathrm{x}) |0\rangle^{\otimes n}.$$

The kernel function then coincides with applying measurements on the prepared quantum states:

$$k(\mathrm{x}^i, \mathrm{x}^j) \to \left|\langle\psi(\mathrm{x}^j)|\psi(\mathrm{x}^i)\rangle\right|^2, \tag{12.2.1}$$

and allows for more expressive models in comparison with the alternative

$$k(\mathrm{x}^i, \mathrm{x}^j) = \phi(\mathrm{x}^i)^\top \phi(\mathrm{x}^j) \to \langle\psi(\mathrm{x}^j)|\psi(\mathrm{x}^i)\rangle. \tag{12.2.2}$$

Huang *et al.* [152] argued that even though the kernel function (12.2.2) seems to be more natural, the quantum kernel (12.2.1) can learn arbitrarily deep quantum neural networks (deep PQC). This is a strong result, especially in combination with the hierarchy of expressive power of parameterised quantum circuits (Chapter 13, Equation (13.2.4)).

Havlíček *et al.* [137] described how a quantum computer can be used to estimate the kernel. The kernel entries are the *fidelities* between different feature vectors (analogous to similarity scores in classical kernel methods). Burnham, Cleve, Watrous, and R. de Wolf [49] and Cincio, Subaşi, Sornborger, and Coles [66] investigated various fidelity estimation methods, such as quantum fingerprinting and a machine learning approach (both relying on the application of a CSWAP gate implementing the swap test). However, by using the fact that the states in the feature space are not arbitrary, the overlap between the quantum states can be estimated from the transition probability:

$$\left| \langle \psi(\mathbf{x}^j) | \psi(\mathbf{x}^i) \rangle \right|^2 = \left| \langle \mathbf{0} | \mathcal{U}^\dagger(\mathbf{x}^j) \mathcal{U}(\mathbf{x}^i) | \mathbf{0} \rangle \right|^2,$$

where, for brevity, we used the notation $|\mathbf{0}\rangle := |0\rangle^{\otimes n}$. The first step is the application of a composition of two consecutive feature map circuits (representing the operators $\mathcal{U}(\mathbf{x}^i)$ and $\mathcal{U}^\dagger(\mathbf{x}^j)$) to the initial state $|\mathbf{0}\rangle$. The second step is the measurement of the final state in the computational basis K times and counting the number κ of all-zero strings $|\mathbf{0}\rangle$. The frequency κ/K of the all-zero string is the estimate of the transition probability (the "similarity score").

The rest of the supervised learning protocol is classical, allowing for the natural embedding of quantumly computed kernels into the overall framework: the algorithm remains essentially classical with only the classically hard task outsourced to the quantum chip.

12.3   Quantum Circuits for Feature Maps

Figure 12.1 displays a schematic representation of the feature map circuit. In this example, we work with an 8-dimensional dataset with features encoded in the rotation angles such that there is a direct mapping of a sample $\mathbf{x}^i := (x_1^i, \ldots, x_8^i)$ into the vector of adjustable circuit parameters $\theta^i := (\theta_1^i, \ldots, \theta_8^i)$. The first section of the circuit implements the operator $\mathcal{U}(\mathbf{x}^i)$, creating an entangled state due to the layer of fixed two-qubit CZ gates, whereas the second section of the circuit implements $\mathcal{U}^\dagger(\mathbf{x}^j)$. Here, we use the fact that

$$R_X^\dagger(\theta) = R_X(-\theta), \quad R_Y^\dagger(\theta) = R_Y(-\theta), \quad CZ^\dagger = CZ.$$

It is easy to see that if the samples x^i and x^j are identical (so that $\theta^i = \theta^j$), then $\mathcal{U}(x^i)\mathcal{U}^\dagger(x^j) = \mathcal{I}$ and all K measurements will return the all-zero string $|0\rangle$.

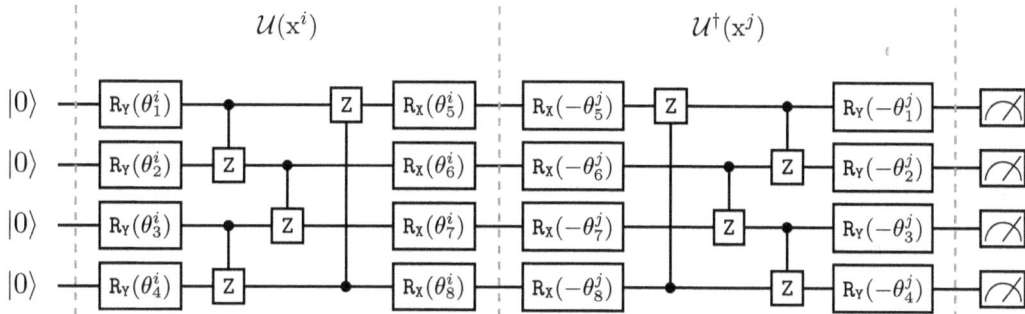

Figure 12.1: Schematic quantum kernel circuit.

The rest of the protocol is classical – the quantum computer is used to assist the classifier with the calculation of a kernel function, which would not be feasible if only classical computational resources were available.

Let us now apply the quantum kernel method to the Australian Credit Approval dataset (introduced in Chapter 8) in order to estimate the degree of similarity between samples drawn from the same class and samples drawn from two different classes. The ACA dataset consists of 690 samples, with 383 samples labelled as Class 0 and 307 samples labelled as Class 1, so the dataset is reasonably well balanced. Each sample consists of 14 features (continuous, integer, binary). In Chapter 8, we built a QNN classifier and tested its performance on the ACA dataset, employing the *angle encoding* scheme as explained in Section 7.2. We would like to build a feature map that is consistent with the angle encoding scheme and does not require the construction of a too deep PQC. In fact, we would like to build a feature map using the PQC that is as close as possible to the one shown in Figure 12.1. The proposed scheme can be embedded into all existing NISQ systems we considered earlier in this book. For example, we can use IBM's Melbourne system shown in Figure 12.2.

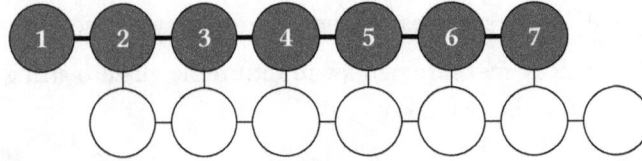

Figure 12.2: Embedding of the quantum kernel circuit into IBM's Melbourne system.

We know that 7 quantum registers (shown as shaded qubits connected by the thick lines in Figure 12.2) can encode a 14-feature data sample if we follow the angle encoding scheme. The corresponding circuit is shown in Figure 12.3. The linear sequential connectivity between the physical qubits makes the choice of the two-qubit gates straightforward (and, in fact, similar to the one in Figure 12.1).

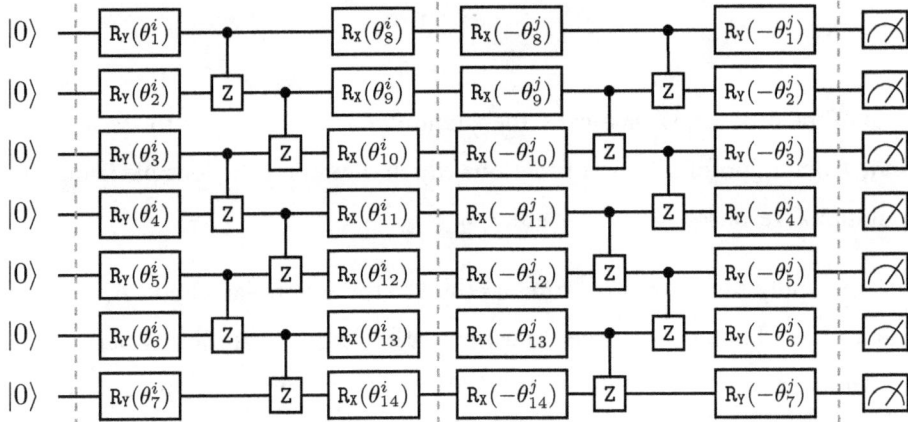

Figure 12.3: Quantum kernel circuit for the ACA dataset.

In the circuit shown in Figure 12.3, the angles θ^i and θ^j encode the data samples x^i and x^j, which can be drawn either from the same class or from two different classes. Running the circuit K times and calculating the number κ of all-zero bitstrings (after measurement) gives us the measure of similarity between samples x^i and x^j (estimated as the ratio κ/K). Figure 12.4 displays the mean values of the transition probabilities (similarity scores) obtained using the quantum kernel (12.2.1) by running the quantum circuit on the Qiskit simulator $K = 10,000$ times for each pair of data samples. The mean values were calculated across all possible pairs of samples from the corresponding classes.

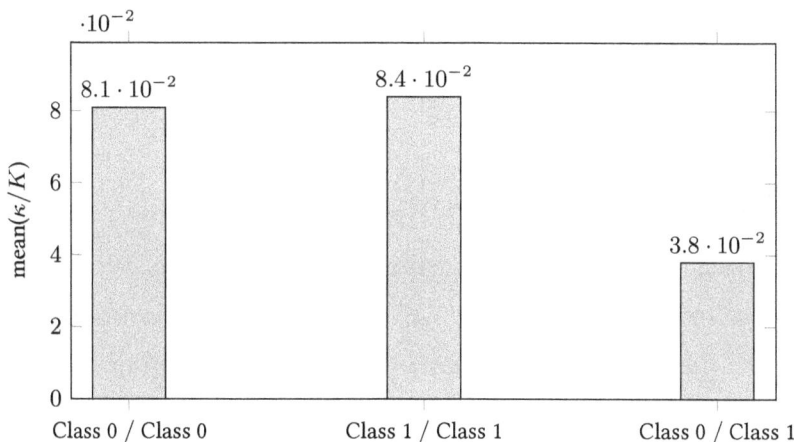

Figure 12.4: Mean values of the quantum kernel (12.2.1) for the ACA dataset.

As expected, samples drawn from the same class have, on average, significantly larger similarity scores given by the quantum kernel compared with samples drawn from two different classes.

Figure 12.5 displays similar results obtained with the help of the classical Gaussian kernel (12.1.2). The ACA dataset features were normalised by mapping their values to the interval $[0, 1]$, and the scaling parameter was set to $\sigma = 1$.

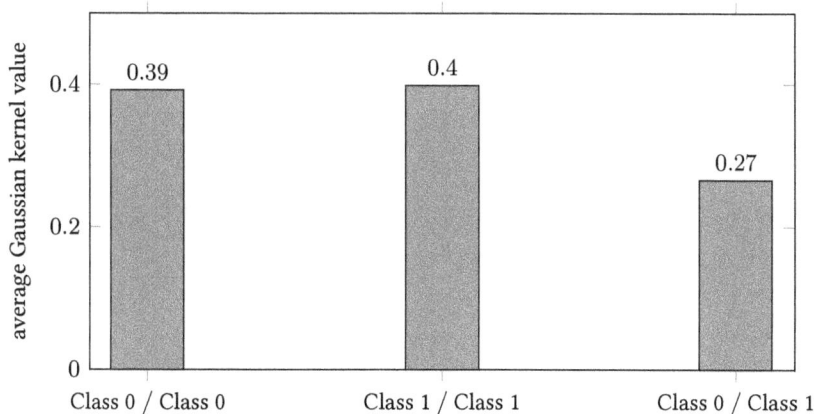

Figure 12.5: Mean values of the Gaussian kernel for the ACA dataset.

> Quantum kernels that can be efficiently calculated on quantum computers
> have the potential to improve the performance of hybrid quantum-classical
> machine learning models.

12.4   Classical Two-Sample Test

The classical and quantum kernels considered above can be used to calculate the closeness
of two samples from the given dataset. We can even do it systematically for all pairs of
samples in the dataset. However, in many situations, we are interested in comparing the
closeness of (or the distance between) whole datasets rather than individual samples.

In other words, given two datasets A and B, with samples drawn from some unknown
multivariate probability distributions, the task is to measure the distance between these
datasets in order to decide whether the null hypothesis of both sets of samples being drawn
from the same probability distribution can be rejected or not. This is a two-sample test
problem, which we consider below. We start with describing a popular classical two-sample
test before introducing its quantum counterpart.

When it comes to the multivariate distribution classification problem, one of the most
widely used measures of similarity is the Maximum Mean Discrepancy (MMD), a kernel-
based two-sample statistical test used to determine whether two distributions are the same.
MMD is widely used as a cost function in various generative machine learning models [122].

MMD is defined by the idea of representing distances between distributions as distances
between mean embeddings of features. If we have distributions P and Q over a set $\mathcal{F}$,
the MMD is defined by a feature map $\varphi : \mathcal{F} \to \mathcal{H}$, where $\mathcal{H}$ is a reproducing kernel
Hilbert space.

In general, the MMD is defined as

$$MMD(P, Q) = \|\mathbb{E}_{x \sim P}[\varphi(x)] - \mathbb{E}_{y \sim Q}[\varphi(y)]\|_{\mathcal{H}}.$$

Example: $\mathcal{F} = \mathcal{H} = \mathbb{R}^d$ and $\varphi(\mathrm{x}) = \mathrm{x}$. In that case, MMD is just the distance between the means of the two distributions:

$$MMD(P, Q) = \|\mathbb{E}_{\mathrm{x} \sim P}[\mathrm{x}] - \mathbb{E}_{\mathrm{y} \sim Q}[\mathrm{y}]\|_{\mathbb{R}^d} = \|\mu_P - \mu_Q\|_{\mathbb{R}^d}.$$

Example: $\mathcal{F} = \mathbb{R}^d$ and $\mathcal{H} = \mathbb{R}^p$ with $\varphi(\mathrm{x}) = A^\top \mathrm{x}$, where A is a $d \times p$ matrix:

$$MMD(P, Q) = \|\mathbb{E}_{\mathrm{x} \sim P}[A^\top \mathrm{x}] - \mathbb{E}_{\mathrm{y} \sim Q}[A^\top \mathrm{y}]\|_{\mathbb{R}^p}$$

$$= \|A^\top \mathbb{E}_{\mathrm{x} \sim P}[\mathrm{x}] - A^\top \mathbb{E}_{\mathrm{y} \sim Q}[\mathrm{y}]\|_{\mathbb{R}^p} = \|A^\top (\mu_P - \mu_Q)\|_{\mathbb{R}^p}.$$

If $p < d$, then we get a *weaker* distance, and if $p > d$, then we get a *stronger* distance than in the previous example.

Example: $\mathcal{F} = \mathbb{R}^1$ and $\mathcal{H} = \mathbb{R}^2$ with $\varphi(x) = (x, x^2)$:

$$MMD(P, Q) = \sqrt{(\mathbb{E}_{x \sim P}[x] - \mathbb{E}_{y \sim Q}[y])^2 + (\mathbb{E}_{x \sim P}[x^2] - \mathbb{E}_{y \sim Q}[y^2])^2)}.$$

This is a stronger distance that can distinguish not only distributions with different means but different variances as well.

If φ maps to a general reproducing kernel Hilbert space, then we can apply the kernel trick to compute the MMD:

$$\langle \varphi(\mathrm{x}), \varphi(\mathrm{y}) \rangle_{\mathcal{H}} = k(\mathrm{x}, \mathrm{y}),$$

where $k(\cdot, \cdot)$ is some *kernel* function. Many kernels, including the Gaussian kernel, lead to the MMD being 0 if, and only if, the distributions are identical.

$$MMD^2(P, Q) = \|\mathbb{E}_{\mathrm{x} \sim P}[\varphi(\mathrm{x})] - \mathbb{E}_{\mathrm{y} \sim Q}[\varphi(\mathrm{y})]\|_{\mathcal{H}}^2 =$$

$$= \langle \mathbb{E}_{\mathrm{x} \sim P}[\varphi(\mathrm{x})], \mathbb{E}_{\mathrm{x}' \sim P}[\varphi(\mathrm{x}')] \rangle_{\mathcal{H}} - \langle \mathbb{E}_{\mathrm{x} \sim P}[\varphi(\mathrm{x})], \mathbb{E}_{\mathrm{y} \sim Q}[\varphi(\mathrm{y})] \rangle_{\mathcal{H}}$$

$$- \langle \mathbb{E}_{\mathrm{y} \sim Q}[\varphi(\mathrm{y})], \mathbb{E}_{\mathrm{x} \sim P}[\varphi(\mathrm{x})] \rangle_{\mathcal{H}} + \langle \mathbb{E}_{\mathrm{y} \sim Q}[\varphi(\mathrm{y})], \mathbb{E}_{\mathrm{y}' \sim Q}[\varphi(\mathrm{y}')] \rangle_{\mathcal{H}}$$

$$= \mathbb{E}_{\mathsf{x},\mathsf{x}'\sim P}k(\mathsf{x},\mathsf{x}') - 2\mathbb{E}_{\mathsf{x}\sim P,\mathsf{y}\sim Q}k(\mathsf{x},\mathsf{y}) + \mathbb{E}_{\mathsf{y},\mathsf{y}'\sim Q}k(\mathsf{y},\mathsf{y}').$$

We can estimate the MMD with the Gaussian kernel directly from the data. For two datasets, a dataset $\mathcal{X}$ of length N and a dataset $\mathcal{Y}$ of length M, we have the following:

a) A Gaussian kernel for samples i and j:

$$k(\mathsf{x}_i, \mathsf{x}_j) = \exp\left(-\frac{\|\mathsf{x}_i - \mathsf{x}_j\|^2}{2\sigma^2}\right) = \exp\left(-\frac{\mathsf{x}_i^\top \mathsf{x}_i - 2\mathsf{x}_i^\top \mathsf{x}_j + \mathsf{x}_j^\top \mathsf{x}_j}{2\sigma^2}\right). \qquad (12.4.1)$$

b) MMD:

$$MMD^2 = \frac{1}{N^2}\sum_{i,j=1}^{N} k(\mathsf{x}_i, \mathsf{x}_j) - \frac{2}{NM}\sum_{i=1}^{N}\sum_{j=1}^{M} k(\mathsf{x}_i, \mathsf{y}_j)$$
$$+ \frac{1}{M^2}\sum_{i,j=1}^{M} k(\mathsf{y}_i, \mathsf{y}_j). \qquad (12.4.2)$$

With all the useful properties outlined above and the ease of computation, it is no wonder that MMD with the Gaussian kernel is one of the most popular two-sample tests for multivariate distributions. This also motivates our choice of MMD as a classical benchmark for the proposed quantum algorithm.

12.5 Quantum Two-Sample Test

Let A and B be two datasets, sampled, respectively, from the probability distributions $P(\mathsf{x})$ and $Q(\mathsf{x})$, $\mathsf{x} \in \mathcal{D} \subseteq \mathbb{R}^m$. The cardinality of A and B will be indicated as $[A]$ and $[B]$. The samples $\mathsf{a}_i \in A$ and $\mathsf{b}_j \in B$ are of the form $\mathsf{a}_i = (a_i(1), \ldots, a_i(m))$ and $\mathsf{b}_j = (b_j(1), \ldots, b_j(m))$. Next, we define a transformation φ that maps the samples a_i and b_j into vectors $\varphi(\mathsf{a}_i)$ and $\varphi(\mathsf{b}_j)$ of a Hilbert space $\mathcal{H}$. On a quantum computer, the feature map φ is realised via a unitary transformation $\mathcal{U}$ (parameterised quantum circuit) applied to the state $|\mathbf{0}\rangle := |0\rangle^{\otimes n}$. We will use the following notations:

$$\varphi(\mathsf{x}) \rightarrow \mathcal{U}(\mathsf{x})|\mathbf{0}\rangle = |\mathsf{x}\rangle.$$

Note that if the transformation φ is specified on n quantum registers, the dimensionality of the Hilbert space $\mathcal{H}$ is 2^n.

Then we can construct two density matrices encoding the probability distributions $P(\mathrm{x})$ and $Q(\mathrm{x})$ as

$$p = \sum_{i=1}^{[A]} \frac{1}{[A]}|\mathrm{a}_i\rangle\langle \mathrm{a}_i| \quad \text{and} \quad q = \sum_{j=1}^{[B]} \frac{1}{[B]}|\mathrm{b}_j\rangle\langle \mathrm{b}_j|.$$

The density matrices p and q and the projectors $|\mathrm{x}\rangle\langle\mathrm{x}|$ that form them are all vectors in the space $\mathcal{H}_{\mathcal{O}}$ of linear operators on $\mathcal{H}$ (note that projectors – the "measurement" operators – are linear but not unitary operators). We can introduce a distance d on $\mathcal{H}_{\mathcal{O}}$ and use it as measure of closeness of p and q, and use it as a proxy of whether A and B are sampled from the same probability distribution.

In the limit case of infinitely many samples, it is easy to see how $d(p, q) = 0$ iff $P(\mathrm{x}) = Q(\mathrm{x})$. In fact, in such case

$$\begin{aligned} p - q &= \int_{\mathcal{D}} d\mathrm{x} P(\mathrm{x})|\mathrm{x}\rangle\langle\mathrm{x}| - \int_{\mathcal{D}} d\mathrm{x} Q(\mathrm{x})|\mathrm{x}\rangle\langle\mathrm{x}| \\ &= \int_{\mathcal{D}} d\mathrm{x}(P(\mathrm{x}) - Q(\mathrm{x}))|\mathrm{x}\rangle\langle\mathrm{x}|, \end{aligned}$$

which is equal to 0 iff $P(\mathrm{x}) - Q(\mathrm{x}) = 0, \forall\mathrm{x}$, given the assumed injectivity of $\mathcal{U}$ (the injective quantum feature map is the one that does not map different data points to the same point in the target space) or, equivalently, iff $P(\mathrm{x}) = Q(\mathrm{x}), \forall\mathrm{x}$. Given d is a distance, then $d(p, q) = 0$ iff $P(\mathrm{x}) = Q(\mathrm{x}), \forall\mathrm{x}$, since $d(p, q) = 0$ iff $p = q$ for any distance d.

In the case of finite A and B, a distance $d > 0$ would in general be measured even if A and B contained elements sampled from $P(\mathrm{x})$ and $Q(\mathrm{x})$ with $P = Q$, and this discrepancy from $d = 0$ is proportional to the variance of the distribution and inversely proportional to the cardinality of the datasets.

The key metric we consider is the Frobenius distance:

$$d_F(p, q) = \frac{1}{2}\|p - q\| = \frac{1}{2}\sqrt{\mathrm{Tr}((p - q)^2)}.$$

The proposed quantum two-sample test is analogous to the MMD test. In fact, the distance between density matrices can be thought of as a specific type of (quantum) MMD. Note that the association of datasets to density matrices can also be defined with classical transformations and projectors in Euclidean spaces. The quality of the method thus depends on the "span" of the mapped points in the target space or, more precisely, on the degree of linear independence of the mapped points in the target space.

> The quantum feature map, especially in the presence of entanglement, can produce a large degree of linear independence, which corresponds to comparing a larger number of components of the probability distributions in a functional space.

To summarise, our task is two-fold:

1. We need to construct the injective quantum feature map that will encode the classical datasets into the density matrices.

2. We need to calculate the Frobenius distance between the given density matrices.

And the proposed solution is as follows:

1. Classical samples are mapped into quantum states with the help of a PQC organised as a quantum neural network where the layers of one-qubit gates (encoding data) alternate with the layers of two-qubit gates (creating entanglement).

2. The Frobenius distance between two density matrices is calculated using the approximate quantum state technique (see, e.g., [247]). The exact step-by-step algorithm is described in the next section [109].

12.6 Estimation of the Frobenius Distance on Quantum Computer

In the space of linear operators $\mathcal{H}_\mathcal{O}$ with dimension $D_\mathcal{O} = D^2 = (2^n)^2$ acting in the Hilbert space $\mathcal{H}$ of a quantum system with dimension $D = 2^n$, let us consider a distance squared d^2 between two Hermitian operators p and q derived from a linear product $(\cdot, \cdot)$ defined in $\mathcal{H}_\mathcal{O}$:

$$d^2(p, q) = \|p - q\|^2 = (p - q, p - q).$$

If $\{\mathcal{B}_i\}$ is a complete orthonormal set in $\mathcal{H}_\mathcal{O}$, then we can express p and q as vectors in $\mathbb{C}^{D_\mathcal{O}}$ with components $p_i := (\mathcal{B}_i, p)$ and $q_i := (\mathcal{B}_i, q)$, respectively, and d^2 can be expressed as the Euclidean distance in $\mathbb{C}^{D_\mathcal{O}}$:

$$d^2(p, q) = \sum_{i=1}^{D_\mathcal{O}} |p_i - q_i|^2 = \|\mathrm{p} - \mathrm{q}\|_E^2,$$

with column vectors $\mathrm{p} := (p_1, \ldots, p_{D_\mathcal{O}})^\top$ and $\mathrm{q} := (q_1, \ldots, q_{D_\mathcal{O}})^\top$. The choice of the linear product

$$(\mathcal{L}, \mathcal{M}) = \mathrm{Tr}(\mathcal{L}^\dagger \mathcal{M})$$

induces the Frobenius norm. The Pauli operators $\{\mathrm{V}_i := \mathrm{V}_i^1 \otimes \ldots \otimes \mathrm{V}_i^n\}$, with $\mathrm{V}_i^j = \{\mathrm{I}, \mathrm{X}, \mathrm{Y}, \mathrm{Z}\}$ corresponding to the j-th single-qubit identity and Pauli operators, are an orthogonal set with this choice of linear product. Next, we fix the normalisation of the operators $\{\mathrm{V}_i\}$, that is, we choose $\mathcal{B}_i = \mathrm{V}_i/\sqrt{D}$. If p and q are representing density matrices, the elements p_i and q_i become real numbers that can be expressed as

$$p_i = \frac{1}{\sqrt{D}}\mathrm{Tr}(\mathrm{V}_i p) = \frac{1}{\sqrt{D}}\langle \mathrm{V}_i \rangle_p \quad \text{and} \quad q_i = \frac{1}{\sqrt{D}}\mathrm{Tr}(\mathrm{V}_i q) = \frac{1}{\sqrt{D}}\langle \mathrm{V}_i \rangle_q.$$

Then we have

$$d^2(p, q) = \frac{1}{D}\sum_{i=1}^{D_\mathcal{O}} |\mathrm{Tr}(\mathrm{V}_i p) - \mathrm{Tr}(\mathrm{V}_i q)|^2 = \frac{1}{D}\sum_{i=1}^{D_\mathcal{O}} |\langle \mathrm{V}_i \rangle_p - \langle \mathrm{V}_i \rangle_q|^2.$$

The Frobenius distance, as defined earlier, can be written as a Euclidean distance between vectors of expectation values of Paulis:

$$d_F^2(p,q) = \frac{1}{4D} \sum_{i=1}^{D_\mathcal{O}} |\langle V_i \rangle_p - \langle V_i \rangle_q|^2 = \frac{1}{4D} \|P - Q\|_E^2,$$

where

$$P := (\langle V_1 \rangle_p, \ldots, \langle V_{D_\mathcal{O}} \rangle_p)^\top \quad \text{and} \quad Q := (\langle V_1 \rangle_q, \ldots, \langle V_{D_\mathcal{O}} \rangle_q)^\top.$$

The complete basis set of operators V_i consists of 4^n elements. This makes working with the complete basis set impractical or even impossible for large n. Therefore, we choose a subset of L operators V_i from the full set and approximate the sum on the expectation values of all Pauli operators with the sum on the subset of L Pauli operators:

$$\frac{1}{4D} \sum_{i=1}^{L} |\langle V_i \rangle_p - \langle V_i \rangle_q|^2 \leq \frac{1}{4D} \sum_{i=1}^{D_\mathcal{O}} |\langle V_i \rangle_p - \langle V_i \rangle_q|^2.$$

Thus, we can define the Frobenius distance estimator as

$$\tilde{d}_F(p,q) := \frac{1}{2} \sqrt{\frac{\sum_{i=1}^{L} |\langle V_i \rangle_p - \langle V_i \rangle_q|^2}{D}}.$$

If we consider the subspace of $\mathcal{H}_\mathcal{O}$ spanned by the L Pauli operators then

$$\tilde{d}_F(p,q) := \frac{1}{2} \sqrt{\frac{\|P_L - Q_L\|_E^2}{D}},$$

where

$$P_L := (\langle V_1 \rangle_p, \ldots, \langle V_L \rangle_p)^\top \quad \text{and} \quad Q_L := (\langle V_1 \rangle_q, \ldots, \langle V_L \rangle_q)^\top.$$

The Algorithm 10 provides detailed instructions for obtaining the Frobenius distance estimate on a quantum computer.

Algorithm 10: Frobenius distance estimator for two datasets

Result: Frobenius distance.

1: Design n-qubit quantum circuit (quantum feature map) with layers of one-qubit gates (encoding sample) alternating with layers of two-qubit gates (creating entanglement). By default, the range of rotation angles should be set at $[0, \pi/m]$, where m is the number of data encoding layers.

2: Prepare two datasets (A and B) – calculate features, standardise features, etc.

3: For dataset A:

- For each sample $i = 1, \ldots, [A]$, run the quantum circuit N times, measuring Pauli Z_k on all quantum registers $k = 1, \ldots, n$ (+1 if we measure $|0\rangle$ and -1 if we measure $|1\rangle$). Calculate the expectation value $\langle Z_k \rangle_i^A$ as the average value of measured Z_k across all quantum circuit runs.

- For each sample $i = 1, \ldots, [A]$, place the H gate on **all** quantum registers before measurement and run the quantum circuit N times, measuring Pauli X_k on all quantum registers $k = 1, \ldots, n$ (+1 if we measure $|0\rangle$ and -1 if we measure $|1\rangle$). Calculate the expectation value $\langle X_k \rangle_i^A$ as the average value of measured X_k across all quantum circuit runs.

- For each sample $i = 1, \ldots, [A]$, place the $HS^\dagger$ gates on **all** quantum registers before measurement and run quantum circuit N times measuring Pauli Y_k on all quantum registers $k = 1, \ldots, n$ (+1 if we measure $|0\rangle$ and -1 if we measure $|1\rangle$). Calculate the expectation value $\langle Y_k \rangle_i^A$ as the average value of measured Y_k across all quantum circuit runs.

- For each sample $i = 1, \ldots, [A]$, place either H or $HS^\dagger$ gates on **some** quantum registers before measurement and run the quantum circuit N times, measuring either Pauli X, Pauli Y, or Pauli Z on the corresponding quantum registers (+1 if we measure $|0\rangle$ and -1 if we measure $|1\rangle$). If gate H was placed on quantum register k before measurement and gates $HS^\dagger$ were placed on quantum register l before measurement, we will measure X_k and Y_l; the expectation value $\langle X_k Y_l \rangle_i^A$ is then calculated as the average value of the product $X_k Y_l$ of measured values of X_k and Y_l across all quantum circuit runs.

4: For dataset B:

- For each sample $j = 1, \ldots, [B]$, run the quantum circuit N times, measuring Pauli Z_k on all quantum registers $k = 1, \ldots, n$ (+1 if we measure $|0\rangle$ and -1 if we measure $|1\rangle$). Calculate the expectation value $\langle Z_k \rangle_j^B$ as the average value of measured Z_k across all quantum circuit runs.

- For each sample $j = 1, \ldots, [B]$, place the H gate on **all** quantum registers before measurement and run the quantum circuit N times, measuring Pauli X_k on all quantum registers $k = 1, \ldots, n$ (+1 if we measure $|0\rangle$ and -1 if we measure $|1\rangle$). Calculate the expectation value $\langle X_k \rangle_j^B$ as the average value of measured X_k across all quantum circuit runs.

- For each sample $j = 1, \ldots, [B]$, place $HS^\dagger$ gates on **all** quantum registers before measurement and run the quantum circuit N times, measuring Pauli Y_k on all quantum registers $k = 1, \ldots, n$ (+1 if we measure $|0\rangle$ and -1 if we measure $|1\rangle$). Calculate the expectation value $\langle Y_k \rangle_j^B$ as the average value of measured Y_k across all quantum circuit runs.

- For each sample $j = 1, \ldots, [B]$, place either H or $HS^\dagger$ gates on **some** quantum registers before measurement and run the quantum circuit N times, measuring either Pauli X, Pauli Y, or Pauli Z on the corresponding quantum registers (+1 if we measure $|0\rangle$ and -1 if we measure $|1\rangle$). If gate H was placed on quantum register k before measurement and gates $HS^\dagger$ were placed on quantum register l before measurement, we will measure X_k and Y_l; the expectation value $\langle X_k Y_l \rangle_j^B$ is then calculated as the average value of the product $X_k Y_l$ of measured values of X_k and Y_l across all quantum circuit runs.

5: Without loss of generality, let us assume that we measured $3n$ second order Paulis (i.e., the same number as all first order Paulis). We have now two arrays: $[A] \times 6n$ array of expectation values of first and second order Paulis for dataset A and $[B] \times 6n$ array of expectation values of first and second order Paulis for dataset B. For every column of these arrays we calculate the average expectation values across all samples in the datasets: $\overline{\langle X_1 \rangle^A} = \frac{1}{[A]} \sum_{i=1}^{[A]} \langle X_1 \rangle_i^A , \ldots$

6: We have now two $6n$-dimensional vectors for datasets A and B:

$$\left(\overline{\langle X_1\rangle^A}, \overline{\langle X_2\rangle^A}, \ldots\right), \quad \left(\overline{\langle X_1\rangle^B}, \overline{\langle X_2\rangle^B}, \ldots\right).$$

The next step is to calculate the vector of differences:

$$c = \left(\overline{\langle X_1\rangle^A} - \overline{\langle X_1\rangle^B}, \overline{\langle X_2\rangle^A} - \overline{\langle X_2\rangle^B}, \ldots\right).$$

7: Calculate the estimate of the Frobenius distance:

$$d_F(A, B) = \frac{1}{2}\sqrt{\frac{c^\top c}{2^n}}.$$

12.7   Feature Map for Two-Sample Test

Figure 12.6 displays another (deeper) quantum circuit that implements the quantum feature map for the ACA dataset. Similar to the one shown in Figure 12.3, it maps each classical sample from the dataset into the corresponding quantum state. The circuit consists of seven quantum registers, four data encoding layers and three entanglement layers (CZ gates).

We encode data through one-qubit gate operations, which are rotations of the individual qubit states around the x and y axes. Each sample is a vector of 14 features, and each feature is mapped into the corresponding rotation angle. This, effectively, dictates to us the feature mapping scheme where each quantum register encodes two features and all rotation angles are defined on the interval $[0, \pi/4]$. To be more specific, the proposed angle encoding scheme is given by the following expression:

$$\theta_j^i = \frac{\pi}{4}\left(\frac{x_j^i - \min(x_j)}{\max(x_j) - \min(x_j)}\right), \quad i = 1, \ldots, N_{samples}, \quad j = 1, \ldots, 14.$$

where $\min(x_j)$ and $\max(x_j)$ are the minimum and maximum values of feature x_j across all samples.

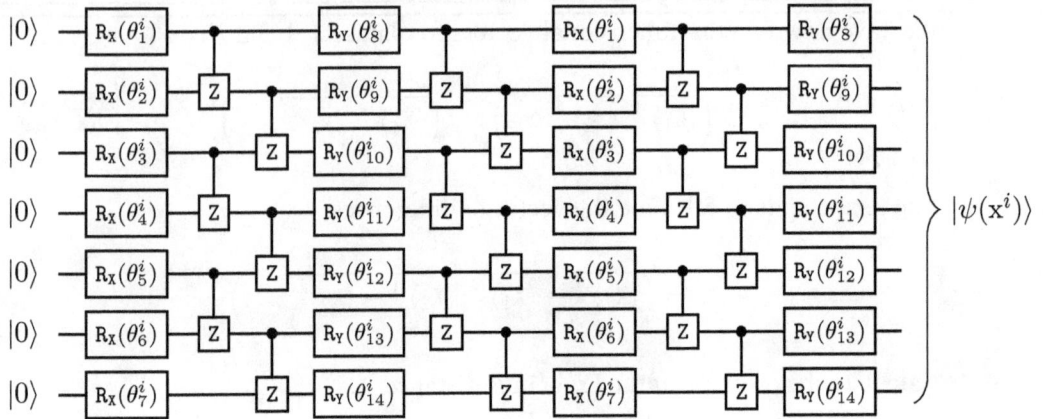

Figure 12.6: Quantum feature map circuit for the ACA dataset with repeat feature encoding.

Figure 12.7 shows the results of the MMD and Frobenius distance calculations for the ACA dataset. MMD was calculated as per the expressions (12.4.1) and (12.4.2). The features were normalised by mapping their values to the interval $[0, 1]$ and the Gaussian kernel scaling parameter was set at $\sigma = 1$. The Frobenius distance was calculated using the Qiskit simulator as per the algorithm 10 with the quantum feature map given by Figure 12.6.

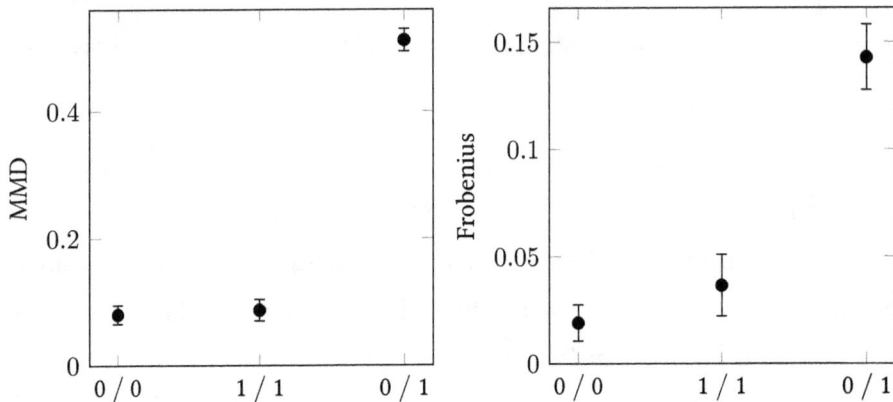

Figure 12.7: Classical (MMD) and quantum (Frobenius) distances.

The MMD and Frobenius distances were calculated for 100 scenarios consisting of random splits of Class 0 and Class 1 datasets. In each scenario, Class 0 dataset (383 samples) was split randomly into datasets A and B of roughly equal sizes and Class 1 dataset (307 samples) was split randomly into datasets C and D, also of roughly equal sizes. In the charts of Figure 12.7, "0/0" corresponds to distances between datasets A and B (both of Class 0), "1/1" corresponds to distances between datasets C and D (both of Class 1), and "0/1" corresponds to distances between A and C as well as between B and D (i.e., of different classes). MMD and Frobenius distances were calculated for all scenarios: in the figure, mean values and standard deviations are shown as, respectively, dots and error bars.

We see that both classical and quantum two-sample tests can clearly differentiate between Class 0 and Class 1 datasets, with the distance between them being much larger than the distances between subsets of Class 0 or subsets of Class 1.

12.8 , Density Matrix Classifier

In this section, we extend the range of available discriminative QML models by adding a new classifier based on two key concepts introduced earlier: the quantum feature map and the Frobenius distance between the density matrices. A quantum feature map consists of encoding each sample from a given dataset into the corresponding quantum state with the help of a PQC. A low-dimensional sample can be mapped into a quantum state – a state vector in a high-dimensional Hilbert space. It can also be described by the corresponding density matrix. All samples belonging to a particular class can be represented by the class density matrix, which is an equally weighted linear combination of the individual sample density matrices (equally weighted statistical ensemble of all samples belonging to the class). The classification of a new sample is then performed by measuring the distances between the density matrix representing the sample and density matrices representing all classes in the dataset. There are many possible ways in which these distances can be measured, the most suitable being the Frobenius distance.

Let D be a dataset consisting of samples belonging to N non-overlapping subsets (classes) $C_1, \ldots, C_N$:

$$C_1 \cup \ldots \cup C_N = D, \quad C_i \cap C_j = \emptyset \quad \forall\, i \neq j\,. \tag{12.8.1}$$

Each sample in the dataset, $x \in D$, can be represented as a vector of features, $x := (x_1, \ldots, x_m)^\top$, where m is the number of features, or the *dimensionality* of the dataset. We define the feature map φ as a function that maps samples x into vectors $\varphi(x)$ of a Hilbert space $\mathcal{H}$. The feature map φ is realised via a unitary transformation $\mathcal{U}(x)$ applied to the initial state $|0\rangle \equiv |0\rangle^{\otimes n}$:

$$\varphi(x) := \mathcal{U}(x) |0\rangle \langle 0| \mathcal{U}(x)^\dagger \equiv |x\rangle \langle x| . \tag{12.8.2}$$

Note that if the transformation φ is specified on n qubits, then the dimension of the Hilbert space $\mathcal{H}$ of linear operators is $4^n = 2^n \times 2^n$.

The quantum state $|x\rangle := \mathcal{U}(x) |0\rangle$ is a *pure* quantum state encoding sample x, and the feature map $\varphi(x)$ is the corresponding density matrix. The density matrix formalism allows us to describe not only pure but also *mixed* quantum states, i.e., statistical ensembles of the pure quantum states (Section 1.3). Thus, the density matrix approach can be used to encode not just a single sample but the whole subset of samples that belong to a particular class.

For each class $C_i \subset D, i = 1, \ldots, N$, we construct a density matrix ρ_i defined as an equally weighted statistical ensemble of all samples in C_i:

$$\rho_i = \frac{1}{[C_i]} \sum_{k=1}^{[C_i]} |x^k\rangle \langle x^k| , \tag{12.8.3}$$

where $[C_i]$ denotes the cardinality of C_i and the sum goes over all samples in class C_i.

We can estimate the *closeness* of any two classes, C_i and C_j, as the distance between the corresponding density matrices ρ_i and ρ_j. There are many possible ways to define the distance between density matrices [296, 329]. For example:

- Jensen-Shannon distance:

$$d_J(\rho_i, \rho_j) = \sqrt{S\left(\frac{\rho_i + \rho_j}{2}\right) - \frac{1}{2}S(\rho_i) - \frac{1}{2}S(\rho_j)} ,$$

where $S(\rho) = -\text{Tr}(\rho \ln \rho)$ is the von Neumann entropy.

- Bures distance:

$$d_B(\rho_i, \rho_j) = \sqrt{1 - \sqrt{F(\rho_i, \rho_j)}} \,,$$

where $F(\rho_i, \rho_j) = (\text{Tr}\sqrt{\sqrt{\rho_i}\rho_j\sqrt{\rho_i}})^2$ is the fidelity of two quantum states.

- Hellinger distance:

$$d_H(\rho_i, \rho_j) = \sqrt{1 - \text{Tr}(\sqrt{\rho_i}\sqrt{\rho_j})} \,.$$

However, the distance that we will consider in the experiments that will follow is the Frobenius distance:

$$d_F(\rho_i, \rho_j) = \frac{1}{2}\sqrt{\text{Tr}((\rho_i - \rho_j)^2)} \,. \tag{12.8.4}$$

The primary motivation for choosing the Frobenius distance over other alternatives is the fact that it can be efficiently estimated on a quantum computer as explained in Section 12.6.

12.8.1 DMC definition

We now have all the necessary ingredients to construct a classifier. The Density Matrix Classifier (DMC) [111] is a lazy classifier in the spirit of the K-nearest neighbours model [264] since it does not require training and all information about the dataset is encoded in the density matrices. It also has some similarities with SVMs due to the feature map of low-dimensional samples into vectors in a high-dimensional Hilbert space.

The DMC classifier can be formulated as the following algorithm:

1) We start with the dataset consisting of N non-overlapping subsets (classes), $C_1, \ldots, C_N$, as described in (12.8.1). Each sample x in the dataset is an m-dimensional vector of real-valued features.

2) We specify an n-qubit quantum circuit that implements the unitary transformation $\mathcal{U}(x)$ applied to the initial all-zero state $|0\rangle^{\otimes n}$. For example, the quantum circuit may consist of several layers of adjustable one-qubit gates encoding components of x into qubit rotation angles and fixed two-qubit gates creating entanglement.

3) For each class $C_i, i = 1, \ldots, N$, and for each sample in class $C_i, |x^k\rangle, k = 1, \ldots, [C_i]$, we construct the feature $\varphi(x^k)$ given by the expression (12.8.2) using the unitary transformation $\mathcal{U}(x^k)$.

4) For each class $C_i, i = 1, \ldots, N$, we construct a density matrix ρ_i defined as an equally weighted statistical ensemble of all samples in C_i as per the expression (12.8.3).

5) For each new sample to be classified, x, we construct its own feature map $\varphi(x)$ and density matrix ρ using the unitary transformation $\mathcal{U}(x)$. For each class $C_i, i = 1, \ldots, N$, we calculate the Frobenius distance $d_F(\rho_i, \rho)$ given by the expression (12.8.4). The new sample is assigned a class label corresponding to the class with the smallest Frobenius distance.

12.8.2   The GI dataset and multiclass classification

We illustrate the performance of the density matrix classifier on a multiclass classification problem – the Glass Identification (GI) dataset [112]. The dataset consists of 214 samples, with each glass sample described by nine continuous features (Refractive Index, Sodium, Magnesium, Aluminium, Silicon, Potassium, Calcium, Barium, and Iron) and four class labels (Window, Container, Tableware, and Headlamp). The dataset is heavily imbalanced, which makes it harder to find the right balance between precision and recall and prompts us to choose the F1 score (`sklearn.metrics.f1_score` with the "average" parameter set equal to "macro") as the preferred metric. Table 12.1 shows the distribution of samples by classes.

Class label	Number of samples
Window	163
Container	13
Tableware	9
Headlamp	29

Table 12.1: Distribution of samples by classes.

Feature importance analysis conducted with the help of the random forest classifier suggests the ranking of GI features in terms of their relative predicting power, as shown in Table 12.2.

Feature	Importance
Magnesium (Mg)	21.5%
Barium (Ba)	18.9%
Aluminium (Al)	13.6%
Sodium (Na)	12.5%
Potassium (K)	10.7%
Calcium (Ca)	8.1%
Refractive Index (RI)	6.9%
Silicon (Si)	6.4%
Iron (Fe)	1.4%

Table 12.2: GI feature importances.

We perform classification using all nine features as well as the top three features as per the feature importance analysis. The reason for experimenting with only the three most important features will become clear later. As always, we use the Qiskit quantum simulator and benchmark the DMC performance against several standard classical models.

12.8.3   Feature encoding and quantum feature map

Feature encoding is performed using the angle encoding scheme. Each feature $x_i, i = 1, \ldots, m$ of a sample $\mathrm{x} := (x_1, \ldots, x_m)^\top$ is mapped into the corresponding rotation angle – the adjustable parameter of a one-qubit gate. Given the dataset C with the total number of samples equal to the dataset cardinality $[C]$, for each sample $\mathrm{x}^k, k = 1, \ldots, [C]$, the mapping of the feature x_i^k to the corresponding rotation angle θ_i^k is performed as follows:

$$\theta_i^k = \alpha\pi \frac{x_i^k - \min_{j=1,\ldots,[C]} x_i^j}{\max_{j=1,\ldots,[C]} x_i^j - \min_{j=1,\ldots,[C]} x_i^j}, \tag{12.8.5}$$

where α is the range parameter. In our experiments we set $\alpha = 1/4$ and, therefore, we have $\theta_i^k \in [0, \pi/4]$, $\forall\, i, k$.

However, before the feature is mapped into a rotation angle of a one-qubit gate, it should be pre-processed to even out its distribution. If the feature is distributed more or less evenly, the mapping scheme (12.8.5) works well without any pre-processing. But if the feature values are concentrated around the minimum or maximum values, we need to stretch the distribution up from the minimum or down from the maximum in order to enhance the discriminatory power of the quantum feature map. In our experiments, we adopt the following simple feature preprocessing rule:

- **If** median $-$ min $< \frac{1}{4}$(max $-$ min) **Then** $x \to x^\gamma$, $\gamma = \frac{1}{2}$.

- **If** median $-$ min $> \frac{3}{4}$(max $-$ min) **Then** $x \to x^\gamma$, $\gamma = \frac{3}{2}$.

Figures 12.8 and 12.9 display the quantum feature maps for, respectively, the nine features and the top three features scenarios. Figures 12.10 and 12.11 show the corresponding feature embeddings on the QPU graph. In both scenarios, we have repeat encoding of the features in the same qubits between the layers of two-qubit gates creating entanglement.

The QPU graph is a 3×3 square grid of computational qubits connected by tunable couplers. This graph can either represent a section of an 84-qubit *Ankaa-2* system or a full 9-qubit *Novera* system – a 9-qubit version of the *Ankaa* architecture [268], the 4th generation of Rigetti's QPUs.

The top-three-features scenario differs from the nine-features scenario in two important respects. First, the top-three-features dataset is a harder to classify problem since we work with limited information. Second, the same 9-qubit QPU graph can be used for the multiple qubit encoding: each feature is encoded in three qubits. This should provide the DMC with an opportunity to demonstrate the expressive power of quantum feature map in outperforming the classical benchmarks.

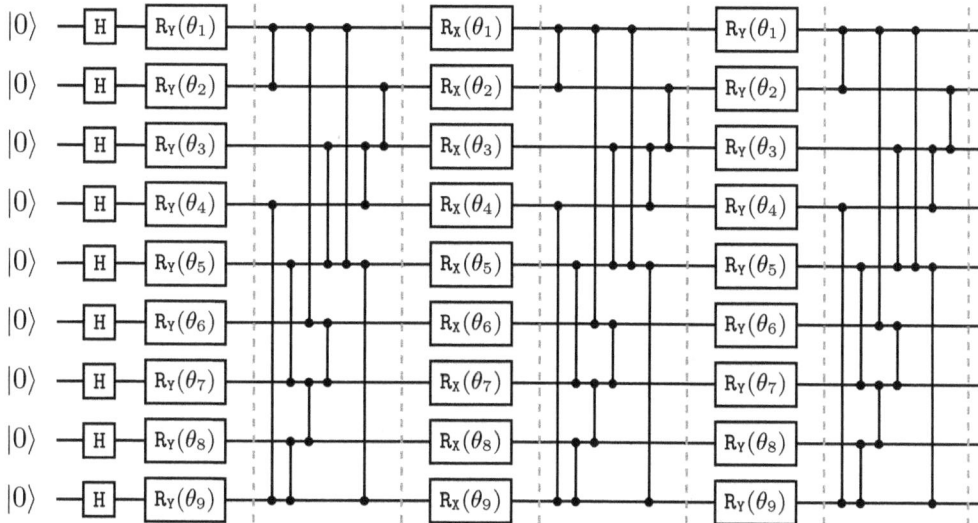

Figure 12.8: *9-qubit quantum circuit (quantum feature map) with single qubit feature encoding repeated three times.*

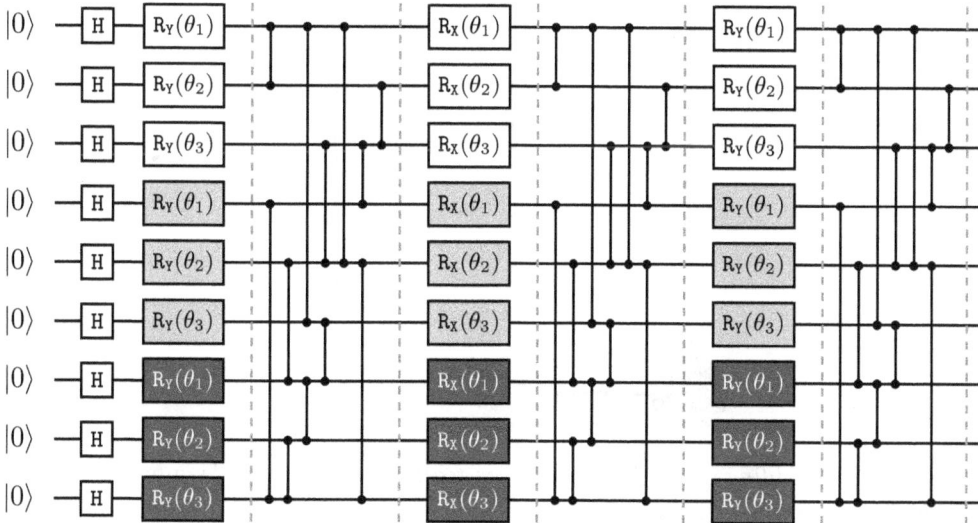

Figure 12.9: *9-qubit quantum circuit (quantum feature map) with triple qubit feature encoding repeated three times.*

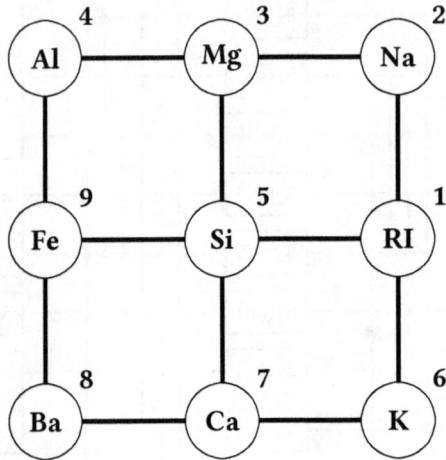

Figure 12.10: *Embedding of a 9-qubit quantum feature map with single qubit feature encoding on the square-grid QPU graph (Ankaa/Novera). Circles indicate computational qubits and lines indicate tunable couplers. Qubits are indexed from 1 to 9 and feature names are shown inside the circles.*

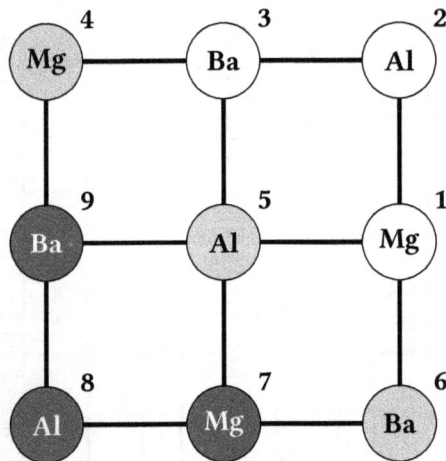

Figure 12.11: *Embedding of a 9-qubit quantum feature map with triple qubit encoding of the most important features on the square-grid QPU graph (Ankaa/Novera). Circles indicate computational qubits and lines indicate tunable couplers. Qubits are indexed from 1 to 9 and feature names are shown inside the circles.*

12.8.4 Classical benchmarks

The classical benchmarks of choice are four standard classical classifiers that are widely used in data science applications: Multi-Layer Perceptron (MLP, `sklearn.neural_network.-MLPClassifier`), Random Forest (RF, `sklearn.ensemble.RandomForestClassifier`), Support Vector Machine (SVM, `sklearn.svm.SVC`), and and Logistic Regression (LR, `sklearn.-linear_model.LogisticRegression`). We use the scikit-learn implementation of these classifiers with the set of optimised parameters shown in Table 12.3.

scikit-learn classifier	parameter value
Logistic Regression Classifier	solver = 'lbfgs'
	penalty = 'l2'
	C = 10
	max_iter = 5000
	random_state = 0
MLP Classifier	solver = 'adam'
	hidden_layer_sizes = (50, 50)
	alpha = 0.1
	max_iter = 5000
	random_state = 0
Random Forest Classifier	criterion = 'entropy'
	n_estimators = 500
	random_state = 0
Support Vector Classifier	kernel = 'rbf'
	C = 100
	random_state = 0

Table 12.3: scikit-learn classifiers – optimised model parameters. All other model parameters were fixed at their default values.

Similar to the feature standardisation for DMC, where all features are mapped to the same rotation angle interval $[0, \alpha\pi]$ as per the expression (12.8.5), classical classifiers such as MLP, SVM, and LR also work with the standardised features. For our experiments, we use the `sklearn.preprocessing.StandardScaler` class, which standardises features by removing their mean values and scaling them to unit variance.

12.8.5   Experimental results

Figure 12.12 displays F1 score results for classical benchmarks and DMC executed on the `Qiskit` quantum simulator. All F1 scores are out-of-sample: we used the `sklearn.model_-selection.train_test_split` class to generate ten random 80:20 splits into the train and test datasets.

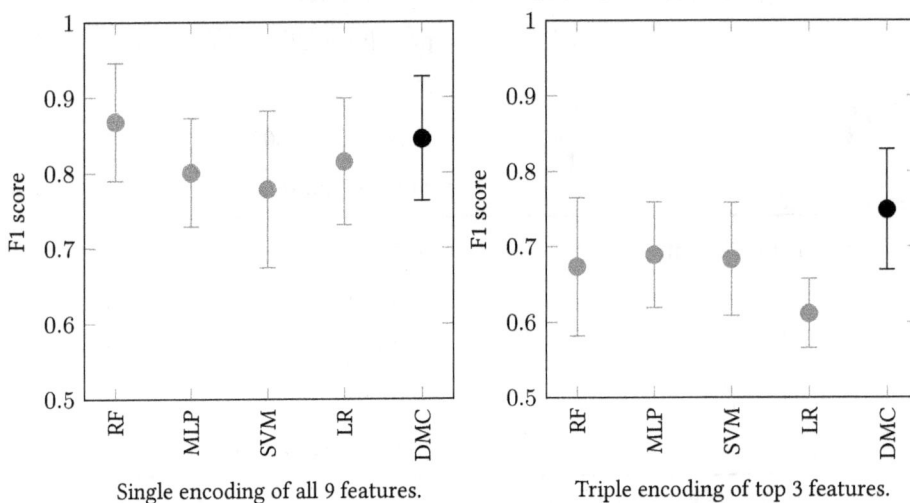

Single encoding of all 9 features.   Triple encoding of top 3 features.

Figure 12.12: Out-of-sample F1 scores for quantum and classical classifiers. All F1 scores were calculated with the "macro" averaging across classes. Error bars indicate ±1 standard deviation.

In Figure 12.12, the dots indicate the mean values and the error bars indicate the uncertainty in F1 scores (±1 standard deviation). The nine-features scenario with single qubit encoding has a higher average F1 score across all classifiers in comparison with the top-three-features scenario with triple qubit encoding, which is not surprising since nine features contain significantly more information as per the feature importance analysis presented in Table 12.2.

DMC performs at par with classical benchmarks on the nine-features scenario. At the same time, it outperforms all classical benchmarks on the top-three-features scenario with triple qubit feature encoding. We can interpret this result as the evidence of the expressive power of the quantum feature map and the ability of DMC to preserve more valuable information. While it may not be clearly visible when we work with all nine features, dealing with the reduced set of features puts more pressure on the classifiers and suppresses the performance of the classical models more than quantum DMC.

Summary

In this chapter, we learned about quantum kernels, which can replace classical kernels in hybrid quantum-classical protocols.

We also introduced the quantum two-sample test and the density matrix classifier based on the calculation of the Frobenius distance between two density matrices that encode the datasets.

The key element in both cases is the quantum feature map, which can be efficiently realised on a quantum computer with the help of a parameterised quantum circuit. In the next chapter, we will explore the sources of power of PQCs.

13

The Power of Parameterised Quantum Circuits

As we have seen in the previous chapters, there is a wide range of QML models based on parameterised quantum circuits. One reason for this is their tolerance to noise [237], which is important when we work with the NISQ hardware. However, this does not fully explain the popularity of PQCs or why they are considered strong competitors to classical ML models. There must be some fundamental properties of PQCs that make them superior to their classical counterparts. In this chapter, we discuss two such properties: resistance to overfitting and larger expressive power.

Resistance to overfitting is a direct consequence of the fact that a typical PQC – one without mid-circuit measurement – can be represented by a linear unitary operator. Linear models impose strong regularisation, thus preventing overfitting. At the same time, the model remains powerful due to the mapping of the input into the higher-dimensional Hilbert space where it may be easier to perform classification if the PQC is trained as a discriminative model (QNN).

Expressive power is related to the model's ability to express different relationships between variables, i.e., its ability to learn complex data structures. It appears that PQCs trained as generative models (QCBM) have strictly larger expressive power than their equivalent classical versions (such as RBM).

13.1   Strong Regularisation

Parameterised quantum circuits trained as classifiers face the same challenge as classical models: the need to generalise well to unseen data points. Classically, we have a wide range of supervised learning models and regularisation techniques to choose from. These regularisation techniques that fight overfitting are model specific. For example, we can try to restrict the depth of the decision trees or to impose a penalty term in the cost function when training neural networks.

Consider a conventional feedforward neural network as, arguably, the most direct classical counterpart of a quantum classifier. In both classical and quantum cases, the signal travels through the network in one direction and the layers of quantum gates can be compared to the layers of classical activation units. Regardless of whether we apply L_1 (Lasso) or L_2 (Ridge) penalty terms, or use dropout techniques, we would like to have a measure of regularisation present in the network. This is an interesting theoretical problem as well as an important practical task that allows us to develop an optimal strategy for fighting overfitting. Ideally, such a measure should be applicable to both classical and quantum neural networks to provide a meaningful comparison of their respective regularisation properties.

Very often, relatively small network weights are associated with a high degree of regularisation and relatively high network weights are symptoms of overfitting. However, it would be highly desirable to have a formal mathematical tool for quantifying the network capacity to overfit. One such possible well-defined measure that captures the degree of regularisation is the Lipschitz constant.

13.1.1   Lipschitz constant

Following Gouk [120], given two metric spaces $(\mathcal{X}, d_{\mathcal{X}})$ and $(\mathcal{Y}, d_{\mathcal{Y}})$, a function $f : \mathcal{X} \to \mathcal{Y}$ is said to be Lipschitz continuous if there exists a constant $k \geq 0$ such that

$$d_{\mathcal{Y}}(f(x_1), f(x_2)) \leq k d_{\mathcal{X}}(x_1, x_2), \quad \text{for all } x_1, x_2 \in \mathcal{X}.$$

The value of k is known as the Lipschitz constant, and the function is referred to as k-Lipschitz. We are interested in the smallest possible Lipschitz constant or, at least, its upper bound. To obtain the upper bound estimate, we should note some useful properties of feedforward neural networks.

In the case of a j-th layer of a feedforward neural network, x_1 and x_2 are the n-dimensional sample outputs of the previous layer, $j - 1$, and $f(x_1)$ and $f(x_2)$ are the m-dimensional outputs of layer j. The metrics $d_{\mathcal{X}}$ and $d_{\mathcal{Y}}$ can be, for example, L_1 or L_2 norms.

A feedforward neural network consisting of l fully connected layers can be expressed as a series of function compositions:

$$f(x) = (\phi_l \circ \phi_{l-1} \circ \ldots \circ \phi_1)(x),$$

where each ϕ_j implements the j-th layer affine transformation of the n-dimensional input x, parameterised by an $m \times n$ weight matrix W_j and an m-dimensional bias vector b_j:

$$\phi_j(x) = W_j x + b_j.$$

The composition of a k_1-Lipschitz function with a k_2-Lipschitz function is a $k_1 k_2$-Lipschitz function [120]. Therefore, we can compute the Lipschitz constants for each layer separately and combine them together to obtain an upper bound on the Lipschitz constant for the entire network.

Choose $d_{\mathcal{X}}$ and $d_{\mathcal{Y}}$ to be the L_2 norms $\|\cdot\|_2$. In this case, we obtain the following relationship from the definition of Lipschitz continuity for the fully connected network layer j:

$$\|(\mathbf{W}_j\mathbf{x}_1 + \mathbf{b}_j) - (\mathbf{W}_j\mathbf{x}_2 + \mathbf{b}_j)\|_2 \le k\|\mathbf{x}_1 - \mathbf{x}_2\|_2.$$

Introducing $\mathbf{a} = \mathbf{x}_1 - \mathbf{x}_2$ and assuming that $\mathbf{x}_1 \ne \mathbf{x}_2$ we arrive at the estimate

$$\frac{\|\mathbf{W}_j\mathbf{a}\|_2}{\|\mathbf{a}\|_2} \le k. \qquad (13.1.1)$$

The smallest Lipschitz constant of the fully connected network layer, $L(\phi_j)$, is equal to the supremum of the left-hand side of inequality (13.1.1):

$$L(\phi_j) := \sup_{\mathbf{a}\ne 0} \frac{\|\mathbf{W}_j\mathbf{a}\|_2}{\|\mathbf{a}\|_2}. \qquad (13.1.2)$$

The operator norm (13.1.2) is given by the largest singular value of the weight matrix $\mathbf{W}_j$, which corresponds to the spectral norm – the maximum scale by which the matrix can stretch a vector. It is straightforward to calculate using any of the suitable open-source packages, for example, `sklearn.decomposition.TruncatedSVD` from the `scikit-learn` machine learning package.

In the case of quantum neural networks, any parameterised quantum circuit operating on n qubits, regardless how complex and deep, can be represented by a $2^n \times 2^n$ unitary matrix. Since all singular values of the unitary matrix are equal to one, this gives us a natural benchmark for comparison of regularisation capabilities of various networks.

13.1.2   Regularisation example

The Australian Credit Approval (ACA) dataset [262, 263] we analysed in Chapter 8 can serve as a good illustrative example. We can compare the performance of classical and quantum neural networks while monitoring regularisation as measured by the Lipschitz constant.

The classical neural network is an MLP Classifier with two hidden layers. Each hidden layer holds the same number of activation units as the number of features in the ACA dataset (14), so that we have to calculate the largest singular values for two 14×14 square matrices. The features are standardised with `sklearn.preprocessing.StandardScaler`. We also use `sklearn.neural_network.MLPClassifier` to construct the classifier with the set of hyperparameters shown in Table 13.1:

Hyperparameter	Value
Number of hidden layers:	2
Number of activation units in each layer:	14
Activation function:	tanh
Solver:	adam
Intial learning rate:	0.01
Number of iterations:	5000
Random state:	0
Regularisation parameter, α:	variable

Table 13.1: MLP Classifier hyperparameters.

The MLP Classifier regularisation parameter α is our control variable. It controls the L_2 regularisation term in the network cost function: the larger this parameter, the more large network weights are penalised. All other parameters were set at their default values.

The quantum neural network is shown in Figure 8.5. The parameterised quantum circuit consists of just 7 fixed two-qubit gates (CZ) and 15 adjustable one-qubit gates (R_X and R_Y). Table 13.2 compares the MLP and the QNN classifiers on the in-sample and out-of-sample datasets (the ACA dataset was split 50:50 into training and test datasets using `sklearn.-preprocessing.StandardScaler`).

We observe that QNN provides strong regularisation with similar performance on the in-sample and out-of-sample datasets as expected from the network represented by the unitary matrix.

Classifier	Average F_1 score (in-sample)	Average F_1 score (out-of-sample)	Lipschitz Constant (upper bound)
MLP, $\alpha = 0.001$	1.00	0.78	36.2
MLP, $\alpha = 0.01$	1.00	0.79	33.5
MLP, $\alpha = 0.1$	1.00	0.80	18.6
MLP, $\alpha = 1$	0.99	0.83	7.4
MLP, $\alpha = 10$	0.90	0.86	1.3
MLP, $\alpha = 40$	0.85	0.86	0.5
MLP, $\alpha = 50$	0.35	0.37	1e-05
QNN	0.86	0.85	1.0

Table 13.2: F_1 scores and Lipschitz constants for MLP and QNN classifiers trained on the ACA dataset.

Further, we observe that the equivalent degree of regularisation can be achieved by MLP only with exceptionally large values of the regularisation parameter α. Making α any larger completely destroys the learning abilities of the network. For the chosen MLP configuration, the critical value of α is between 40 and 50.

> Parameterised quantum circuits can be represented as (high-dimensional) norm-preserving unitary matrices. This ensures strong regularisation properties of the quantum neural networks.

Now we can move on to the next feature of the parameterised quantum circuits: their expressive power. We can define the expressivity of a PQC as the circuit's ability to generate pure quantum states that are well representative of the Hilbert space [290]. In other words, from the QML point of view, the expressive power of a PQC is its ability to learn ("express") complex data structures. In the following section, we will try to quantify the degree of expressivity inherent in different PQC types.

13.2 Expressive Power

We saw in previous chapters how PQCs can be applied to solving optimisation problems (QAOA and VQE) as well as to various machine learning tasks covering both discriminative (QNN classifier) and generative (QCBM market generator) use cases. In general, the PQCs we used for quantum machine learning tasks can be divided into two types [88]: tensor network PQC (similar to the QNN circuit in Figure 8.4) and multilayer PQC (similar to the QCBM circuit in Figure 9.1). What is their expressive power and how can we rank them? Before trying to answer this question, let us have a look at a simple illustrative example: quantum circuits specified on a single quantum register.

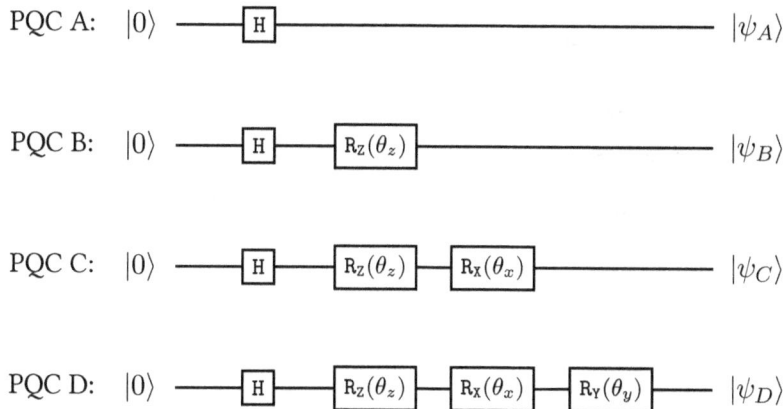

$$\theta_x \sim U[-\pi, \pi], \quad \theta_y \sim U[-\pi, \pi], \quad \theta_z \sim U[-\pi, \pi],$$

Figure 13.1: PQCs with different expressive powers.

Figure 13.1 displays four one-qubit circuits with dramatically different expressive powers, where $U[-\pi, \pi]$ denotes the Uniform distribution over the closed interval $[-\pi, \pi]$. Let us go through them one by one.

PQC A starts with the qubit state initialised as $|0\rangle$ – North Pole on the Bloch sphere (Figure 7.3). The only gate is the Hadamard gate H that moves $|0\rangle$ to $(|0\rangle + |1\rangle)/\sqrt{2}$. Thus, state $|\psi_A\rangle$ can only be a single point on the Bloch sphere.

PQC B also starts with the qubit state initialised as $|0\rangle$ and applies the Hadamard gate transforming the initial state into $(|0\rangle + |1\rangle)/\sqrt{2}$ before applying the rotation R_z around the z axis by an angle θ_z drawn from the Uniform distribution on $[-\pi, \pi]$. The final state $|\psi_B\rangle$ can be any point on the equator, all reached with equal probability.

PQC C adds a rotation R_X to PQC B, by an angle θ_x drawn from the Uniform distribution on $[-\pi, \pi]$. With two rotations around two orthogonal axes we can reach any point on the Bloch sphere. However, with angles θ_z and θ_x drawn from the Uniform distribution on $[-\pi, \pi]$ we do not have a Uniform distribution of points on the Bloch sphere for state $|\psi_C\rangle$. We observe the highest density around points $(|0\rangle + |1\rangle)/\sqrt{2}$ and $(|0\rangle - |1\rangle)/\sqrt{2}$, which are the points where the equator crosses the $0°$ and $180°$ meridians, and the lowest density along the $90°$ and $270°$ meridians.

Finally, PQC D adds one more rotation R_Y around the y axis by an angle θ_y drawn from the Uniform distribution on $[-\pi, \pi]$. This rotation results in spreading the previously clustered points more evenly around the Bloch sphere, thus making all points on the Bloch sphere equally accessible.

Therefore, in terms of our ability to explore the Hilbert space, we have the following hierarchy of the expressive power of the PQCs introduced above:

$$\text{PQC D} > \text{PQC C} > \text{PQC B} > \text{PQC A}.$$

We can now return to the PQCs developed in the previous chapters.

13.2.1   Multilayer PQC

A multilayer PQC (MPQC) consists of multiple blocks of quantum circuits in which the arrangement of quantum gates in each block is identical [27, 204]. Figure 13.2 shows a schematic representation of the MPQC.

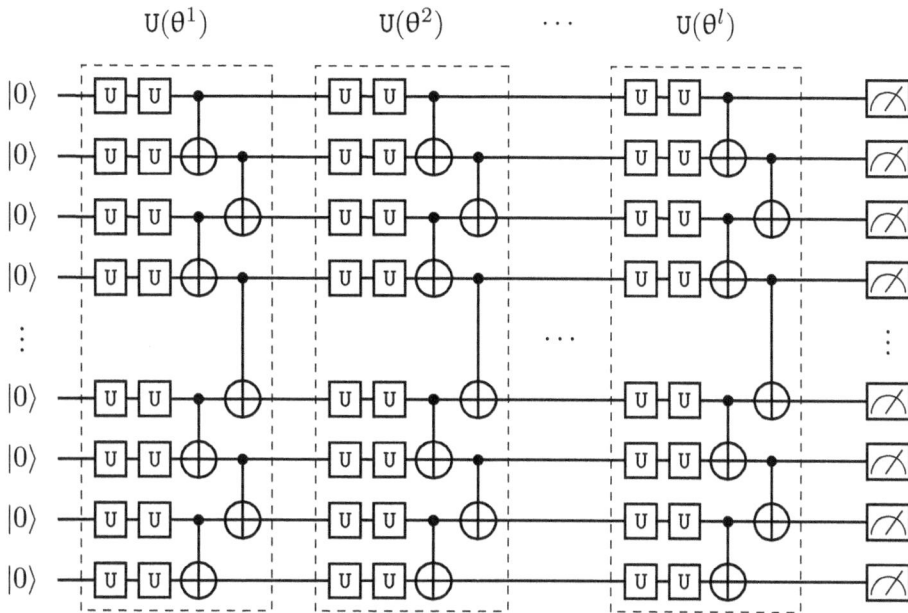

$$U(\theta^1) \qquad\qquad U(\theta^2) \qquad \cdots \qquad U(\theta^l)$$

Figure 13.2: Schematic representation of a multilayer PQC.

The following mathematical formalism can be used to describe MPQC. The input n-qubit quantum state with all qubits initialised as $|0\rangle$ in the computational basis is $|0\rangle^{\otimes n}$, the total number of blocks is denoted l, and the i-th block is denoted $U(\theta^i)$, where the number of parameters is proportional to the number of qubits, and n is logarithmically proportional to the dimension of the generated data (this reflects our assumption about the data encoding scheme). The generated output state of the circuit thus reads

$$|\psi\rangle = \prod_{i=l}^{1} U(\theta^i)|0\rangle^{\otimes n}.$$

13.2.2   Tensor network PQC

A tensor network PQC (TPQC) treats each block as a local tensor. The arrangement of the blocks follows a particular network structure, such as matrix product states or tree tensor networks [153]. A schematic representation of TPQC is shown in Figure 13.3.

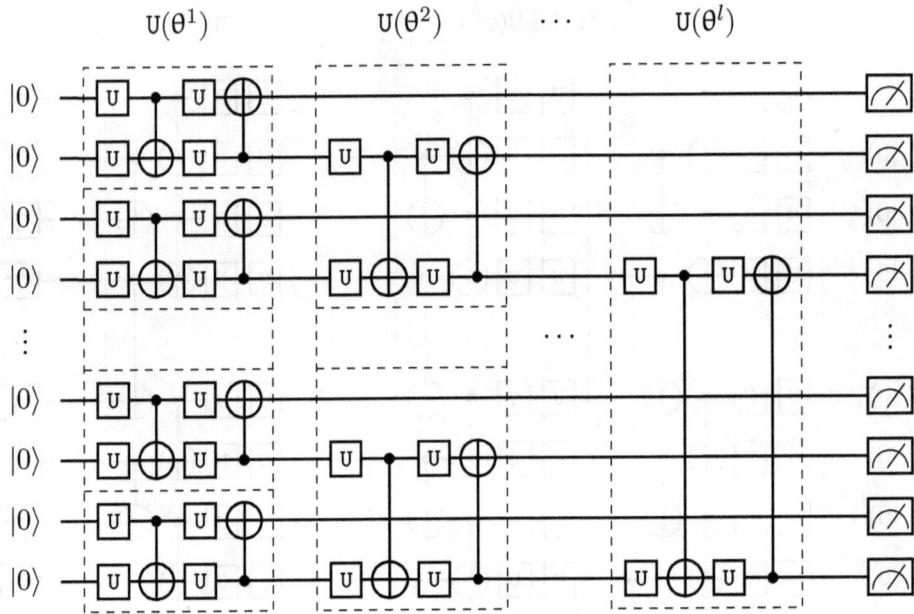

Figure 13.3: Schematic representation of a tensor network PQC.

Mathematically, the i-th block $\mathsf{U}(\theta^i)$ is composed of M_i local tensor blocks, with $M_i \propto n/2^i$, denoted as $\mathsf{U}(\theta^i) = \bigotimes_{j=1}^{M_i} \mathsf{U}(\theta_j^i)$. Note that many of these tensor blocks may be identity operators. The generated state is thus of the form

$$|\psi\rangle = \prod_{i=l}^{1} \bigotimes_{j=1}^{M_i} \mathsf{U}(\theta_j^i)|0\rangle^{\otimes n}.$$

13.2.3   Measures of expressive power

The main question to answer is whether MPQC and TPQC have larger expressive power in comparison with their classical counterparts, such as classical neural networks. The expressive power of a model can be defined in many different ways, for example, as a model capacity to express different relationships between variables [21]. Deep neural networks serve as a good example of powerful models capable of learning complex data structures [94]. Therefore, the power of a model can be quantified by its complexity, with the *Vapnik-Chervonenkis dimension* being a complexity measure of choice [318]. The objective is to provide an estimate of how well a model generalises to the unseen data.

Another popular approach is the *Fisher information*, which describes the geometry of a model parameter space [269]. Arguably, the *effective dimension* based on Fisher information, rather than Vapnik-Chervonenkis dimension, is a better measure to study the power of quantum and classical neural networks [1].

However, one of the most natural metrics of expressive power is *entanglement entropy*, which allows us to establish a well-defined ranking of quantum and classical machine learning models. In this chapter, we will present the expressive power estimates obtained in [88] for TPQC and MPQC based on entanglement entropy.

Let us recall the definitions of entropy in statistical mechanics (the Gibbs entropy S) and in information theory (the Shannon entropy H) introduced in Chapter 6:

$$S := -k_B \sum_i p_i \log(p_i) \quad \text{and} \quad \text{H} := - \sum_i p_i \log_2(p_i). \tag{13.2.1}$$

Here, p_i is the probability that the microstate i is taken from an equilibrium ensemble in the case of the Gibbs entropy, and that the message i is picked from the message space in the case of the Shannon entropy.

These definitions of entropy can be extended to the quantum case. In Chapter 1, we introduced the density matrix as a universal tool for describing pure and mixed quantum states:

$$\rho := \sum_{i=1}^{N} \sum_{j=1}^{N} \rho_{ij} |i\rangle \langle j|,$$

where $(|i\rangle)_{i=1,\ldots,N}$ are the basis vectors of a given quantum system. The *von Neumann entropy* S is then defined as

$$S(\rho) := -\text{Tr}(\rho \log(\rho)).$$

Since the density matrix is Hermitian, it is *diagonalisable*, so that there exists a basis $(|k\rangle)_{k=1,\dots,N}$ such that

$$\rho = \sum_{k=1}^{N} \rho_{kk} |k\rangle \langle k| =: \sum_{k=1}^{N} p_k |k\rangle \langle k|, \quad \text{where } \sum_{k=1}^{N} p_k = 1.$$

The eigenvalues of the operator $\rho \log(\rho)$ are thus $(p_k \log(p_k))_{k=1,\dots,N}$, and we obtain the following expression for the von Neumann entropy:

$$S(\rho) = -\text{Tr}(\rho \log(\rho)) = -\sum_{k} p_k \log(p_k). \tag{13.2.2}$$

From (13.2.1) and (13.2.2) we see that for the orthogonal mixture of quantum states, the quantum and classical entropies coincide.

If the system has two component parts, A and B, we can define the *reduced density matrix* as the *partial trace* of the density matrix over the subspace of the Hilbert space we are not interested in. Let $(|a_i\rangle)_{i=1,\dots,N}$ be the standard orthonormal basis of the Hilbert space $\mathbb{H}_A$ of system A, and $(|b_j\rangle)_{j=1,\dots,M}$ be the standard orthonormal basis of the Hilbert space $\mathbb{H}_B$ of system B. The density matrix ρ_{AB} of the bipartite system AB on the tensor product Hilbert space $\mathbb{H}_A \otimes \mathbb{H}_B$ can then be represented as

$$\rho_{AB} = \sum_{i=1}^{N} \sum_{j=1}^{M} \sum_{k=1}^{N} \sum_{l=1}^{M} c_{ijkl} |a_i\rangle \langle a_k| \otimes |b_j\rangle \langle b_l|,$$

for some coefficients c_{ijkl}. The partial traces then read

$$\text{Tr}_B(\rho_{AB}) = \sum_{i=1}^{N} \sum_{j=1}^{M} \sum_{k=1}^{N} \sum_{l=1}^{M} c_{ijkl} |a_i\rangle \langle a_k| \langle b_l | b_j\rangle,$$

which is a reduced density matrix ρ_A on $\mathbb{H}_A$, and

$$\text{Tr}_A(\rho_{AB}) = \sum_{i=1}^{N} \sum_{j=1}^{M} \sum_{k=1}^{N} \sum_{l=1}^{M} c_{ijkl} |b_j\rangle \langle b_l| \langle a_k | a_i\rangle,$$

which is a reduced density matrix ρ_B on $\mathbb{H}_B$. Note that $\mathrm{Tr}(|a_i\rangle \langle a_k|) = \langle a_k|a_i\rangle$ and $\mathrm{Tr}(|b_j\rangle \langle b_l|) = \langle b_l|b_j\rangle$.

Example: Consider the two-qubit system in the state

$$|\psi\rangle = \frac{1}{\sqrt{2}}(|01\rangle + |10\rangle),$$

which is one of the four maximally entangled Bell states (Section 6.5.2). We assume that the first qubit is system A and the second qubit is system B. This state corresponds to the following density matrix:

$$\rho_{AB} := |\psi\rangle \langle\psi| = \frac{1}{2}\Big(|01\rangle \langle 01| + |01\rangle \langle 10| + |10\rangle \langle 01| + |10\rangle \langle 10| \Big).$$

Let us now act on this state with the partial trace $\mathrm{Tr}_B(\cdot)$:

$$\rho_A := \mathrm{Tr}_B(\rho_{AB}) = \frac{1}{2}\Big(|0\rangle \langle 0| \langle 1|1\rangle + |0\rangle \langle 1| \langle 0|1\rangle + |1\rangle \langle 0| \langle 1|0\rangle + |1\rangle \langle 1| \langle 0|0\rangle \Big)$$

$$= \frac{1}{2}\Big(|0\rangle \langle 0| + |1\rangle \langle 1| \Big) = \frac{1}{2}\begin{bmatrix} 1 & 0 \\ 0 & 1 \end{bmatrix}.$$

$$(13.2.3)$$

The reduced density matrix ρ_A in (13.2.3) is the same as the density matrix ρ in (1.3.8), which describes a statistical ensemble of states $|0\rangle$ and $|1\rangle$ (mixed state), i.e., a physical system prepared to be either in state $|0\rangle$ or state $|1\rangle$ with equal probability.

The *entanglement entropy* of a bipartite system AB is then defined as

$$\mathcal{S}(\rho_A) := -\mathrm{Tr}(\rho_A \log(\rho_A)) = -\mathrm{Tr}(\rho_B \log(\rho_B)) =: \mathcal{S}(\rho_B),$$

and can be used as a measure of expressive power of a model in the following way. First, note that TPQC, MPQC, and classical neural networks have a close connection with *tensor networks*, such as matrix product states (MPS) [88]. The key question is then whether the given quantum system can be efficiently represented by the MPS.

A quantum system that satisfies the *area law* (its entanglement entropy grows proportionally with the boundary area) has an efficient MPS representation. At the same time, a quantum system that satisfies the *volume law* (its entanglement entropy grows proportionally with the volume) cannot be efficiently represented by MPS [88].

13.2.4   Expressive power of PQC

In Chapter 5, we introduced the Restricted Boltzmann Machine (RBM) – a neural network operating on stochastic binary activation units, which is a natural classical counterpart of parameterised quantum circuits. We considered two types of RBM:

- a shallow two-layer network where the activation units in the visible layer are connected to the activation units in the hidden layer with no connections between the activation units within the same layer;
- a deeper multi-layer network of stacked RBMs where the hidden layer of the k-th RBM serves as the visible layer of the $(k + 1)$-th RBM. Such stacked RBMs (trained sequentially) are called Deep Boltzmann Machines (DBMs).

It is also possible to impose further restrictions on the connections between the RBM layers. In *short-range* RBMs, we restrict the connectivity of the hidden layer activation units such that they are allowed to connect to the limited number of activation units in the visible layer that are in close proximity to each other (local connectivity) [84]. In *long-range* RBMs, we allow connections between the hidden layer activation units and the visible layer activation units that are not necessarily local.

It has been established by Deng, Li, and Sarma [85] that the entanglement entropy of all short-range RBM states satisfies an area law for arbitrary dimensions and bipartition geometry. For long-range RBM states, such states could exhibit volume law entanglement. Therefore, long-range RBMs are capable of representing quantum states with large entanglement.

It is probably not surprising that a DBM would have even larger expressive power than a single RBM. However, using entanglement entropy as a measure of expressive power, Du,

Hsieh, Liu, and Tao have proven in [88] that MPQCs have strictly larger expressive power than DBMs. The main result can be formulated as the following theorem:

Theorem 10 (Expressive Power Theorem). *The expressive power of MPQC and TPQC with* $\mathcal{O}(poly(n))$ *single qubit gates and* CNOT *gates, and classical neural networks with* $\mathcal{O}(poly(n))$ *trainable parameters, where* n *refers to the number of qubits or visible units, can be ordered as*

$$MPQC > DBM > long\text{-}range\ RBM > TPQC > short\text{-}range\ RBM. \qquad (13.2.4)$$

Theorem 10 provides a solid theoretical foundation for experimental works aimed at establishing quantum advantage of PQC-based QML models. The larger expressive power of PQCs in comparison with their classical counterparts prompted the development of many such models in recent years. For example, a hybrid quantum-classical approach, suitable for NISQ devices and harnessing the greater expressive power of quantum entanglement, was proposed in [59]. It was shown through numerical simulations that the Quantum Long Short Term Memory (QLSTM) model learns faster than the equivalent classical LSTM with a similar number of network parameters. In addition, the convergence of QLSTM was shown to be more stable than that of its classical counterpart. A Quantum Convolutional Neural Network (QCNN) was proposed in [58] which, due to its larger expressive power, achieved greater test accuracy compared to classical CNNs. The source of expressive power was the replacement of the classical convolutional filters with quantum convolutional kernels based on variational quantum circuits.

Multi-layer parameterised quantum circuits such as QCBM have strictly more expressive power than classical models such as RBM when only a polynomial number of parameters is allowed. For systems that exhibit quantum supremacy, a classical model cannot learn to reproduce the statistics unless it uses exponentially scaling resources [28].

Summary

In this chapter, we learned where the power of parameterised quantum circuits comes from. We started with the observation that quantum neural networks enjoy strong regularisation inherent in their architecture. This is due to the fact that any PQC, however wide and deep, is a unitary linear operator.

Next, we considered the expressive power of parameterised quantum circuits and established the concept of the expressive power hierarchy. The main result (Theorem 10) supports the experimental findings indicating the presence of the elements of quantum advantage in various QML models compatible with the main characteristics of NISQ devices.

In the next chapter, we will go deeper into the less explored territory of new quantum algorithms and those algorithms that are beyond the reach of NISQ devices.

14

Advanced QML Models

The first generation of quantum algorithms appeared in the 1990s when quantum computers existed only as a concept. On the one hand, the absence of actual quantum hardware was a huge disadvantage since it made direct experiments impossible; on the other hand, it stimulated theoretical research not inhibited by the limitations and constraints of the imperfect early quantum computers. Researchers focused on devising algorithms that would achieve quadratic or even exponential speedup, assuming that powerful, error-free quantum computers would be available one day. It was the time when Shor's prime factorisation algorithm [289] and Grover's search algorithm [124] were discovered. Incidentally, as the book was about to be released, Peter Shor was named one of the four recipients of the 2022 Breakthrough Prize in Fundamental Physics (along with C. H. Bennett, G. Brassard, and D. Deutsch) for their foundational work in quantum information. Many such algorithms, in turn, relied on basic building blocks such as Quantum Phase Estimation and Quantum Fourier Transform [302]. These algorithms played an important role in demonstrating the capabilities of universal gate model quantum computers – if only they were available! A quarter of a century later, we are facing a different problem: the development of practical quantum computing algorithms and techniques that would allow us to extract value from NISQ computers. While quantum computing hardware is improving at a breathtaking pace, it is still too far from a state where it can break RSA encryption. What are existing quantum computers capable of doing? What is their relative strength in comparison with classical computers? In this chapter, we look at several new, NISQ-friendly algorithms that bring us one step closer to achieving the quantum advantage.

14.1   Quantum Generative Adversarial Networks

Generative Adversarial Networks (GANs) are powerful statistical techniques to generate (as much as needed) data close enough (in some sense) to given samples. They were introduced in [119] and originally tested on image data. Since then, they have seen wide applications in finance for time series generation [326, 327], tuning of trading models [190], portfolio management [211], synthetic data generation [17], and diverse types of fraud detection [285]. The gist of it is to have a generator and a discriminator compete against each other in order to improve themselves: the generator improves by becoming better at generating good samples (i.e., close to real data) from random noise, whereas the discriminator improves by being able to recognise real data from "fake" (namely generated) data. Both the generator and the discriminator are usually built as neural networks with hyperparameters over which to optimise. Mathematically, given a generator $\mathfrak{G}(\cdot, \theta^{\mathfrak{G}})$ and a discriminator $\mathfrak{D}(\cdot, \theta^{\mathfrak{D}})$, where $\theta^{\mathfrak{G}}$ and $\theta^{\mathfrak{D}}$ represent the hyperparameters, the problem reads as follows:

$$\min_{\theta^{\mathfrak{G}}} \max_{\theta^{\mathfrak{D}}} \left\{ \mathbb{E}_{\mathrm{x} \sim \mathbb{P}_{\text{data}}} \left[\log(\mathfrak{D}\left(\mathrm{x}; \theta^{\mathfrak{D}}\right)) \right] + \mathbb{E}_{\mathrm{z} \sim \mathbb{P}_{\mathfrak{G}(\cdot, \theta^{\mathfrak{G}})}} \left[\log \left(1 - \mathfrak{D}\left(\mathfrak{G}\left(\mathrm{z}; \theta^{\mathfrak{G}}\right); \theta^{\mathfrak{D}}\right) \right) \right] \right\},$$

where $\mathrm{x} \sim \mathbb{P}_{\text{data}}$ means some sample x generated from the original dataset, whereas $\mathrm{z} \sim \mathbb{P}_{\mathfrak{G}}$ refers to a sample generated from the generator $\mathfrak{G}$. We refer the interested reader to [95] for an overview of the advantages and the pitfalls of GANs in finance. Given this popularity and the existence of quantum neural networks (Chapter 8), it is thus natural to explore the question of whether GANs can be extended to the quantum world, and whether there is any advantage in doing so.

The main principles of a Quantum Generative Adversarial Network (QGAN) – introduced simultaneously by Lloyd and Weedbrook [207] and by Dallaire-Demers and Killoran [77] – remain the same, relying on two actors, a generator and a discriminator, competing against each other. In [207], the authors translated the classical problem in the language of density matrices (described in Section 1.3.1): Given some data represented by a density matrix σ (not necessarily describing a pure state) and a generator $\mathfrak{G}$ generating some output density matrix ρ, the discriminator is tasked with identifying the true data from the fake data. More precisely, it makes a positive operator-valued measurement (Section 1.2.3) with outcomes T (for True) or F (for False).

The probability that the measurement yields a positive answer given the true data is $\mathbb{P}(T|\sigma) = \mathrm{Tr}(T\sigma)$, while the probability that it yields a positive answer given generated data is $\mathbb{P}(T|\mathfrak{G}) = \mathrm{Tr}(T\rho)$. The adversarial game, similarly to the classical case, therefore reads

$$\min_{\mathfrak{G}} \max_{T} \left\{ \mathrm{Tr}(T\rho) - \mathrm{Tr}(T\sigma) \right\}. \tag{14.1.1}$$

Note that both the set of positive measurement operators T (with 1-norm less than one) and the set of density matrices ρ are convex, ensuring that the optimisation problem (14.1.1) admits at least one optimum. However, these two sets are infinite dimensional, making the optimisation problem hard to solve. Following similar arguments, Dallaire-Demers and Killoran [77] further proposed to model both the generator and the discriminator as variational quantum circuits parameterised by some vector of parameters describing, for example, the rotation angles of all the gates. A natural question then is whether some optimal architecture of variational quantum circuit might exist. While there is no clear answer at this stage – as far as we know – recent developments have improved our understanding and the power of such circuits.

Starting from n qubits, a quantum generator $\mathfrak{G} : \mathbb{C}^{2^n} \to \mathbb{C}^{2^n}$ takes the form of a multi-layer quantum neural network, for example, of the following form:

$$\mathfrak{G} := \prod_{l=L}^{1} U_l(\theta_l). \tag{14.1.2}$$

For each layer $l \in \{1, \ldots, L\}$, the unitary gate $U_l(\theta_l)$ acts on all n qubits at the same time, and depends on a vector of parameters (or hyperparameters) θ_l. In order to avoid (too expensive) high-order qubit gates, entanglement takes the form of pairwise controlled unitary gates, and we therefore assume that, for each $l \in \{1, \ldots, L\}$, U_l is composed of one- or two-qubit gates only. One possible (though not the only one) way to parameterise U_l is with the following principles in mind:

- any one-qubit unitary gate can be decomposed into a sequence of three rotation gates R_Z, R_X, and R_Y, as proved in [238, Theorem 4.1];

- following [280], *imprimitive* two-qubit gates (i.e., two-qubit gates that map product states to non-product states), together with one-qubit gates, ensure quantum universality [47]. In particular the decomposition $R_X(\theta)Q(\phi)$ is universal [47, Corollary 9.2], for $\theta, \phi \in [0, 2\pi)$, where

$$
Q(\phi) := \begin{bmatrix} 1 & 0 & 0 & 0 \\ 0 & 1 & 0 & 0 \\ 0 & 0 & 1 & 0 \\ 0 & 0 & 0 & e^{i\phi} \end{bmatrix}.
$$

The general form of the L-layer neural network is therefore (14.1.2), where each layer gate $U_l(\theta_l)$ takes the form

$$
U_l(\theta_l) = \left\{ \bigotimes_{i=1}^{n} R_X(\theta_e^i) Q^{1+(i \bmod n)}(\theta_{imp}^i) \right\}
$$
$$
\left\{ \left(\bigotimes_{i=1}^{n} R_Z(\theta_{Z,l}^i) \right) \left(\bigotimes_{i=1}^{n} R_X(\theta_{X,l}^i) \right) \left(\bigotimes_{i=1}^{n} R_Y(\theta_{Y,l}^i) \right) \right\},
$$

where Q^i means that qubit i is the control qubit and the gate acts on qubit $(i+1)$. Note that $1 + (i \bmod n) = 1 + i$ when $i \in \{1, \ldots, n-1\}$ and is equal to 1 when $i = n$. The total number of hyperparameters is therefore $5n$ per layer, thus $5nL$ in total. The discriminator itself may or may not be of a quantum nature (following a construction similar to the generator), depending on the problem at hand (it is in [18] but not in [292], for example), and the nature of the problem – in particular the potential need to encode/decode data from quantum to classical (with a high cost) may influence this choice.

The finite-dimensional optimisation (14.1.1) is usually carried out via some gradient descent method; the gradients themselves are computed via separate quantum circuits in [77], or rather – more efficiently – using the parameter-shift rule, explained in Section 8.3.3 (see also [279]), which allows for an exact computation of the gradient from the original circuit.

QGANs are a very new and active research area, and promise to be one where NISQ-based algorithms will be particularly fruitful. They are intimately linked with developments of QNNs as a whole, and current advances in the field relate to the following, which we encourage the reader to follow closely over the next few years:

- QGAN to generate probability distributions: we refer the interested reader to [18, 292, 341] for univariate distributions, mostly in the context of finance, and to [5, 339] for multivariate distributions;

- Quantum Convolutional Neural Networks: In [173], the authors show how to handle non-linearities in (quantum) deep neural networks; [69, 325] explain how to reduce the number of required gates (equivalently, the number of rotation parameters) in the circuit, and [151] highlights the importance and sufficiency of two-qubit interactions, more amenable to NISQ devices;

- Quantum Wasserstein GAN: In [55] – mimicking recent results in classical Wasserstein GANs [13, 128] – the authors introduced a Wasserstein semimetric between quantum data, which they use to reduce the number of required quantum gates.

14.2 Bayesian Quantum Circuit

Parameterised quantum circuits can be used to construct a quantum state with desired properties and to modify it in a controlled way. Measuring the final state is then equivalent to drawing a sample from a probability distribution in the form of a bitstring. This is the key concept behind the Quantum Circuit Born Machine (QCBM) we considered in Chapter 9.

The Bayesian Quantum Circuit (BQC) is another quantum generative machine learning model that extends the capabilities of QCBM [88]. Unlike QCBM, which operates only on *data* qubits, encoding the desired probability distribution, BQC has additional *ancillary* qubits encoding the *prior* distribution. The BQC circuit is shown in Figure 14.1.

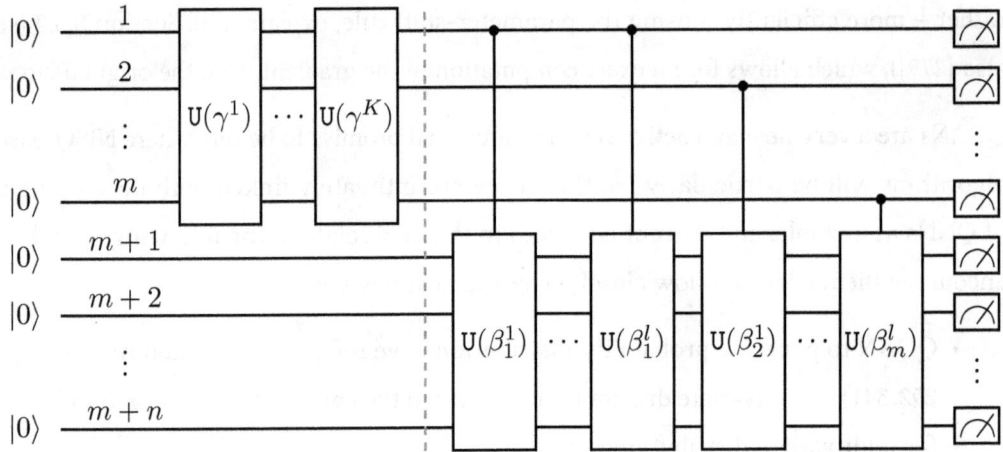

Figure 14.1: Schematic representation of BQC.

The first m quantum registers in the circuit are ancillary qubits. After applying K operator blocks $\mathtt{U}(\gamma^i)_{i=1,\dots,K}$, to the initial state $|0\rangle^{\otimes m}$, we construct the state $|\psi\rangle$,

$$|\psi\rangle = \prod_{i=K}^{1} \mathtt{U}(\gamma^i) |0\rangle^{\otimes m} , \tag{14.2.1}$$

and measuring it generates a sample from the prior distribution.

The next n quantum registers are data qubits. Quantum gates operating on them are conditional on the states of the ancillary qubits. Conditionally applying $l \times m$ operator blocks to n data qubits, we obtain a state that is conditional on $|\psi\rangle$. Measuring it will generate a sample from the conditional distribution, which is exactly what is needed to realise a Bayesian model. Bayesian modelling allows us to infer a *posterior* distribution over the parameters θ of the model given some observed data D using Bayes' theorem [57],

$$\mathbb{P}(\theta|D) = \frac{\mathbb{P}(D|\theta)\mathbb{P}(\theta)}{\mathbb{P}(D)} = \frac{\mathbb{P}(D|\theta)\mathbb{P}(\theta)}{\int \mathbb{P}(D|\theta)\mathbb{P}(\theta)d\theta},$$

where $\mathbb{P}(D|\theta)$ is the likelihood, $\mathbb{P}(D)$ is the marginal likelihood or evidence, and $\mathbb{P}(\theta)$ is the prior. We obtain $\mathbb{P}(\theta)$ by repeatedly measuring the state $|\psi\rangle$ given by (14.2.1), $\mathbb{P}(D|\theta)$ by repeatedly measuring the final state after applying the conditional operators $U(\beta)$, and $\mathbb{P}(D)$ by repeatedly measuring the final state after applying the operators $U(\beta)$ unconditionally.

In the case of BQC, the prior is parameterised by the parameters $\gamma := (\gamma^1, \ldots, \gamma^K)$. The posterior can be used to model new unseen data, D^*, using the *posterior predictive* [105]:

$$\mathbb{P}(D^*|D) = \int \mathbb{P}(D^*|\theta)\mathbb{P}(\theta|D)d\theta.$$

This integral averages predictions of all plausible models weighted by posterior probability and is called the *Bayesian model average*.

The BQC can be trained by minimising the maximum mean discrepancy cost function described in Chapter 9. In terms of expressive power, Du, Hsieh, Liu, and Tao [88] showed that a better expressive power of BQC is obtained in comparison with MPQC from a computational complexity perspective.

Bayesian networks can be used for financial asset price forecasting [56], predicting dynamics of limit order book market [214], predicting corporate bankruptcy [52], and to model, analyse, and understand trading behaviour [306].

> The Bayesian Quantum Circuit model extends the capabilities of parameterised quantum circuits trained as generative models (QCBM) through the addition of ancillary quantum registers encoding the prior distribution. As a result, it achieves greater expressive power than MPQC.

14.3   Quantum Semidefinite Programming

In Semidefinite Programming (SDP), one optimises a linear function subject to the constraint that an affine combination of symmetric matrices is positive semidefinite. Such a constraint is non-linear and non-smooth, but convex, so semidefinite programs are convex optimisation problems. Semidefinite programming unifies several standard problems (e.g., linear and quadratic programming) and finds many applications in engineering and combinatorial optimisation [317]. Similarly to finding a quantum counterpart to the classical kernel method, we can specify a quantum version of the SDP.

14.3.1   Classical semidefinite programming

The SDP can be generally defined as the following optimisation problem:

$$\max_{X \in \mathcal{M}_N^+(\mathbb{R})} \mathrm{Tr}(CX), \quad \text{subject to} \quad \mathrm{Tr}(A_j X) \le b_j, \quad \text{for all } j \in [\![M]\!], \qquad (14.3.1)$$

where $[\![M]\!] := \{1, \ldots, M\}$, $\mathcal{M}_n^+(\mathbb{R})$ denotes the set of positive semidefinite matrices of size $N \times N$. Here, Hermitian matrices $(A_j)_{j=1,\ldots,M}$ and C in $\mathcal{M}_N(\mathbb{R})$, and $(b_j)_{j \in [\![M]\!]} \in \mathbb{R}^M$ are the inputs of the problem.

SDP can be applied to complex NP-hard optimisation problems [116], such as various portfolio optimisation problems. For example, it is typically an unrealistic assumption that the distribution of asset returns is known exactly. The necessary information may not be complete and estimates are subject to estimation errors as well as modelling errors (e.g., an assumption of stationarity of the distributions).

14.3.2   Maximum risk analysis

The classical maximum risk analysis problem, assuming there is uncertainty in the estimate of the covariance matrix of asset returns, Σ, can be formulated as

$$\max_{\Sigma \in \mathcal{M}_N^+(\mathbb{R})} \mathbf{w}^\top \Sigma \mathbf{w}, \quad \text{subject to} \quad \Sigma_{ij}^L \le \Sigma_{ij} \le \Sigma_{ij}^U, \quad \text{for all } i, j \in [\![N]\!],$$

where w is the fixed vector of weights and Σ is the problem variable. For each $i, j \in [\![N]\!]$, the matrices Σ_{ij}^L and Σ_{ij}^U are fixed constraints in $\mathcal{M}_N^+(\mathbb{R})$. The task is to establish the maximum possible portfolio risk for the known asset allocation, given uncertainty in the estimate of covariance matrix of asset returns. The problem can be expressed as the following SDP [246]:

$$\max_{\Sigma \in \mathcal{M}_N^+(\mathbb{R})} \mathrm{Tr}\left(w^\top \Sigma w\right),$$

$$\text{subject to} \quad \begin{cases} \mathrm{Tr}(-E_{ij}\Sigma) \leq -\Sigma_{ij}^L, \\ \mathrm{Tr}(E_{ij}\Sigma) \leq \Sigma_{ij}^U, \end{cases} \quad \text{for all } (i, j) \in [\![N]\!] \times [\![N]\!],$$

where we denote $(E_{ij})_{\alpha\beta} := \delta_{i\alpha}\delta_{j\beta}$. The maximum risk analysis problem can be expressed in the same form with different risk measures such as VaR or Expected Shortfall.

14.3.3 Robust portfolio construction

The robust portfolio construction problem aims at finding an asset allocation method that would achieve the minimum *estimation error* in the suggested asset allocation weights. This problem has been addressed in [209] using Monte Carlo simulations to determine the most robust asset allocation method with respect to small changes in the input covariance matrix for the given portfolio.

In the most general case, it can be formulated as the Min-Max problem

$$\min_{w \in \mathcal{W}} \max_{\Sigma \in \mathcal{S}} w^\top \Sigma w,$$

with

$$\mathcal{S} := \left\{ \Sigma \in \mathcal{M}_N^+(\mathbb{R}) \ : \ \Sigma_{ij}^L \leq \Sigma_{ij} \leq \Sigma_{ij}^U, \quad \text{for all } i, j \in [\![N]\!] \right\},$$

$$\mathcal{W} := \left\{ w \in \mathbb{R}^N \ : \ \mathbf{1}^\top w = 1, \quad \mu^\top w \geq R_{\min} \right\},$$

where w is the vector of weights, μ is the vector of expected asset returns, and Σ is the covariance matrix of asset returns.

The following theorem (first proven by von Neumann [322] in 1928) establishes the equivalence of the Min-Max and Max-Min optimisation problems [313]:

Theorem 11 (Minimax Theorem). *Let $\mathcal{X} \subset \mathbb{R}^n$ and $\mathcal{Y} \subset \mathbb{R}^m$ be compact convex sets. If the function $f : \mathcal{X} \times \mathcal{Y} \to \mathbb{R}$ is continuous and concave in x for fixed y and continuous and convex in y for fixed x, then*

$$\min_{y \in \mathcal{Y}} \max_{x \in \mathcal{X}} f(x,y) = \max_{x \in \mathcal{X}} \min_{y \in \mathcal{Y}} f(x,y).$$

Therefore, in general, the Min-Max robust portfolio construction problem (which is convex in w and concave in Σ) is equivalent to the Max-Min problem and can be expressed for the constraints above as an SDP in all variables [246].

14.3.4   Quantum semidefinite programming

The key idea behind Quantum Semidefinite Programming (QSDP) is based on the observation that a normalised positive semidefinite matrix can be naturally represented as a quantum state. Operations on quantum states can sometimes be computationally cheaper to perform on a quantum computer than classical matrix operations. This idea prompted the development of quantum algorithms for SDPs [42].

Consider the SDP (14.3.1) and let $\varepsilon > 0$ be small. An algorithm is called an *ε-approximate quantum SDP oracle* [315] if, for all inputs $g \in \mathbb{R}$ and $\zeta \in (0,1)$, it finds, with success probability $1 - \zeta$, a vector $y \in \mathbb{R}^{M+1}$ and a real number z such that for the density matrix

$$\rho = \frac{\exp\left(-\sum_{j=1}^{M} y_j A_j + y_0 C\right)}{\mathrm{Tr}\left(\exp\left(-\sum_{j=1}^{M} y_j A_j + y_0 C\right)\right)}, \tag{14.3.2}$$

we have that $z\rho$ is an ε-feasible solution with objective value at least $g - \varepsilon$, that is

$$\begin{cases} \mathrm{Tr}(z\rho A_j) \leq b_j + \varepsilon, & \text{for all } j \in [\![M]\!], \\ \mathrm{Tr}(z\rho C) \geq g - \varepsilon, \end{cases}$$

or concludes that no such z and y exist even if we set $\varepsilon = 0$.

A general QSDP-solver for sparse matrices was implemented by Brandão and Svore [42] using the Arora-Kale framework [14]. They observed that the density matrix ρ in (14.3.2) is in fact a $\log(N)$-qubit Gibbs state and can be efficiently prepared as a quantum state on a quantum computer.

The reader should already be familiar with the Gibbs state (Gibbs distribution) in the form

$$\rho = \frac{e^{-\beta \mathcal{H}}}{\mathrm{Tr}(e^{-\beta \mathcal{H}})}, \tag{14.3.3}$$

where $\mathcal{H}$ is the problem Hamiltonian and $\mathrm{Tr}(\exp(-\beta \mathcal{H}))$ is the partition function. The Gibbs (Boltzmann) sampling and the Gibbs (Boltzmann) distribution were discussed in Chapter 5 (in (5.4.2) and (5.4.3)). The form of the partition function in (14.3.3) should not be confusing. Recall from (10.1.1), that since the Hamiltonian is a Hermitian operator, its spectral decomposition yields the representation

$$\mathcal{H} = \sum_i E_i \, |\psi_i\rangle \langle\psi_i| ,$$

which gives the following expression for the Gibbs state:

$$\rho = \frac{e^{-\beta \mathcal{H}}}{Z} = \frac{1}{Z} \sum_i e^{-\beta E_i} \, |\psi_i\rangle \langle\psi_i| ,$$

where the partition function Z is given by

$$Z = \mathrm{Tr}\left(e^{-\beta \mathcal{H}}\right) = \sum_i e^{-\beta E_i} .$$

QSDP gives a square-root unconditional speedup over any classical method for solving SDPs both in N and M [42].

> Quantum Semidefinite Programming is yet another example where quantum speedup can be achieved since operations on quantum states performed on a quantum computer are less computationally expensive than the corresponding matrix operations on a classical computer.

14.4   Symmetric Encryption

Quantum computing and cryptography are often mentioned together. To a large extent this is due to the fame of Shor's algorithm [302], designed to factor integer numbers with an exponential speedup with respect to the best known classical algorithms. Integer number factoring is a hard computational problem for classical computers that, due to its hardness, lies at the heart of the Rivest-Shamir-Adleman (RSA) protocol [270], an asymmetric public-private key cryptosystem. The basic idea behind RSA is very simple. There is a public key based on two large prime numbers and some auxiliary value that can be freely shared with external parties and used for encryption. But only the owner of the private key, who has the knowledge of the two prime numbers used to create the public key, can perform decryption. This asymmetry supports a secure and convenient communication protocol.

However, as the invention of Shor's algorithm shows, asymmetric encryption in the spirit of the RSA protocol may be vulnerable to a quantum attack. This prompts us to have another look at much older (probably, as old as writing itself) symmetric encryption [81, 244]. In the symmetric encryption protocol, the same key is used to encrypt and decrypt the message. It means that the key can only be shared with trusted parties. This significantly reduces its utility as a communication tool between arbitrary external parties, but at the same time allows us to make communication between trusted parties almost perfectly secure as long as the key remains unknown to the potential adversary [169].

In this section, we consider a symmetric encryption algorithm executable on a quantum computer [110]. The algorithm is based on utilising the expressive power of parameterised quantum circuits [28, 88].

14.4.1 Stylised example

Alice and Bob, two trusted parties, would like to communicate securely via unsecure channels. They meet at Alice's office, which is a secure place, where Alice generates a random quantum circuit in the form of Python code on her laptop. The circuit looks like the one shown in Figure 14.2.

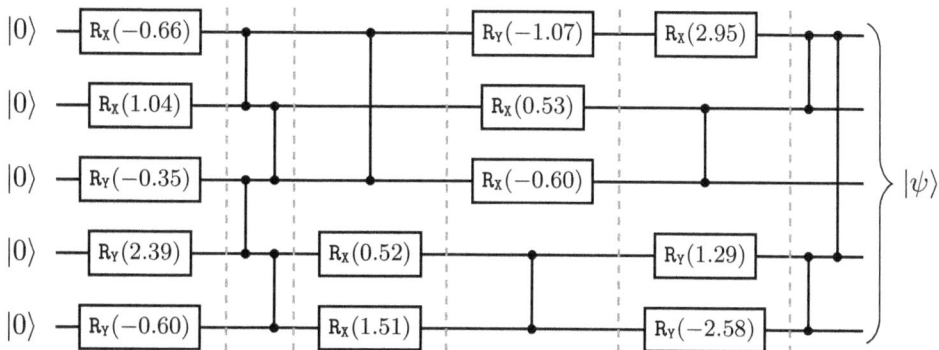

Figure 14.2: Randomly generated quantum circuit. All one-qubit gate parameters (rotation angles) are expressed in radians.

In this stylised example, the randomly generated quantum circuit consists of 5 quantum registers and 6 layers of one-qubit and two-qubit gates. The one-qubit gates are random rotations around the x and the y axes and the two-qubit gates are CZ gates that create entanglement.

Alice shares this circuit with Bob via a secure in-house communication channel. Now both Alice and Bob have the same quantum circuit on their laptops in the form of Python code – completely hardware agnostic. Running this quantum circuit would transform the initial quantum state $|0\rangle^{\otimes n}$ into a unique entangled quantum state $|\psi\rangle$ that can only be fully described by specifying 2^n probability amplitudes, where n is the number of qubits.

Next day, Alice needs some help from Bob. She decides to send him an encrypted message consisting of the letter "V" in binary format:

Letter	Radio Signal	ASCII Binary	ICS Meaning
V	Victor	1010110	"I require assistance"

Therefore, Alice needs to encrypt the bitstring [1010110].

Alice starts with generating a random string of Pauli operators X, Y and Z – the same length as the width of the quantum circuit she shared with Bob:

$$[X, Z, Y, Y, X].$$

This determines the bases in which qubits should be measured: the first and fifth qubits in the x basis, the second qubit in the z basis, the third and fourth qubits in the y basis. The objective is to calculate the expectation values of Pauli operators on various quantum registers. In this stylised example, we have the following possibilities for measuring the expectation value of a tensor product of two Pauli operators P_k and P_l on quantum registers k and l, $\langle P_k P_l \rangle$:

$$\langle X_1 Z_2 \rangle, \langle X_1 Y_3 \rangle, \langle X_1 Y_4 \rangle, \langle X_1 X_5 \rangle, \langle Z_2 Y_3 \rangle, \langle Z_2 Y_4 \rangle, \langle Z_2 X_5 \rangle, \langle Y_3 Y_4 \rangle, \langle Y_3 X_5 \rangle, \langle Y_4 X_5 \rangle.$$

To measure these expectation values, Alice has to add "change of basis" gates to the quantum circuit before measurement, as shown in Figure 14.3:

- Add nothing (or identity gate I) to the second quantum register as the computational basis is the z basis.

- Add H gate to the first and fifth quantum registers to transform the z basis into the x basis.

- Add HS† gates to the third and fourth quantum registers to transform the z basis into the y basis (S† gate should be applied first).

Next, Alice runs this quantum circuit 100,000 times and saves the measurement results in a $100,000 \times 5$ array. If "0" is measured, the eigenvalue of the corresponding Pauli operator is +1. If "1" is measured, the eigenvalue of the corresponding Pauli operator is −1. Therefore, the array entries are either +1 or −1. The number of rows is equal to the number of quantum circuit runs and the number of columns is equal to the number of quantum registers.

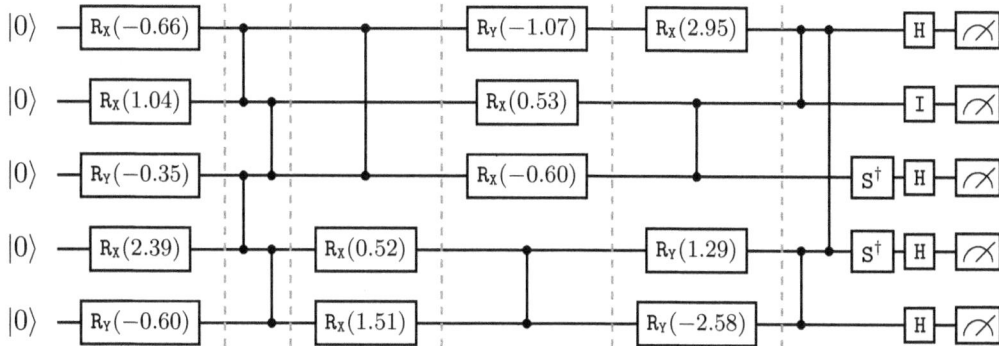

Figure 14.3: Measurement in different bases.

Then Alice computes the expectation values for the pairs of Pauli operators (tensor products of two Pauli operators). The expectation value $\langle P_k P_l \rangle$ is the dot product of columns k and l, normalised by the number of rows. She gets:

$$\langle X_1 Z_2 \rangle = -0.06580$$
$$\langle X_1 Y_3 \rangle = -0.03182$$
$$\langle X_1 Y_4 \rangle = 0.14572$$
$$\langle X_1 X_5 \rangle = 0.02212$$
$$\langle Z_2 Y_3 \rangle = 0.73397$$
$$\langle Z_2 Y_4 \rangle = 0.02899$$
$$\langle Z_2 X_5 \rangle = -0.00624$$
$$\langle Y_3 Y_4 \rangle = 0.01529$$
$$\langle Y_3 X_5 \rangle = -0.03990$$
$$\langle Y_4 X_5 \rangle = -0.27681$$

If expectation value $\langle P_k P_l \rangle$ is larger than ϵ, the pair $P_k P_l$ can be used to encode "1".

If expectation value $\langle P_k P_l \rangle$ is smaller than $-\epsilon$, the pair $P_k P_l$ can be used to encode "0".

The value of ϵ depends on the circuit configuration and the number of runs. Alice sets it at $\epsilon = 0.01$.

Therefore, Alice can use the following pairs of Pauli operators for encrypting her message:

$$\langle X_1 Z_2 \rangle \Rightarrow 0$$
$$\langle X_1 Y_3 \rangle \Rightarrow 0$$
$$\langle X_1 Y_4 \rangle \Rightarrow 1$$
$$\langle X_1 X_5 \rangle \Rightarrow 1$$
$$\langle Z_2 Y_3 \rangle \Rightarrow 1$$
$$\langle Z_2 Y_4 \rangle \Rightarrow 1$$
$$\langle Z_2 X_5 \rangle \Rightarrow - \quad \text{(expectation value falls within } [-\epsilon, \epsilon] \text{ interval)}$$
$$\langle Y_3 Y_4 \rangle \Rightarrow 1$$
$$\langle Y_3 X_5 \rangle \Rightarrow 0$$
$$\langle Y_4 X_5 \rangle \Rightarrow 0$$

Alice sends Bob a text message that reads: XZYYX.

Then Alice sends Bob an email that reads: (2,3) (4,5) (1,4) (1,2) (2,4) (1,5) (3,5).

After Bob receives the text message, he knows that he must change the basis on quantum registers 1, 3, 4, and 5 (add H to the first and fifth quantum registers and add HS† to the third and fourth quantum registers).

After Bob receives the email, he knows what expectation values he needs to calculate and in what order. Bob runs the quantum circuit (the same as Alice's) and gets the following expectation values (they would differ slightly from Alice's values due to the finite number of quantum circuit runs and some hardware noise):

$$\langle Z_2 Y_3 \rangle = 0.73666 \Rightarrow 1$$
$$\langle Y_4 X_5 \rangle = -0.27318 \Rightarrow 0$$
$$\langle X_1 Y_4 \rangle = 0.14722 \Rightarrow 1$$
$$\langle X_1 Z_2 \rangle = -0.06782 \Rightarrow 0$$
$$\langle Z_2 Y_4 \rangle = 0.02876 \Rightarrow 1$$
$$\langle X_1 X_5 \rangle = 0.02092 \Rightarrow 1$$
$$\langle Y_3 X_5 \rangle = -0.03930 \Rightarrow 0$$

Bob correctly decrypts the message, reads the bitstring [1010110], and learns that Alice requires his assistance.

Two unsecure channels (text and email) were used to transmit the message. Even if both channels are compromised, it is impossible to decipher the message without the knowledge of the quantum circuit that creates the quantum state in which Pauli operators are measured. In other words, the quantum circuit plays the role of a symmetric key used to encrypt and decrypt messages between two trusted parties.

In this stylised example, there are two additional important elements of the security protocol that provide additional protection:

- No Pauli pairs are repeated.

- Only some of the possible Pauli pairs are used.

In practice, when dealing with the task of encrypting a large amount of data rather than a single ASCII symbol, we would face additional challenges. This means that the general encryption protocol based on measuring expectation values of second order Pauli operators should provide additional degrees of security.

In fact, we cannot use ASCII encoding for the plain text symbols. To see why, let us have a look at how ASCII encodes letters. First, we notice that all upper-case letters (from "A" to

"Z") start with "10" as leading digits in the 7-digit binary encoding, while all lower-case letters (from "a" to "z") start with "11" as the two leading digits. Secondly, letter encodings have an unequal number of "0"s and "1"s. For example, upper-case "A" is encoded as 5 "0"s and 2 "1"s: 1000001. This means that a relatively simple frequency analysis can help to break the encryption given a sufficiently large amount of text. Therefore, we need to generate a random mapping of the plain text symbols to the bitstrings of fixed length simultaneously with the generation of the quantum circuit. The key condition is that the number of "0"s and "1"s in each randomly generated bitstring should be the same to prevent any possible attempt at frequency analysis. A sample random mapping that illustrates this principle is shown in Table 14.1. There, each plain text symbol is represented by a 12-bit bitstring with six "0"s and six "1"s. The table is shared with the trusted parties together with the generated quantum circuit.

Another consideration concerns the architecture of the quantum circuit. Security can be dramatically improved if the quantum circuit is modified by making the one-qubit gates in the first layer adjustable, as shown in Figure 14.4. Making the first layer adjustable turns the generated quantum circuit into a de facto family of quantum circuits, where the configuration of adjustable parameters plays the role of a "seed" that can be set according to some rules. For example, it can be derived from the time stamp of some process as explained below. Without this functionality, we will be limited in how much text we can encode before we start reusing the same pairs of Pauli operators with the same expectation values. With this functionality, assuming that we perform a regular reset of the seed, the same pairs of Pauli operators will have different expectation values for different seeds. This should prevent any meaningful attempt at analysis of large amounts of intercepted encrypted text as long as the seed is changed before the Pauli pairs are repeated.

In Figure 14.4, the one-qubit gates in the first layer are rotations around the x, y, and z axes by angle $\alpha_i \theta$, where i indicates the quantum register number. Coefficients $\alpha_i \in [-1, 1]$ are fixed. Parameter θ is a function of the time stamp of the generated vector of Pauli operators. For example, we can adopt the following scheme for setting the value of θ.

Symbol	Encoding	Symbol	Encoding	Symbol	Encoding
A	111100100010	a	100110001011	0	101010110100
B	111101010000	b	100001100111	1	001001101011
C	011101101000	c	100000101111	2	001010110101
D	011011001010	d	101100101010	3	100010101101
E	110001110010	e	101000001111	4	101000110011
F	110100010101	f	011110101000	5	010100011011
G	110100111000	g	100111101000	6	011010001110
H	010010101101	h	001011000111	7	011111000010
I	001110011010	i	000100101111	8	011100001011
J	011110010100	j	001110001101	9	110100001101
K	001011010101	k	100101010011	+	010101100101
L	100001110110	l	110010011010	-	011001011010
M	110101010010	m	100010101011	*	001011110001
N	011011000110	n	010101101100	/	011001001110
O	001110001011	o	011101011000	=	001101000111
P	101011000011	p	101011010100	(	110000110011
Q	011110010001	q	101000011101	)	010100111001
R	100100011101	r	101101010100	.	001110101001
S	101111100000	s	010101010011	,	101001110010
T	001111011000	t	011001101001	:	010011000111
U	101100010101	u	111000010101	;	110010101010
V	110100011010	v	111101001000	?	001101100011
W	111110000100	w	110111010000	!	110010110100
X	111011000001	x	001101001011	%	100110110100
Y	101101001001	y	000111001011	'	101010011010
Z	100001101101	z	101110110000	*space*	000110010111

Table 14.1: Sample encoding scheme (randomly generated).

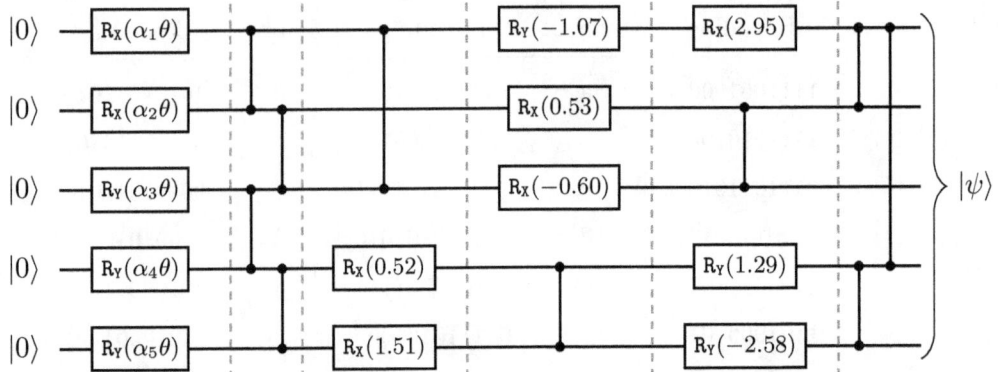

Figure 14.4: Quantum circuit with adjustable gates in the first layer.

Let us write down the time stamp in the format: YYMMDDhhmmss. For example, 13:03:46 on 27 November 2023 would be represented by the following integer number: 231127130346. Then the value of parameter θ is set equal to YYMMDDhhmmss modulo 2π. In our example, it would be $\theta = 1.32392129$. The value of θ is unique for each time stamp. At the same time, the knowledge of θ will be of no use unless the whole quantum circuit is known. It means that the time stamp can also be shared between the trusted parties using the unsecure channel – the same channel that is used for communicating the vector of Pauli operators.

Making the parameters of the first layer's gates depend on the time stamp ensures that it is fruitless to try to establish the mapping between the pairs of Pauli operators and their binary values: each random vector of Pauli operators that are used to encode the message will be accompanied by a unique quantum circuit given by the unique time stamp. This would allow us to reuse the same quantum circuit multiple times since every configuration of adjustable parameters in the first layer will lead to the corresponding unique quantum state in which Pauli operators are measured.

14.4.2 General algorithm

We can now formulate the general symmetric encryption/decryption algorithm [110].

Algorithm 11: Symmetric Encryption

1: Generation of a random quantum circuit on N quantum registers with M layers of one-qubit and two-qubit gates (e.g., $M = \text{int}(2\sqrt{N})$). The one-qubit gates are random rotations around the x, y, and z axes; the two-qubit gates are suitable native gates (e.g., `CZ` or `iSWAP`). The first layer of the quantum circuit consists of one-qubit gates with parameters (rotation angles) of the form $\alpha_i \theta$, where all $\alpha_i \in [-1, 1]$, $i = 1, \ldots, N$, are randomly generated coefficients and the parameter θ is the same for all one-qubit gates in the first layer.

2: Generation of a random mapping scheme that would provide a one-to-one mapping between any desired plain text symbol and a $2n$-digit bitstring with exactly n "0"s and exactly n "1"s.

3: Sharing of the generated quantum circuit and the plain text symbol mapping scheme with the trusted party. The quantum circuit is a symmetric key that can be used by the trusted parties to encrypt and decrypt data in binary format.

4: When one trusted party (Alice) wants to communicate with another trusted party (Bob), the protocol is as follows:

    a) Alice converts the plain text she wants to encrypt into the bitstring using the generated mapping scheme.

    b) Alice generates a random vector of Pauli operators $\{X, Y, Z\}$. The length of the vector should be equal to the width of the quantum circuit, with one-to-one mapping between the elements of the vector of Pauli operators and the quantum registers (e.g., $[Z_1, Y_2, X_3, \ldots, Y_N]$).

    c) Alice saves the time stamp of the generated random vector of Pauli operators as an integer number in the format YYMMDDhhmmss. This number modulo 2π (double precision) is the value of parameter θ in the first layer of the generated quantum circuit.

4:   d) Alice adds "change of basis" gates to each quantum register in accordance with the Pauli operator which is assigned to the quantum register:

- nothing to quantum registers with Pauli Z (z basis);
- H to quantum registers with Pauli X (transformation from the z basis to the x basis);
- HS† to quantum registers with Pauli Y (transformation from the z basis to the y basis).

e) Alice executes L runs of the quantum circuit and saves the measurement results in an $L \times N$ array:

- **if** "0" is measured

  then the eigenvalue of the corresponding Pauli operator is $+1$;
- **if** "1" is measured

  then the eigenvalue of the corresponding Pauli operator is -1.

Therefore, the array entries are either $+1$ or -1. The number of rows is equal to the number of quantum circuit runs and the number of columns is equal to the number of quantum registers.

f) Alice randomly selects two quantum registers i and j (without replacement). She calculates the expectation value of Pauli operators $\langle P_i P_j \rangle$ as the dot product of columns i and j, normalised by the number of rows:

- **if** the value of $\langle P_i P_j \rangle$ is larger than ϵ

  then this pair of quantum registers can be used to encode "1";
- **if** the value of $\langle P_i P_j \rangle$ is smaller than $-\epsilon$

  then this pair of quantum registers can be used to encode "0".

The value of ϵ is one of the algorithm's parameters. The broad condition is $\epsilon \geq m\sigma$, where σ is the estimated average standard deviation of expectation values. The value of m can be chosen from some general considerations.

4:    g) Alice takes the first bit from the bitstring she wants to encrypt:

if the bit value is "1" and $\langle P_i P_j \rangle > \epsilon$ or the bit value is "0" and $\langle P_i P_j \rangle < -\epsilon$

then Alice encrypts this bit with the pair (i, j)

else the pair (i, j) is either discarded (if $-\epsilon \leq \langle P_i P_j \rangle \leq \epsilon$) or put on hold (for the next suitable bit), and Alice tries another random pair of Pauli operators (without replacement).

h) This process continues (steps f) and g) are repeated) until Alice encrypts the first K bits with the pairs of quantum register indices. The broad condition is $K < N(N-1)/2$.

i) After that, Alice generates a new random vector of Pauli operators, saves its time stamp as an integer number in the format YYMMDDhhmmss, and repeats the above procedure for the next K bits.

j) This process continues until Alyce encrypts the whole bitstring (the whole dataset in binary format). If the bitstring length is B, then we end up with D vectors of Pauli operators, D time stamps, and D cycles of quantum circuit runs: $(D-1)K < B \leq DK$.

5: The encrypted dataset consists of three arrays:

a) $D \times N$ array of Pauli operators (D rows, N columns), where each kth row is the vector of Pauli operators used to encrypt the $[(k-1)K+1, kK]$ section of the bitstring.

b) $D \times 1$ array of time stamps (a time stamp for each vector of Pauli operators).

c) $D \times K$ array of pairs of quantum register indices (D rows, K columns), where each kth row is the vector of quantum register index pairs used to encrypt the $[(k-1)K+1, kK]$ section of the bitstring.

Alice sends the first and second arrays to Bob via a preferred unsecure communication channel (e.g., email). Then Alice sends the third array to Bob via the same or a different unsecure channel (e.g., second email).

6: After receiving the arrays from Alice:

a) Bob takes the first value from the time stamp array as an integer number in the format YYMMDDhhmmss. This number modulo 2π is the value that should be assigned to the parameter θ in the first layer of the quantum circuit.

b) Bob takes the first row from the Pauli operator array and adds the corresponding basis transformation gates to the quantum circuit – exactly the same procedure as done by Alice. Bob runs the quantum circuit L times and saves the measurement results $(+1, -1)$ in an $L \times N$ array.

c) Bob takes the first element (i, j) from the index pairs array and calculates the dot product of columns i and j from the $L \times N$ array:

if the value of the dot product of columns i and j is positive

then Bob decrypts (i, j) as "1"

else Bob decrypts (i, j) as "0".

Bob continues this procedure until he reaches the end of the first row of the index pairs array (i.e., until he decrypts the first K bits).

d) Bob switches to the second row of the Pauli operators array and repeats steps a) and b). He continues until the whole bitstring is decrypted.

e) Bob translates the decrypted bitstring into plain text using the same mapping scheme as Alice.

Remark: With $N = 2,500$, $M = 100$, $L = 500,000$, $K = 2,500,000$, $\epsilon = 0.01$, $m = 6$, and $n = 6$, Algorithm 11 would allow encryption of 65 full pages of dense normal text per single vector of Pauli operators (1 page $= 40$ lines $\times$ 80 symbols per line $\times$ 12 bits per symbol). The 6-sigma threshold is equivalent to 1 typo per about 26,000 pages of dense normal text.

It is possible to transform the proposed algorithm into a multiple keys encryption protocol. For example, Alice splits the quantum circuit into two parts: the first M_1 layers and the second M_2 layers ($M_1 + M_2 = M$). Then, Alice shares the first M_1 layers with Bob and the second M_2 layers with Charlie. It would only be possible to perform decryption if both Bob and Charlie combine their quantum sub-circuits into a single complete quantum circuit.

> Although we have not analysed the quantum symmetric encryption protocol introduced in this section from the computational theory point of view, some general considerations suggest that it can be made as secure as a one-time pad system, which is the only theoretically secure classical cipher [203].

14.4.3 Encryption and Decryption Example

The following example illustrates the application of Algorithm 11 to the encryption and decryption of a meaningful amount of plain text. The sample message is taken from a short story by Edgar Allan Poe, *The Gold-Bug* [256]:

```
A good glass in the bishop's hostel in the devil's seat.

Forty-one degrees and thirteen minutes northeast and by north.

Main branch seventh limb east side.

Shoot from the left eye of the death's-head.

A bee line from the tree through the shot fifty feet out.
```

In the novel, this message was encrypted by Scottish privateer Captain Kidd using a simple substitution cipher and successfully decrypted more than a century later by the novel's protagonist, William Legrand, using frequency analysis[1]. The infamous privateer had a strong reason to encrypt the message as it led to the discovery of a treasure trove worth millions of dollars in mid-nineteenth century money!

---

[1]The punctuation has been changed from its original version to make the message more readable.

We will encrypt and then decrypt this message on the Qiskit quantum simulator. The message is 258 characters long (counting spaces) and should be translated into the 3,096-bit string using the mapping Table 14.1.

Figure 14.5 displays embedding of a 20-qubit quantum circuit on the square-grid QPU graph. Even though we can easily realise the "all-to-all" qubit connectivity on a quantum simulator, our objective is to stay within the restrictions imposed by the limited connectivity of existing quantum processors.

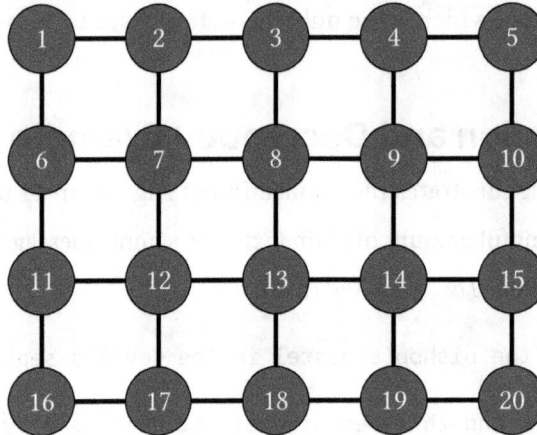

Figure 14.5: Embedding of a 20-qubit quantum circuit on the square-grid QPU graph (Rigetti's Ankaa system).

The corresponding quantum circuit is shown in Figure 14.6. It consists of three layers of one-qubit gates with adjustable parameters (R_X, R_Y) and three layers of fixed two-qubit gates (CZ). In total, the quantum circuit is configured by 100 adjustable parameters. Each one-qubit rotation angle ϕ_i^j is defined on the interval $[-\pi/2, \pi/2]$, thus attempting to provide sufficiently uniform coverage of the qubit states on the Bloch sphere.

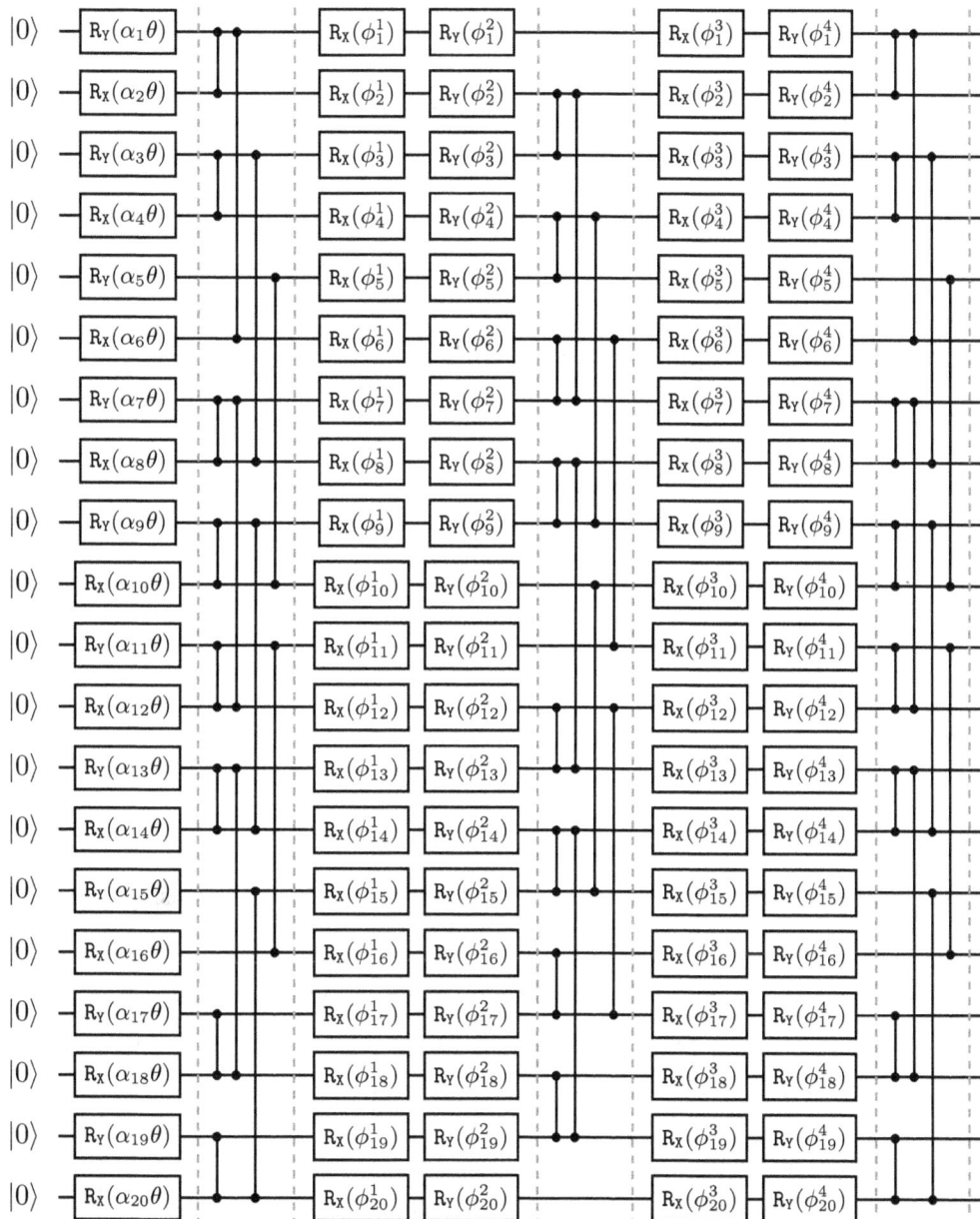

Figure 14.6: The 20-qubit quantum circuit for the execution of symmetric encryption/decryption protocol on a quantum simulator.

Since our task is to encode 3,096 bits and a single cycle (single set of basis gates) can produce at most $(20 \cdot 19)/2 = 190$ unique Pauli pairs, we have to run multiple cycles when working with the 20-qubit circuit. Also, not all Pauli pairs will be eligible for encoding purposes due to the finite number of quantum circuit runs. The proportion of eligible unique Pauli pairs is regulated by the value of the threshold parameter. Numerical experiments indicate that with 10,000 quantum circuit runs, we need to set the expectation value threshold parameter ϵ to 0.05 in order to ensure the 6 standard deviations confidence level [110]. This is a demanding requirement that may push the number of cycles up significantly unless we increase the number of quantum circuit runs and, therefore, decrease the threshold value.

However, we can decrease the value of ϵ without increasing the number of quantum circuit runs and without sacrificing the accuracy by applying an error detection and error correction technique suggested by the very nature of the proposed encryption protocol. The encryption algorithm requires every plain text character to be represented by a bitstring with an equal number of "0"s and "1"s. For example, six "0"s and six "1"s, as per the mapping Table 14.1. If the decrypted bitstring has an unequal number of "0"s and "1"s, then we immediately know that the bitstring has been encrypted/decrypted incorrectly. As long as the error rate remains low, it is highly unlikely that two errors (two bit-flips) would happen for the same bitstring and, therefore, all errors that occurred can be detected.

The encryption protocol also suggests an obvious error correction mechanism. For example, if the decrypted bitstring has seven "0"s and five "1"s, we can try to flip each "0" into "1" one by one and test the obtained seven new bitstrings with an equal number of "0"s and "1"s as to whether they encode a plain text character. This is the exact approach we are taking in our experiment with the 20-qubit circuit given in Figure 14.6.

The numerical experiment consists of executing the encryption and decryption protocol end to end, as specified by Algorithm 11 with 10,000 quantum circuit runs while varying the values of the threshold parameter ϵ.

In our experiment, we observed error-free encryption and decryption for $\epsilon > 0.035$.

Once the value of ϵ was reduced below 0.035, we started to see instances of incorrect encryption and decryption. A typical decrypted message would then look as follows:

```
A good glass in the bishop's hostel in the devil's seat.

Forty-one degrees and thirteen minutes northeast and by north.

Main branch seventh limb east side.

Shoot from the left eye of the death's $ hea $ .

A bee line from the tree through the shot fifty feet out.
```

Here, the symbol $ indicates detected incorrectly encrypted/decrypted character. The application of the error correction mechanism outlined above restores the message to its original form:

```
A good glass in the bishop's hostel in the devil's seat.

Forty-one degrees and thirteen minutes northeast and by north.

Main branch seventh limb east side.

Shoot from the left eye of the death's - hea d .

A bee line from the tree through the shot fifty feet out.
```

The first incorrect bitstring has been error-corrected to the bitstring representing character "-" and the second incorrect bitstring has been error-corrected to the bitstring representing character "d". In some cases, the proposed error correction method can suggest several valid substitutions. To ensure uniqueness, the mapping between plain text characters and the corresponding bitstrings should be organised in such a way that the Hamming distance between any two bitstrings is > 2. This may require longer bitstrings (e.g., 14 bits long) but would significantly increase the number of eligible Pauli pairs – often by a substantial multiple for a relatively small number of quantum circuit runs.

For $\epsilon < 0.03$, we observe instances of more than one error per bitstring and, therefore, can no longer rely on the error correction. In this case, the solution is to increase the number of quantum circuit runs.

Summary

In this chapter, we introduced several new promising quantum algorithms. First, we learned about quantum generative adversarial networks – powerful generative models that are based on the same principle as their classical counterparts: a generator and a discriminator competing against each other.

Next, we introduced the Bayesian quantum circuit model that expands the concept of a Bayesian neural network to parameterised quantum circuits. BQC is a promising generative model with larger expressive power than QCBM/MPQC (covered in Chapters 9 and 13).

Then we looked at quantum SDP and its potential to outperform classical SDP. This is a topic of active research.

Finally, with so much attention attracted to Shor's algorithm and the possibility of undermining public key cryptography (asymmetric encryption based on public key-private key pairing), we considered a possible quantum symmetric encryption protocol which is resistant to both classical and quantum attacks.

In the next chapter, we will cover the algorithms designed for fault tolerant quantum computers. As the NISQ era draws to a close, these algorithms will become progressively more relevant to many real-world applications of quantum computers.

15

Beyond NISQ

We would like to complete this book with a glance beyond the capabilities of NISQ computers. This chapter presents several important algorithms that, one day, will become the main building blocks of many quantum computing applications. We start with describing the workhorse of many important quantum algorithms, the Quantum Fourier Transform (QFT), before moving to its flagship application, the Quantum Phase Estimation (QPE), and then discussing the possibility of achieving quantum speedup with the Quantum Monte Carlo (QMC) and the Quantum Linear Solver (QLS) algorithms.

15.1 Quantum Fourier Transform

In the classical setting, the discrete Fourier transform maps a vector

$$\mathbf{x} := (x_0, \ldots, x_{2^n-1}) \in \mathbb{C}^{2^n}$$

to a vector

$$\mathbf{y} := (y_0, \ldots, y_{2^n-1}) \in \mathbb{C}^{2^n},$$

the components of which read

$$y_k = \frac{1}{\sqrt{2^n}} \sum_{j=0}^{2^n-1} \exp\left(\frac{2\pi i j k}{2^n}\right) x_j, \quad \text{for each } k = 0, \ldots, 2^n - 1.$$

Similarly, the QFT is the linear map

$$|k\rangle \longmapsto \frac{1}{\sqrt{2^n}} \sum_{j=0}^{2^n-1} \exp\left(\frac{2\pi i k j}{2^n}\right) |j\rangle,$$

and the operator

$$_q\mathcal{F} := \frac{1}{\sqrt{2^n}} \sum_{k,j=0}^{2^n-1} \exp\left(\frac{2\pi i k j}{2^n}\right) |j\rangle \langle k|$$

represents the Fourier transform matrix which is unitary as $_q\mathcal{F}_q\mathcal{F}^\dagger = \mathcal{I}$. In an n-qubit system with basis $(|0\rangle, \ldots, |2^n - 1\rangle)$, for a given state $|j\rangle$, we use the binary representation

$$j := \overline{j_1 \cdots j_n},$$

with $(j_1, \ldots, j_n) \in \{0, 1\}^n$ so that

$$|j\rangle = |j_1 \cdots j_n\rangle = |j_1\rangle \otimes \ldots \otimes |j_n\rangle.$$

Likewise, the notation

$$\overline{0.j_1 j_2 \ldots j_n}$$

represents the binary fraction $\sum_{i=1}^{n} 2^{-i} j_i$. Elementary algebra (see [238, Section 5.1] for details) then yields

$$_q\mathcal{F}|j\rangle = \frac{1}{\sqrt{2^n}} \left(|0\rangle + e^{2\pi i \overline{0.j_n}} |1\rangle\right) \otimes \left(|0\rangle + e^{2\pi i \overline{0.j_{n-1}j_n}} |1\rangle\right) \otimes \cdots$$
$$\cdots \otimes \left(|0\rangle + e^{2\pi i \overline{0.j_1 \ldots j_n}} |1\rangle\right). \tag{15.1.1}$$

15.2   Quantum Phase Estimation

The goal of QPE is to estimate the unknown phase $\varphi \in [0, 1)$ for a given unitary operator $\mathcal{U}$ with an eigenvector $|u\rangle$ and eigenvalue $\exp(2\pi i \varphi)$. Consider a register of size m and define

$$b^* := \sup_{j \leq 2^m \varphi} \left\{ j = 2^m \, \overline{0.j_1 \cdots j_m} \right\}.$$

Thus with $b^* = \overline{b_1 \cdots b_m}$, we obtain that

$$2^{-m} b^* = \overline{0.b_1 \cdots b_m}$$

is the best m-bit approximation of φ from below. The QPE procedure uses two registers, with the first containing m qubits initially in the state $|0\rangle$. Selecting m relies on the number of digits of accuracy for the estimate of φ, and the probability with which we wish to obtain a successful phase estimation procedure.

QPE allows us to implement a measurement for any Hermitian operator. Note that we always measure individual qubits. If we want to measure a more complex observable, we can use a QPE that implements the von Neumann measurement scheme [227]. The routine prepares an eigenstate of the Hermitian operator in one register and stores the corresponding eigenvalue in a second register.

Up to a SWAP transformation, the quantum phase circuit [238, Section 5.2] gives the output

$$|\psi\rangle = \frac{1}{\sqrt{2^m}} \left(|0\rangle + e^{2\pi i \, \overline{0.\varphi_m}} \, |1\rangle \right) \otimes \left(|0\rangle + e^{2\pi i \, \overline{0.\varphi_{m-1} \varphi_m}} \, |1\rangle \right) \otimes \cdots$$

$$\cdots \otimes \left(|0\rangle + e^{2\pi i \, \overline{0.\varphi_1 \cdots \varphi_m}} \, |1\rangle \right),$$

which is exactly equal to the QFT for the state $|2^m \varphi\rangle = |\varphi_1 \varphi_2 \ldots \varphi_m\rangle$ as in (15.1.1), and therefore $|\psi\rangle = {}_q\mathcal{F} |2^m \varphi\rangle$. Since the QFT is a unitary transformation, we can invert the process to retrieve $|2^m \varphi\rangle$. Algorithm 12 below provides pseudocode for the QPE procedure, and we refer the interested reader to [238, Chapter 5.2] for detailed explanations.

Algorithm 12: Quantum Phase Estimation

Input:

- Unitary matrix (gate) U with $U \ket{u} = e^{2\pi i \varphi} \ket{u}$;

- m ancilla qubits initialised at $\ket{0}$.

1: Prepare the initial state with $\ket{0}^{\otimes m}$ being the m-qubit ancilla register and $\ket{u}$ being the n-qubit eigenstate register.

2: Map to

$$\frac{1}{\sqrt{2^m}} \sum_{j=0}^{2^m - 1} \ket{j} \ket{u}$$

with Hadamard gates applied to the ancilla register.

3: Map to

$$\frac{1}{\sqrt{2^m}} \sum_{j=0}^{2^m - 1} \ket{j} U^j \ket{u} = \frac{1}{\sqrt{2^m}} \sum_{j=0}^{2^m - 1} \ket{j} e^{2\pi i j \varphi} \ket{u}$$

with controlled U^j gates applied to the eigenstate register.

4: Compute

$$\ket{\widetilde{\varphi}} \ket{u}$$

using the inverse QFT, where $\widetilde{\varphi}$ is an m-qubit approximation of φ.

5: Measure to deduce $\widetilde{\varphi}$.

Result: Phase estimate $\widetilde{\varphi}$.

15.3   Monte Carlo speedup

Leveraging on the speedup provided by the QPE, Montanaro [231] devised a Monte Carlo scheme providing quantum speedup compared to the classical one.

15.3.1   Classical Monte Carlo

Monte Carlo techniques represent a wide array of methods to simulate statistics of random processes. We refer the interested reader to the excellent monograph [115] for a full description and analysis. Consider a one-dimensional random variable X and a function $\phi : \mathbb{R} \to [0, 1]$ such that both $\mathfrak{p} := \mathbb{E}[\phi(X)]$ and $\sigma^2 := \mathbb{V}[\phi(X)]$ are well defined. By the Central Limit Theorem, given an iid collection of random variables $(X_1, \ldots, X_N)$ distributed as X, then

$$\sqrt{N} \frac{\widehat{\mathfrak{p}}_N - \mathfrak{p}}{\sigma}$$

converges to a centered Gaussian with unit variance $\mathcal{N}(0, 1)$ as N tends to infinity, where

$$\widehat{\mathfrak{p}}_N := \frac{1}{N} \sum_{i=1}^{N} X_i$$

is the empirical mean. This implies that, for any $\varepsilon > 0$, we can estimate

$$\mathbb{P}\left(\left|\widehat{\mathfrak{p}}_N - \mathfrak{p}\right| \leq \varepsilon\right) = \mathbb{P}\left(|\mathcal{N}(0, 1)| \leq \frac{\varepsilon \sqrt{N}}{\sigma}\right),$$

so that, for any $z > 0$ and $\delta \in (0, 1)$, in order to get an estimate of the form

$$\mathbb{P}\left(\left|\widehat{\mathfrak{p}}_N - \mathfrak{p}\right| \leq z\right) = 1 - \delta,$$

we need $N = \mathcal{O}(1/\varepsilon^2)$ samples.

15.3.2   Quantum Monte Carlo

Consider now an operator $\mathcal{A}$ of the form

$$\mathcal{A} |0\rangle^{\otimes n} = \sum_{x \in \{0,1\}^k} \alpha_x |\psi_x\rangle |x\rangle,$$

for some $k \leq n$, where each $|\psi_x\rangle$ is a quantum state with $n - k$ qubits and $|x\rangle$ is a quantum state with k qubits, and $\alpha_x \in \mathbb{C}$ is some amplitude, the meaning of which will be made clear below. We simply assume that $\{|\psi_x\rangle\}_{x \in \{0,1\}^k}$ forms an orthogonal family and are in fact "garbage qubits", i.e., qubits that are, for example, used as controlled to build the solution vector $|x\rangle$ from the data. Given the encoded data $|x\rangle$, assume further the existence of the operator $\mathcal{W}$:

$$\mathcal{W} |\mathrm{x}\rangle |0\rangle = |\mathrm{x}\rangle \left(\sqrt{1 - \phi(\mathrm{x})} \, |0\rangle + \sqrt{\phi(\mathrm{x})} \, |1\rangle \right).$$

This can be achieved for example by using the following lemma.

Lemma 9 (Conditional Rotation. Theorem 3.5 in [199]). *Given a quantum state* $|\psi_a\rangle$, *encoding* $a \in [-1, -1]$ *in q qubits, there exists a quantum circuit performing the unitary mapping* $|\psi_a\rangle |0\rangle \longmapsto |\psi_a\rangle \left(a |0\rangle + \sqrt{1 - a^2} |1\rangle \right).$

Consider now the operator $\mathcal{M}$:

$$\mathcal{M} := \left(\mathcal{I}^{n-k} \otimes \mathcal{W} \right) \left(\mathcal{A} \otimes \mathcal{I} \right),$$

where $\mathcal{I}^{n-k}$ means the identity operator acting on $n - k$ qubits, so that

$$
\begin{aligned}
|\psi\rangle &:= \mathcal{M} |0\rangle^{\otimes(n+1)} \\
&= \left(\mathcal{I}^{n-k} \otimes \mathcal{W} \right) \left(\sum_{\mathrm{x} \in \{0,1\}^k} \alpha_\mathrm{x} |\psi_\mathrm{x}\rangle |\mathrm{x}\rangle \right) |0\rangle \\
&= \sum_{\mathrm{x} \in \{0,1\}^k} \alpha_\mathrm{x} \left(\mathcal{I}^{n-k} \otimes \mathcal{W} \right) |\psi_\mathrm{x}\rangle |\mathrm{x}\rangle |0\rangle \\
&= \sum_{\mathrm{x} \in \{0,1\}^k} \alpha_\mathrm{x} |\psi_\mathrm{x}\rangle |\mathrm{x}\rangle \left(\sqrt{1 - \phi(\mathrm{x})} \, |0\rangle + \sqrt{\phi(\mathrm{x})} \, |1\rangle \right) \\
&=: |\Psi_B\rangle |0\rangle + |\Psi_G\rangle |1\rangle,
\end{aligned}
\tag{15.3.1}
$$

where $|\Psi_B\rangle$,

$$|\Psi_B\rangle := \sum_{\mathrm{x} \in \{0,1\}^k} \alpha_\mathrm{x} \sqrt{1 - \phi(\mathrm{x})} \, |\psi_\mathrm{x}\rangle |\mathrm{x}\rangle,$$

stands for the 'bad' state, and $|\Psi_G\rangle$,

$$|\Psi_G\rangle := \sum_{\mathrm{x} \in \{0,1\}^k} \alpha_\mathrm{x} \sqrt{\phi(\mathrm{x})} \, |\psi_\mathrm{x}\rangle |\mathrm{x}\rangle, \tag{15.3.2}$$

stands for the 'good' state.

Consider now the projector $\mathcal{P} := \mathcal{I}^n \,|1\rangle\langle1|$ and measure the probability of the last qubit of $|\psi\rangle$ to be in state $|1\rangle$, namely

$$\langle\psi|\,\mathcal{P}^\dagger\mathcal{P}\,|\psi\rangle = \langle\psi|\,\mathcal{P}\,|\psi\rangle$$

$$= \Big(\,\langle0|\,\langle\Psi_B| + \langle1|\,\langle\Psi_G|\,\Big)\mathcal{P}\Big(\,|\Psi_B\rangle\,|0\rangle + |\Psi_G\rangle\,|1\rangle\,\Big)$$

$$= \Big(\,\langle0|\,\langle\Psi_B| + \langle1|\,\langle\Psi_G|\,\Big)\Big(\,|\Psi_B\rangle\,|1\rangle\,\langle1|0\rangle + |\Psi_G\rangle\,|1\rangle\,\langle1|1\rangle\,\Big)$$

$$= \Big(\,\langle0|\,\langle\Psi_B| + \langle1|\,\langle\Psi_G|\,\Big)\,|\Psi_G\rangle\,|1\rangle$$

$$= \langle0|\,\langle\Psi_B|\Psi_G\rangle\,|1\rangle + \langle1|\,\langle\Psi_G|\Psi_G\rangle\,|1\rangle$$

$$= \langle\Psi_B|\Psi_G\rangle\,\langle0|1\rangle + \langle\Psi_G|\Psi_G\rangle\,\langle1|1\rangle = |\Psi_G|^2.$$

Now, since the family $\{|\psi_x\rangle\}_x$ is orthogonal, it is easy to see from (15.3.2) that

$$|\Psi_G|^2 = \langle\Psi_G|\Psi_G\rangle$$

$$= \left(\sum_{x\in\{0,1\}^k}\alpha_x^*\sqrt{\phi(x)}\,\langle x|\,\langle\psi_x|\right)\left(\sum_{y\in\{0,1\}^k}\alpha_y\sqrt{\phi(y)}\,|\psi_y\rangle\,|y\rangle\right)$$

$$= \sum_{x,y\in\{0,1\}^k}\alpha_x^*\alpha_y\sqrt{\phi(x)}\sqrt{\phi(y)}\,\langle x|\,\langle\psi_x|\psi_y\rangle\,|y\rangle$$

$$= \sum_{x\in\{0,1\}^k}|\alpha_x|^2\phi(x),$$

which corresponds precisely to the expectation $\mathbb{E}[\phi(X)]$ where the random variable X is discretised over the set with labels $\{0,1\}^k$, and where each $|\alpha_x|^2$ corresponds to the discrete probability of X being in x.

In order to retrieve the expectation we are after, we therefore simply need to run the circuit corresponding to $\mathcal{M}$, measure the output in the computational basis, and determine the probability of observing the state $|1\rangle$.

15.3.3   QMC speedup

The actual speedup of QMC resides in a subtle application of the *Amplitude Estimation* theorem and the *Powering Lemma*, which we present now.

Theorem 12 (Amplitude Estimation. Theorem 12 in [43]). *Assume that we have access to a quantum unitary operator $\mathcal{U}$ such that*

$$\mathcal{U}\ket{0} = \sqrt{1-\mathfrak{p}}\ket{\Psi_B}\ket{0} + \sqrt{\mathfrak{p}}\ket{\Psi_G}\ket{1},$$

for some states $\ket{\Psi_B}, \ket{\Psi_G}$. Then, for any $N \in \mathbb{N}$, the amplitude estimation algorithm outputs the estimate $\widehat{\mathfrak{p}}$ such that

$$\left|\widehat{\mathfrak{p}} - \mathfrak{p}\right| \le 2\pi \frac{\sqrt{\mathfrak{p}(1-\mathfrak{p})}}{N} + \frac{\pi^2}{N^2}$$

with probability at least $8/\pi^2$. Achieving this takes exactly N iterations.

Lemma 10 (Powering Lemma. Lemma 6.1 in [162]). *Let $\mathfrak{p}$ be a quantity to estimate and $\mathcal{U}$ an algorithm that outputs $\widehat{\mathfrak{p}}$ such that $\left|\widehat{\mathfrak{p}} - \mathfrak{p}\right| \le \varepsilon$ except with probability smaller than $1/2$. Then, for any $\delta \in (0,1)$, it suffices to repeat $\mathcal{U}$ about $\mathcal{O}(\log(1/\delta))$ times and to take the median to obtain $\left|\widehat{\mathfrak{p}} - \mathfrak{p}\right| \le \varepsilon$ with probability at least $1 - \delta$.*

In light of (15.3.1), the Amplitude Estimation theorem, combined with the Powering Lemma, shows that in order to obtain an estimate of the empirical mean

$$\bra{\psi}\mathcal{P}^\dagger\mathcal{P}\ket{\psi} = |\Psi_G|^2$$

with probability at least $1 - \delta$ (for any $\delta \in (0,1)$), i.e.,

$$\mathbb{P}\left(\left|\widehat{\mathfrak{p}} - \mathfrak{p}\right| \le \varepsilon\right) \ge 1 - \delta,$$

it suffices to apply the operators $\mathcal{M}$ and $\mathcal{P}$ about $\mathcal{O}(N\log(1/\delta))$ times, with

$$\varepsilon = 2\pi \frac{\sqrt{\mathfrak{p}(1-\mathfrak{p})}}{N},$$

so that, for any fixed $\delta \in (0,1)$, the computational cost is of order $\mathcal{O}(1/\varepsilon)$, achieving quadratic speedup compared to classical Monte Carlo.

15.4 Quantum Linear Solver

Harrow, Hassidim and Lloyd (HHL) [133] devised a quantum algorithm to solve linear systems, beating classical computation times. Linear systems are ubiquitous in applications, and many aspects of quantitative finance rely on being able to solve such (low- or high-dimensional) systems. We highlight below two key examples of fundamental importance in finance: solving Partial Differential Equations (PDEs) and portfolio optimisation.

15.4.1 Theoretical aspects

The problem can be stated as follows: given a matrix $A \in \mathcal{M}_N(\mathbb{C})$ and a vector $b \in \mathbb{C}^N$, find the vector $x \in \mathbb{C}^N$ such that

$$Ax = b. \tag{15.4.1}$$

In order for the algorithm to work, the matrix A needs to be Hermitian. If A is not Hermitian, we can nevertheless consider the augmented system

$$\begin{bmatrix} \mathbf{0}_{N,N} & A \\ A^\dagger & \mathbf{0}_{N,N} \end{bmatrix} \begin{bmatrix} \mathbf{0}_{N,1} \\ x \end{bmatrix} = \begin{bmatrix} b \\ \mathbf{0}_{N,1} \end{bmatrix},$$

similarly to the Hamiltonian embedding in Section 7.6. We assume from now now that A is indeed Hermitian. The first step of the algorithm is to assume that the vector b can be encoded into a quantum state $|b\rangle$ and to then rewrite (15.4.1) as

$$A |x\rangle = |b\rangle, \tag{15.4.2}$$

where we now look for the solution, not as an element of $\mathbb{C}^N$, but as a quantum state.

Since A is Hermitian, it admits the spectral decomposition (Section 1.1.5)

$$A = \sum_{j=0}^{N-1} \lambda_j |\phi_j\rangle \langle\phi_j|,$$

where $\lambda_0, \ldots, \lambda_{N-1}$ are its (not necessarily distinct) strictly positive eigenvalues with corresponding eigenstates $|\phi_0\rangle, \ldots, |\phi_{N-1}\rangle$, and we immediately obtain its inverse:

$$A^{-1} = \sum_{j=0}^{N-1} \frac{1}{\lambda_j} |\phi_j\rangle \langle\phi_j|.$$

We can also decompose $|b\rangle$ into the $(|\phi_j\rangle)_{j=0,\ldots,N-1}$ basis as

$$|b\rangle = \sum_{j=0}^{N-1} b_i |\phi_j\rangle,$$

and therefore the solution to (15.4.2) reads

$$|x\rangle = A^{-1} |b\rangle = \sum_{j=0}^{N-1} \frac{b_j}{\lambda_j} |\phi_j\rangle.$$

The goal of the QLS algorithm is thus to construct such a state, and we summarise it as Algorithm 13 below. Note that, since A is Hermitian, then, for any $t \in \mathbb{R}$, $U := \exp(iAt)$ is unitary with decomposition

$$U = \sum_{j=0}^{N-1} e^{i\lambda_j t} |\phi_j\rangle \langle\phi_j|.$$

In total, the QLS algorithm requires $n_l + n_b + 1$ qubits, where n_l is the number of qubits used to encode the n_l-bit binary representation of $(\lambda_j)_{j=0,\ldots,N-1}$ and n_b is the number of qubits used to convert b into $|b\rangle$ (and also the number of qubits to write the solution state). HHL runtime is of order $\mathrm{poly}(\log(N), \kappa)$ assuming that A is sparse with condition number κ, which yields an exponential speedup compared to the classical $\mathcal{O}(N\sqrt{\kappa})$ runtime.

Algorithm 13: HHL Quantum Linear Solver

Input: Hermitian matrix A and $n_l + n_b + 1$ qubits initialised at $|0\rangle^{\otimes n_l} |0\rangle^{\otimes n_b} |0\rangle$.

1: Load the data b into $|b\rangle$ using n_b qubits (with $N = 2^{n_b}$).

2: Apply QPE with

$$U := \exp(iAt),$$

after which the quantum state of the register is

$$\sum_{j=0}^{N-1} b_j |\lambda_j\rangle_{n_l} |\phi_j\rangle_{n_b} |0\rangle.$$

3: Rotate the ancillary qubit $|0\rangle$ controlled by $|\lambda_j\rangle_{n_l}$ to obtain

$$\sum_{j=0}^{N-1} b_j |\lambda_j\rangle_{n_l} |\phi_j\rangle_{n_b} \left(\sqrt{1 - \frac{C^2}{\lambda_j^2}} |0\rangle + \frac{C}{\lambda_j} |1\rangle \right),$$

for some normalising constant C (with $|C| < \min_j \lambda_j$).

4: Apply the inverse QPE to obtain

$$\sum_{j=0}^{N-1} b_j |0\rangle_{n_l} |\phi_j\rangle_{n_b} \left(\sqrt{1 - \frac{C^2}{\lambda_j^2}} |0\rangle + \frac{C}{\lambda_j} |1\rangle \right).$$

5: Measure the ancillary qubit in the computational basis. If the outcome is $|1\rangle$, the register is in the post-measurement state

$$C \sum_{j=0}^{N-1} \frac{b_i}{\lambda_i} |0\rangle_{n_l} |\phi_j\rangle_{n_b},$$

which, up to a normalisation factor, corresponds to the solution.

Result: Solution $|x\rangle$:

$$|x\rangle = A^{-1} |b\rangle = \sum_{j=0}^{N-1} \frac{b_j}{\lambda_j} |\phi_j\rangle.$$

15.4.2   Solving PDEs

One important example is finite-difference schemes for partial differential equations; standard tools can be consulted in [293], for example, and specific applications to finance can be found in [89].

Consider, for example, the Black-Scholes parabolic PDE

$$\partial_t V_t + rS\partial_S V_t + \frac{\sigma^2}{2}S^2\partial_{SS}^2 V_t = rV_t,$$

with boundary condition $V_T(S)$ (for instance, for a European call option with maturity $T > 0$ and strike $K > 0$, we have $V_T(S) = (S_T - K)_+ := \max(S_T - K, 0)$). Before trying to solve it, it is standard to simplify it. Let $\tau := T - t$ and define $g_\tau(S) := V_t(S)$, then $\partial_t V_t(S) = -\partial_\tau g_\tau(S)$, and hence

$$-\partial_\tau g_\tau + rS\partial_S g_\tau + \frac{\sigma^2}{2}S^2\partial_{SS}^2 g_\tau = rg_\tau,$$

with boundary condition $g_0(S)$. Now introduce $f_\tau(S) := e^{r\tau}g_\tau(S)$ so that

$$-\partial_\tau f_\tau + rS\partial_S f_\tau + \frac{\sigma^2}{2}S^2\partial_{SS}^2 f_\tau = 0,$$

with boundary condition $f_0(S)$. The transformation $x := \log(S)$ and the map $\psi_\tau(x) := f_\tau(S)$ yield, after simplifications,

$$-\partial_\tau \psi_\tau + \left(r - \frac{\sigma^2}{2}\right)\partial_x \psi_\tau + \frac{\sigma^2}{2}\partial_{xx}^2 \psi_\tau = 0, \qquad (15.4.3)$$

with boundary condition $\psi_0(x)$. Finally, setting ϕ_τ via $\psi_\tau(x) =: e^{\alpha x + \beta \tau}\phi_\tau(x)$ with

$$\alpha := -\frac{1}{\sigma^2}\left(r - \frac{\sigma^2}{2}\right) \quad \text{and} \quad \beta := -\frac{1}{2\sigma^2}\left(r - \frac{\sigma^2}{2}\right)^2,$$

implies that equation (15.4.3) becomes the heat equation

$$\partial_\tau \phi_\tau(x) = \frac{\sigma^2}{2} \partial^2_{xx} \phi_\tau(x),\tag{15.4.4}$$

for all $x \in \mathbb{R}$ with the (Dirichlet) boundary condition $\phi_0(x) = \mathrm{e}^{-\alpha x} \psi_0(x)$.

We now discretise this PDE using an explicit scheme, where the time derivative ∂_τ is evaluated by forward difference while the space derivative ∂_{xx} is approximated with a central difference scheme (implicit schemes or more general θ-schemes follow a similar logic). We consider (15.4.4) for $\tau > 0$ and x in some interval $[x_L, x_U] \in \mathbb{R}$, with (Dirichlet) boundary conditions $\phi(0, x) = f(x)$ (payoff at maturity), $\phi(\tau, x_L) = f_L(\tau)$, and $\phi(\tau, x_U) = f_U(\tau)$.

We start by constructing the time-space grid for the approximation scheme. For two integers m and n, we consider a uniform grid, i.e., we split the space axis into m intervals and the time axis into n intervals, and we denote $\mathcal{V} := \{0, 1, \ldots, n\}$ and $\mathcal{W} := \{0, 1, \ldots, m\}$. This means that each point on the grid has coordinates $(i\delta_T, x_L + j\delta_x)$ for $i \in \mathcal{V}$ and $j \in \mathcal{W}$, where

$$\delta_T := \frac{T}{n} \quad \text{and} \quad \delta_x := \frac{x_U - x_L}{m}.$$

At each node, we let $\phi_{i,j} := \phi(i\delta_T, x_L + j\delta_x)$ denote the value of the function u. Note in particular that the boundary conditions imply

$$\phi_{0,j} = f(x_L + j\delta_x), \quad \phi_{i,0} = f_L(i\delta_T), \quad \phi_{i,m} = f_U(i\delta_T).$$

More precisely, we consider the following approximations

$$\partial_\tau \phi(\tau, x) = \frac{\phi(\tau + \delta_T, x) - \phi(\tau, x)}{\delta_T} + \mathcal{O}(\delta_T),$$

$$\partial_{xx} \phi(\tau, x) = \frac{\phi(\tau, x + \delta_x) - 2\phi(\tau, x) + \phi(\tau, x - \delta_x)}{\delta_x^2} + \mathcal{O}(\delta_x^2).$$

Ignoring the terms in δ_T and δ_x^2, the heat equation at the node $(i\delta_T, x_L + j\delta_x)$ becomes

$$\frac{\phi_{i+1,j} - \phi_{i,j}}{\delta_T} + \mathcal{O}(\delta_T) = \frac{\sigma^2}{2} \frac{\phi_{i,j+1} - 2\phi_{i,j} + \phi_{i,j-1}}{\delta_x^2} + \mathcal{O}(\delta_x^2),\tag{15.4.5}$$

which we can rewrite as

$$\phi_{i+1,j} = \frac{\delta_T}{\delta_x^2}\frac{\sigma^2}{2}\phi_{i,j+1} + \left(1 - \frac{\delta_T}{\delta_x^2}\sigma^2\right)\phi_{i,j} + \frac{\delta_T}{\delta_x^2}\frac{\sigma^2}{2}\phi_{i,j-1},$$

for all $i = 0, \ldots, n-1$, $j = 1, \ldots, m-1$. To rewrite this in matrix form, introduce for each $i = 0, \ldots, n$, $\phi_i \in \mathbb{R}^{m-1}$, $\mathrm{B}_i \in \mathbb{R}^{m-1}$ and the matrix $\mathrm{A} \in \mathcal{M}_{m-1}(\mathbb{R})$ by

$$\phi_i := (\phi_{i,1}, \ldots, \phi_{i,m-1})^\top,$$
$$\mathrm{B}_i := (\phi_{i,0}, 0, \ldots, 0, \phi_{i,m})^\top,$$
$$\mathrm{A} := \mathrm{T}_{m-1}\left(1 - \alpha\sigma^2, \frac{\alpha\sigma^2}{2}, \frac{\alpha\sigma^2}{2}\right),$$

where

$$\alpha := \frac{\delta_T}{\delta_x^2}$$

and where $\mathrm{T}_{m-1}(\cdot)$ denotes the tridiagonal matrix of dimension $(m-1) \times (m-1)$.

The recursion (15.4.5) thus becomes

$$\phi_{i+1} = \mathrm{A}\phi_i + \frac{\alpha\sigma^2}{2}\mathrm{B}_i, \quad \text{for each } i = 0, \ldots, n-1, \tag{15.4.6}$$

with the time boundary condition

$$\phi_0 = (\phi_{0,1}, \ldots, \phi_{0,m-1})^\top = (f(x_L + \delta_x), \ldots, f(x_L + (m-1)\delta_x))^\top.$$

Leaving the boundary term B_i aside, the recursion (15.4.6) thus is exactly of the form (15.4.1) and can therefore be tackled using the HHL algorithm.

This is the obvious first step to investigate the use of HHL-type algorithms in quantitative finance, and further developments have already been proposed in [104, 108, 202, 337], with or without finance applications in mind.

15.4.3 Application to portfolio optimisation

The second immediate application of QLS in finance is for portfolio optimisation. Indeed, the standard Markowitz-type problem of the form (3.3.1) in Section 3.3 is readily formulated (at least for weights in $\{0, 1\}$) as a linear problem, the constraints only increasing the dimension as Lagrange multipliers. We shall not dive into the details here as this is a rather novel development with huge potential but limited results so far, and instead refer the reader to [201, 332] for promising implementations and details.

Summary

In this chapter, we covered several important quantum algorithms that rely on the existence of a quantum computing hardware with characteristics that exceed the capabilities of currently available NISQ computers. However, the very presence of these algorithms and their potential to achieve quadratic or even exponential speedup provides strong motivation for the rapid development of quantum computers.

This chapter completes the book. Looking ahead, we see a bright future for quantum computing. In the update to their quantum computing development roadmap [154], IBM outlined an exciting vision with the goal of building quantum-centric supercomputers. This will incorporate quantum processors, classical processors, quantum communication networks, and classical networks. The most recent deliverables include 433-qubit *Osprey* processor (released in 2022) and 1,121-qubit *Condor* processor (released in 2023). The next step is the development of ways to link processors together into a modular system capable of scaling without physical limitations. This modular, multi-chip scaling technology has also been envisaged by Rigetti. Rigetti released their 4th generation single-chip 84-qubit *Ankaa* processor in 2023 and plans to release the multi-chip 336-qubit *Lyra* processor in the near future. This new processor is expected to combine the anticipated improvements of the 84-qubit processor with the modular, multi-chip scaling technology of Rigetti's Aspen-M machine. These machines are expected to deliver increased performance across the key dimensions of speed, scale, and fidelity [267].

We also expect to see significant progress in the trapped ion space. IonQ announced several major breakthroughs that may have a major impact on the way quantum algorithms are designed and run on trapped ion quantum computing hardware. This includes, for example, a new family of n-qubit gates, such as the n-qubit Toffoli gate, which flips a select qubit if and only if all the other qubits are in a particular state. Unlike standard two-qubit quantum computing gates, the n-qubit Toffoli gate acts on many qubits at once, leading to more efficient operations [156].

Table 15.1 summarises the outlook from several recent roadmaps (IBM [155], Quantinuum [261], IonQ [157]), all pointing towards achieving improvements in the number of high fidelity quantum logic gates due to hardware development and error correction.

	IBM	Quantinuum	IonQ
2025	156 qubits (x7 = 1092 qubits) 5k gates	96 qubits 2Q gate error $< 5 \cdot 10^{-4}$ logical error $< 10^{-4}$	2Q gate error $< 10^{-3}$ logical error $\sim 10^{-5}$
2027	156 qubits (x7 = 1092 qubits) 10k gates	192 qubits 2Q gate error $< 2 \cdot 10^{-4}$ logical error $\sim 10^{-5}$	
2029	200 qubits 100M gates	1000's qubits 2Q gate error $< 10^{-4}$ logical error $\sim 10^{-10}$–10^{-5}	

Table 15.1: IBM, Quantinuum, and IonQ roadmaps. The emphasis is very much on the quality of individual qubits and error correction.

Quantum annealing is going from strength to strength. In a white paper [39], D-Wave introduced the new *Zephyr* graph with better connectivity than its predecessors, *Chimera* and *Pegasus*. Plans are in place for a 7,000-qubit chip based on *Zephyr* [93]. Early benchmarks with smaller-scale prototype systems consisting of 500+ qubits have demonstrated more compact embedding, lower error rates, improved solution quality, and an increased probability of finding optimal solutions [225].

But, ultimately, it is up to the users to try and test various hardware and software solutions on a variety of use cases. We encourage our readers to experiment and apply the methods of quantum computing to their own spheres of interest and discover new quantum algorithms and applications. This is an exciting journey and a great opportunity to participate in the collective effort of achieving quantum advantage for the benefit of wider society.

Bibliography

[1] A. Abbas, D. Sutter, C. Zoufal, A. Lucchi, A. Figalli and S. Woerner. The power of quantum neural networks. *Nature Computational Science*, 1, 2021.

[2] D.M. Abrams, N. Didier, B.R. Johnson, M.P. da Silva and C.A. Ryan. Implementation of the XY interaction family with calibration of a single pulse. *Nature Electronics*, 3, 2020.

[3] D.H. Ackley, G.E. Hinton and T.J. Sejnowski. A learning algorithm for Boltzmann machines. *Cognitive Science*, 9(1), 1985.

[4] S. Adachi and M. Henderson. Application of quantum annealing to training of deep neural networks. arXiv:1510.06356, 2015.

[5] G. Agliardi and E. Prati. Optimal tuning of quantum generative adversarial networks for multivariate distribution loading. *Quantum Reports*, 4(1), 2022.

[6] A. Agresti. Categorical Data Analysis. Wiley, 3rd Edition, 2013.

[7] D. Aharonov, W. van Dam, J. Kempe, Z. Landau, S. Lloyd and O. Regev. Adiabatic quantum computation is equivalent to standard quantum computation. *SIAM Review*, 50(4), 2008.

[8] D. Aharonov and L. Zhou. Hamiltonian sparsification and gap-simulations. arXiv:1804.11084, 2018.

[9]   O. Akbilgic, H. Bozdogan and M.E. Balaban. Istanbul Stock Exchange Dataset. UCI
      Machine Learning Repository. `http://archive.ics.uci.edu/ml`. UC Irvine, 2013.

[10] O. Akbilgic, H. Bozdogan and M.E. Balaban. A novel Hybrid RBF neural networks
      model as a forecaster. *Statistics and Computing*, 24(3), 2013.

[11] M. Amin and M. Steininger. Adiabatic quantum computation with superconducting
      qubits. US Patent US7135701B2, 2006.

[12] K.P. Anagnostopoulos and G. Mamanis. A portfolio optimization model with three
      objectives and discrete variables. *Computers & Operations Research*, 37(7), 2010.

[13] M. Arjovsky, S. Chintala and L. Bottou. Wasserstein generative adversarial networks.
      International Conference on Machine Learning, 2017.

[14] S. Arora and S. Kale. A combinatorial, primal-dual approach to semidefinite programs.
      Journal of the ACM, 63(2), 2016.

[15] S. Arunachalam, V. Gheorghiu, T. Jochym-O'Connor, M. Mosca and P.V. Srinivasan.
      On the robustness of bucket brigade quantum RAM. *New Journal of Physics*, 17(12),
      2015.

[16] F. Arute, K. Arya, R. Babbush, D. Bacon, J.C. Bardin, R. Barends, R. Biswas, S. Boixo,
      F.G.S.L. Brandao, D.A. Buell, B. Burkett, Y. Chen, Z. Chen, B. Chiaro, R. Collins, W.
      Courtney, A. Dunsworth, E. Farhi, B. Foxen, A. Fowler, C. Gidney, M. Giustina, R.
      Graff, K. Guerin, S. Habegger, M.P. Harrigan, M.J. Hartmann, A. Ho, M. Hoffmann, T.
      Huang, T.S. Humble, S.V. Isakov, E. Jeffrey, Z. Jiang, D. Kafri, K. Kechedzhi, J. Kelly,
      P.V. Klimov, S. Knysh, A. Korotkov, F. Kostritsa, D. Landhuis, M. Lindmark, E. Lucero,
      D. Lyakh, S. Mandrà, J.R. McClean, M. McEwen, A. Megrant, X. Mi, K. Michielsen, M.
      Mohseni, J. Mutus, O. Naaman, M. Neeley, C. Neill, M.Y. Niu, E. Ostby, A. Petukhov,
      J.C. Platt, C. Quintana, E.G. Rieffel, P. Roushan, N.C. Rubin, D. Sank, K.J. Satzinger, V.
      Smelyanskiy, K.J. Sung, M.D. Trevithick, A. Vainsencher, B. Villalonga, T. White, Z.J.
      Yao, P. Yeh, A. Zalcman, H. Neven and J.M. Martinis. Quantum supremacy using a
      programmable superconducting processor. *Nature*, 574, 2019.

[17] S.A. Assefa, D. Dervovic, M. Mahfouz, R.E. Tillman, P. Reddy and M. Veloso. Generating synthetic data in finance: opportunities, challenges and pitfalls. *Proceedings of the First ACM International Conference on AI in Finance*, 2020.

[18] A. Assouel, A. Jacquier and A. Kondratyev. A quantum generative adversarial network for distributions. *Quantum Machine Intelligence*, 4(2), 2022.

[19] J.E. Avron and A. Elgart. Adiabatic theorem without a gap condition. *Communications In Mathematical Physics*, 203(445), 1999.

[20] J.E. Avron, R. Seiler and L.G. Yaffe. Adiabatic theorems and applications to the quantum Hall effect. *Communications In Mathematical Physics*, 110(1), 1987.

[21] P. Baldi and R. Vershynin The capacity of feedforward neural networks. *Neural Networks*, 116, 2019.

[22] V. Bapst, L. Foini, F. Krzakala, G. Semerjian and F. Zamponi. The quantum adiabatic algorithm applied to random optimization problems: the quantum spin glass perspective. *Physics Reports*, 523(127), 2013.

[23] F. Barahona. On the computational complexity of Ising spin glass models. *Journal of Physics A*, 15(10), 1982.

[24] B. Barak, A. Moitra, R. O'Donnell, P. Raghavendra, O. Regev, D. Steurer, L. Trevisan, A. Vijayaraghavan, D. Witmer and J. Wright. Beating the random assignment on constraint satisfaction problems of bounded degree. *Proceedings, International Conference on Randomization and Computation*, 2015.

[25] P. Barkoutsos, G. Nannicini, A. Robert, I. Tavernelli and S. Woerner. Improving Variational Quantum Optimization using CVaR. *Quantum*, 4, 2020.

[26] A.G. Baydin, B.A. Pearlmutter, A.A. Radul and J.M. Siskind. Automatic differentiation in machine learning: a survey. *Journal of Machine Learning Research*, 18, 2018.

[27] M. Benedetti, D. Garcia-Pintos, O. Perdomo, V. Leyton-Ortega, Y. Nam and A. Perdomo-Ortiz. A generative modeling approach for benchmarking and training

shallow quantum circuits. *Quantum Information*, 5(45), 2019.

[28] M. Benedetti, E. Lloyd, S. Sack and M. Fiorentini. Parameterized quantum circuits as machine learning models. *Quantum Science and Technology*, 4(4), 2019.

[29] M. Benedetti, B. Coyle, M. Fiorentini, M. Lubasch and M. Rosenkranz. Variational Inference with a Quantum Computer. *Physical Review Applied*, 16(044057), 2021.

[30] K. Berahmand, F. Daneshfar, E.S. Salehi, Y. Li and Y. Xu. Autoencoders and their applications in machine learning: a survey. *Artificial Intelligence Review*, 57(28), 2024.

[31] J. Berkson. Application of the Logistic Function to Bio-Assay. *Journal of the American Statistical Association*, 39(227), 1944.

[32] C. Bernhardt. Quantum Computing for Everyone. MIT Press, 2019.

[33] M.V. Berry. Quantal phase factors accompanying adiabatic changes. *Proceedings of the Royal Society of London A*, 392(45), 1964.

[34] D. Berry, G. Ahokas, R. Cleve and B.C. Barry. Efficient quantum algorithms for simulating sparse Hamiltonians. *Communications in Mathematical Physics*, 270(2), 2007.

[35] J.D. Biamonte and P.J. Love. Realizable Hamiltonians for universal adiabatic quantum computers. *Physical Review A*, 78(1), 2008.

[36] A. Billionnet and B. Jaumard. A decomposition method for minimizing quadratic pseudo-Boolean functions. *Operations Research Letters*, 8(3), 1989.

[37] C. Bishop. Pattern Recognition and Machine Learning. Springer, 2006.

[38] S. Boixo, V.N. Smelyanskiy, A. Shabani, S. V. Isakov, M. Dykman, V.S. Denchev, M.H. Amin, A.Y. Smirnov, M. Mohseni and H. Neven. Computational multiqubit tunnelling in programmable quantum annealers. *Nature Communications*, 7(10327), 2016.

[39] K. Boothby, A.D. King and J. Raymond. Zephyr Topology of D-Wave Quantum Processors. D-Wave Technical Report, 2021.

[40] V. Bornemann. Homogenization in Time of Singular Perturbed Mechanical Systems. *Lecture Notes in Mathematics*, 1687, Springer, 1998.

[41] M. Born and V. Fock. Beweis des Adiabatensatzes. *Zeitschrift für Physik*, 51(3), 1928.

[42] F.G.S.L. Brandão and K.M. Svore. Quantum speed-ups for semidefinite programming. 58th Annual Symposium on Foundations of Computer Science, IEEE, 2017.

[43] G. Brassard, P. Hoyer, M. Mosca and A. Tapp. Quantum amplitude amplification and estimation. *Contemporary Mathematics*, 305, 2002.

[44] S. Bravyi, A. Kliesch, R. Koenig and E. Tang. Obstacles to Variational Quantum Optimization from Symmetry Protection. *Physical Review Letters*, 125(260505), 2020.

[45] S. Bravyi, D.P. DiVincenzo, R.I. Oliveira and B.M. Terhal. The complexity of stoquastic local Hamiltonian problems. *Quantum Information & Computation*, 8(5), 2008.

[46] C.D. Bruzewicz, J. Chiaverini, R. McConnell and J.M. Sage. Trapped-ion quantum computing: progress and challenges. *Applied Physics Reviews*, 6(2), 2019.

[47] J.-L. Brylinski and R. Brylinski. Universal quantum gates. In *Mathematics of Quantum Computation*. Chapman and Hall/CRC, 2002.

[48] H. Bühler, B. Horvath, T. Lyons, I. Perez Arribaz and B. Wood. A data-driven market simulator of financial time series for small data environments. arXiv:2006.14498, 2020.

[49] H. Buhrman, R. Cleve, J. Watrous and R. de Wolf. Quantum fingerprinting. *Physical Review Letters*, 87(167902), 2001.

[50] Y. Cai. Achieve the minimum width of neural networks for universal approximation. *International Conference on Learning Representations*, 2023.

[51] E. Campbell, A. Khurana and A. Montanaro. Applying quantum algorithms to constraint satisfaction problems. *Quantum*, 3, 2019.

[52] Y. Cao, X. Liu, J. Zhai and S. Hua. A two-stage Bayesian network model for corporate bankruptcy prediction. *International Journal of Finance & Economics*, 27(1), 2022.

[53] M.A. Carreira-Perpiñán and G.E. Hinton. On contrastive divergence learning. *AIS-TATS*, 2005.

[54] M. Cerezo, A. Sone, T. Volkoff, L. Cincio and P.J. Coles. Cost function dependent barren plateaus in shallow parametrized quantum circuits. *Nature Communications*, 12, 2021.

[55] S. Chakrabarti, H. Yiming, T. Li, S. Feizi and X. Wu, Xiaodi. Quantum Wasserstein generative adversarial networks. *NeurIPS*, 32, 2019.

[56] R. Chandra and Y. He. Bayesian neural networks for stock price forecasting before and during COVID-19 pandemic. *PLoS ONE*, 16(7), 2021.

[57] D.T. Chang. Bayesian Neural Networks: Essentials. arXiv:2106.13594, 2021.

[58] S.Y.-C. Chen, T.-C. Wei, C. Zhang, H. Yu and S. Yoo. Quantum Convolutional Neural Networks for High Energy Physics Data Analysis. *Physical Review Research*, 4(1), 2022.

[59] S.Y.-C. Chen, S. Yoo and Y.-L.L. Fang. Quantum Long Short-Term Memory. *International Conference on Acoustics, Speech and Signal Processing*, 2022.

[60] S. Cheng, J. Chen and L. Wang. Information perspective to probabilistic modeling: Boltzmann machines versus Born machines. *Entropy*, 20(583), 2018.

[61] A.M. Childs, E. Farhi and J. Preskill. Robustness of adiabatic quantum computation. *Physical Review A*, 65, 2001.

[62] K. Cho, A. Ilin and T. Raiko. Improved learning of Gaussian-Bernoulli restricted Boltzmann machines. *Proceedings of the 20th International Conference on Artificial Neural Networks*, 2011.

[63] V. Choi. Minor-embedding in adiabatic quantum computation: I. The parameter setting problem. *Quantum Information Processing*, 7(5), 2008.

[64] V. Choi. Minor-embedding in adiabatic quantum computation: II. Minor-universal graph design. *Quantum Information Processing*, 10(3), 2011.

[65] C. Ciliberto, M. Herbster, A.D. Ialongo, M. Pontil, A. Rocchetto, S. Severini and L. Wossnig. Quantum machine learning: a classical perspective. *Proceedings of the Royal Society A*, 474(2209), 2018.

[66] L. Cincio, Y. Subaşi, A.T. Sornborger and P.J. Coles. Learning the quantum algorithm for state overlap. *New Journal of Physics*, 20(11), 2018.

[67] J.I. Cirac, R. Blatt, A.S. Parkins and P. Zoller. Preparation of Fock states by observation of quantum jumps in an ion trap. *Physical Review Letters*, 70(6-8), 1993.

[68] J.I. Cirac and P. Zoller. Quantum Computations with Cold Trapped Ions. *Physical Review Letters*, 74(20), 1995.

[69] I. Cong, S. Choi and M.D. Lukin. Quantum convolutional neural networks. *Nature Physics*, 15(12), 2019.

[70] C. Cortes and V. Vapnik. Support-vector networks. *Machine Learning*, 20(3), 1995.

[71] L. Coslovich, R. Pesenti and W. Ukovich. Large-scale set partitioning problems: Some real-world instances hide a beneficial structure. *Technological and Economic Development of Economy*, 12(1), 2006.

[72] B. Coyle, M. Henderson, J. Chan Jin Le, N. Kumar, M. Paini and E. Kashefi. Quantum versus classical generative modelling in Finance. *Quantum Science and Technology*, 6(2), 2021.

[73] B. Coyle, D. Mills, V. Danos and E. Kashefi. The Born supremacy: Quantum advantage and training of an Ising Born machine. *Quantum Information*, 6(1), 2020.

[74] A. Crespi, R. Ramponi, R. Osellame, L. Sansoni, I. Bongioanni, F. Sciarrino, G. Vallone and P. Mataloni. Integrated photonic quantum gates for polarization qubits. *Nature Communications*, 1570, 2011.

[75] G. Cybenko. Approximation by superpositions of a sigmoidal function. *Mathematics of Control, Signals and Systems*, 2, 1989.

[76] E.D. Dahl. Programming with D-Wave: map coloring problem. D-Wave White Paper,

2013.

[77] P.-L. Dallaire-Demers and N. Killoran. Quantum generative adversarial networks. *Physical Review A*, 98(1), 2018.

[78] G.B. Dantzig and J.H. Ramser. The truck dispatching problem. *Management Science*, 6(1), 1959.

[79] S. Darolles and C. Gouriéroux. Conditionally Fitted Sharpe Performance with an Application to Hedge Fund Rating. *Journal of Banking & Finance*, 34(3), 2010.

[80] B.S. Dees, L. Stanković, A.G. Constantinides and D.P. Mandic. Portfolio Cuts: A Graph-Theoretic Framework to Diversification. *International Conference on Acoustics, Speech and Signal Processing*, 2020.

[81] H. Delfs and H. Knebl. Introduction to Cryptography: Principles and Applications. Springer, 2007.

[82] V. DeMiguel, L. Garlappi and R. Uppal. Optimal Versus Naive Diversification: How Inefficient is the $1/N$ Portfolio Strategy? *The Review of Financial Studies*, 22(5), 2009.

[83] V.S. Denchev, S. Boixo, S.V. Isakov, N. Ding, R. Babbush, V. Smelyanskiy, J. Martinis and H. Neven. What is the computational value of finite-range tunneling? *Physical Review X*, 6(3), 2016.

[84] D.-L. Deng, X. Li and S.D. Sarma. Machine Learning Topological States. *Physical Review B*, 96(195145), 2017.

[85] D.-L. Deng, X. Li and S.D. Sarma. Quantum Entanglement in Neural Network States. *Physical Review X*, 7(021021), 2017.

[86] P.A.M. Dirac. The Principles of Quantum Mechanics. Oxford University Press, 1930.

[87] D.P. DiVincenzo. The physical implementation of quantum computation. *Fortschritte der Physik*, 48(9–11), 2000.

[88] Y. Du, M.-H. Hsieh, T. Liu and D. Tao. Expressive power of parameterized quantum circuits. *Physical Review Research*, 2(033125), 2020.

[89] D.J. Duffy. Finite Difference Methods in Financial Engineering: a Partial Differential Equation Approach. John Wiley & Sons, 2013.

[90] D-Wave Systems. Practical quantum computing. D-Wave Overview, 2020.

[91] D-Wave Systems. D-Wave QPU Architecture: Topologies. D-Wave, 2021.

[92] D-Wave Systems. Advantage Performance Update. D-Wave, 2021.

[93] D-Wave Systems. Ahead of the Game: D-Wave Delivers Prototype of Next-Generation Advantage2 Annealing Quantum Computer. D-Wave, 2022.

[94] G.K. Dziugaite and D.M. Roy. Computing nonvacuous generalization bounds for deep (stochastic) neural networks with many more parameters than training data. *Proceedings of the Uncertainty in Artificial Intelligence Conference*, 2017.

[95] F. Eckerli and J. Osterrieder. Generative Adversarial Networks in finance: an overview. arXiv:2106.06364, 2021.

[96] E. Farhi, J. Goldstone and S. Gutmann. A Quantum approximate optimisation algorithm. arXiv:1411.4028, 2014.

[97] E. Farhi, J. Goldstone, S. Gutmann, J. Lapan, A. Lundgren and D. Preda. A quantum adiabatic evolution algorithm applied to random instances of an NP-Complete problem. *Science*, 292(5516), 2001.

[98] E. Farhi, J. Goldstone, S. Gutmann and M. Sipser. Quantum computation by adiabatic evolution. arXiv:0001106, 2000.

[99] E. Farhi and A.W. Harrow. Quantum Supremacy through the Quantum Approximate Optimization Algorithm. arXiv:1602.07674, 2016.

[100] E. Farhi and H. Neven. Classification with quantum neural networks on near-term processors. arXiv:1802.06002, 2018.

[101] R. Feynman, R. Leighton and M. Sands. The Schrödinger equation in a classical context: a seminar on superconductivity. The Feynman Lectures on Physics. Addison Wesley, 2006.

[102] A. Fischer and C. Igel. An introduction to restricted Boltzmann machines. In *Progress in Pattern Recognition, Image Analysis, Computer Vision, and Applications.* Lecture Notes in Computer Science, 7441, Springer, 2012.

[103] A. Fischer and C. Igel. Training restricted Boltzmann machines: an introduction. *Pattern Recognition*, 47(1), 2014.

[104] F. Fontanela, A. Jacquier and M. Oumgari. A quantum algorithm for linear PDEs arising in finance. *SIAM Journal on Financial Mathematics*, 12(4), 2021.

[105] V. Fortuin. Priors in Bayesian deep learning: a review. *International Statistical Review*, 90(3), 563-591, 2022.

[106] M. Frank. Foundations of generalized reversible computing. *International Conference on Reversible Computation*, 2017.

[107] Y. Freund and R.E. Schapire. A decision-theoretic generalization of on-line learning and an application to boosting. *Journal of Computer and System Sciences*, 55(1), 1997.

[108] P. García-Molina, J. Rodríguez-Mediavilla and J.J. García-Ripoll. Quantum Fourier analysis for multivariate functions and applications to a class of Schrödinger-type partial differential equations. *Physical Review A*, 105, 2022.

[109] D. Garvin, O. Kondratyev, A. Lipton and M. Paini. Quantum two-sample test for investment strategies. SSRN:4789400, 2024.

[110] D. Garvin, O. Kondratyev, A. Lipton and M. Paini. Symmetric encryption on a quantum computer. Cryptology ePrint Archive:2024/1836, 2024.

[111] D. Garvin, O. Kondratyev, A. Lipton and M. Paini. Density matrix classifier. SSRN:5014387, 2024.

[112] B. German. Glass Identification. UCI Machine Learning Repository. http://archive.ics.uci.edu/ml. UC Irvine, 1987.

[113] A. Gilyén, S. Lloyd and E. Tang. Quantum-inspired low-rank stochastic regression with logarithmic dependence on the dimension. arXiv:1811.04909, 2018.

[114] V. Giovannetti, S. Lloyd and L. Maccone. Quantum random access memory. *Physical Review Letters*, 100(16), 2008.

[115] P. Glasserman. Monte Carlo Methods in Financial Engineering. Stochastic Modelling and Applied Probability, 53, Springer, 2003.

[116] M.X. Goemans. Semidefinite programming in combinatorial optimization. *Mathematical Programming*, 79, 1997.

[117] L. Gonon and A. Jacquier. Universal Approximation Theorem and error bounds for quantum neural networks and reservoirs. arXiv:2307.12904, 2023.

[118] I. Goodfellow, Y. Bengio and A. Courville. Deep Learning. MIT Press, 2016.

[119] I. Goodfellow, J. Pouget-Abadie, M. Mirza, B. Xu, D. Warde-Farley, S. Ozair, A. Courville and Y. Bengio. Generative Adversarial Nets. NiPS, 2014.

[120] H. Gouk. Regularisation of neural networks by enforcing Lipschitz continuity. *Machine Learning*, 110(2), 2021.

[121] E. Grant, T.S. Humble and B. Stump. Benchmarking Quantum Annealing Controls with Portfolio Optimization. *Physical Review Applied*, 15(1), 2021.

[122] A. Gretton, K.M. Borgwardt, M.J. Rasch, B. Schölkopf and A. Smola. A Kernel Two-Sample Test. *Journal of Machine Learning Research*, 13(25), 2012.

[123] G. Gripenberg. Approximation by neural networks with a bounded number of nodes at each level. *Journal of Approximation Theory*, 122(2), 2003.

[124] L.K. Grover. A fast quantum mechanical algorithm for database search. *Proceedings of the 28th annual ACM symposium on Theory of computing*. ACM, 1996.

[125] L.K. Grover and T. Rudolph. Creating superpositions that correspond to efficiently integrable probability distributions. arXiv:0208112, 2002.

[126] M. Grundmann. Quantum devices of reduced dimensionality. Encyclopedia of Condensed Matter Physics, Elsevier, 2005.

[127] G.G. Guerreschi. Solving quadratic unconstrained binary optimization with divide-and-conquer and quantum algorithms. arXiv:2101.07813, 2021.

[128] I. Gulrajani, F. Ahmed, M. Arjovsky, V. Dumoulin and A.C. Courville. Improved training of Wasserstein GANs. *NeurIPS*, 30, 2017.

[129] S. Hadfield, Z. Wang, B. O'Gorman, E.G. Rieffel, D. Venturelli and R. Biswas. From the quantum approximate optimization algorithm to a quantum alternating operator ansatz. *Algorithms*, 12(2), 2019.

[130] B.C. Hall. Quantum Theory for Mathematicians. *Graduate Texts in Mathematics*, 267, Springer, 2013.

[131] R. Hamerly, T. Inagaki, P.L. McMahon, D. Venturelli, A. Marandi, E. Ng, C. Langrock, K. Inaba, T. Honjo, K. Enbutsu, T.Umeki, R. Kasahara, S. Utsunomiya, S. Kako, K. Kawarabayashi, R.L. Byer, M.M. Fejer, H. Mabuchi, D. Englund, E. Rieffel, H. Takesue and Y. Yamamoto. Experimental investigation of performance differences between coherent Ising machines and a quantum annealer. *Science Advances*, 5(5), 2019.

[132] J.M. Hammersley and P. Clifford. Markov fields on finite graphs and lattices. Unpublished, 1971.

[133] A.W. Harrow, A. Hassidim and S. Lloyd. Quantum algorithm for linear systems of equations. *Physical Review Letters*, 103(150502), 2009.

[134] R. Hassan, B. Cohanim, O. de Weck and G. Venter. A comparison of particle swarm optimization and the genetic algorithm. *46th AIAA/ASME/ASCE/AHS/ASC Structures, Structural Dynamics and Materials Conference*, 2005.

[135] T. Hastie, R. Tibshirani and J. Friedman. The Elements of Statistical Learning. Springer Series in Statistics, 2009.

[136] M.B. Hastings. Classical and quantum bounded depth approximation algorithms. *Quantum Information & Computation*, 19(13-14), 2019.

[137] V. Havlíček, A.D. Córcoles, K. Temme, A.W. Harrow, A. Kandala, J.M. Chow and J.M.

Gambetta. Supervised learning with quantum-enhanced feature spaces. *Nature*, 567, 2019.

[138] J. He and L. Kang. On the convergence rates of genetic algorithms. *Theoretical Computer Science*, 229, 1999.

[139] L.-P. Henry, S. Thabet, C. Dalyac and L. Henriet. Quantum evolution kernel: Machine learning on graphs with programmable arrays of qubits. *Physical Review A*, 104(3), 2021.

[140] A.D. Hill, M.J. Hodson, N. Didier and M.J. Reagor. Realization of arbitrary doubly-controlled quantum phase gates. arXiv:2108.01652, 2021.

[141] G. Hinton. Training products of experts by minimizing contrastive divergence. *Neural Computation*, 14(8), 2002.

[142] G. Hinton. A practical guide to training restricted Boltzmann machines. In *Neural Networks: Tricks of the Trade*. Lecture Notes in Computer Science, 7700, Springer, 2012.

[143] G. Hinton and R. Salakhutdinov. Reducing the dimensionality of data with neural networks. *Science*, 313, 2006.

[144] T.K. Ho. Random Decision Forests. *Proceedings of the 3rd International Conference on Document Analysis and Recognition*, 1995.

[145] C.A.R. Hoare. Algorithm 64: Quicksort. *Communications of the ACM*, 4(7), 1961.

[146] M. Hodson, B. Ruck, H. Ong, D. Garvin and S. Dulman. Portfolio rebalancing experiments using the Quantum Alternating Operator Ansatz. arXiv:1911.05296, 2019.

[147] Z. Holmes, K. Sharma, M. Cerezo and P.J. Coles. Connecting ansatz expressibility to gradient magnitudes and barren plateaus. *PRX Quantum*, 3(1), 2022.

[148] F.-Y. Hong, Y. Xiang, Z.-Y. Zhu, L.-Z. Jiang and L.-N. Wu. Robust quantum random access memory. *Physical Review A*, 86(1), 2012.

[149] K. Hornik. Approximation capabilities of multilayer feedforward networks. *Neural*

Networks, 4(2), 1991.

[150] K. Hornik, M. Stinchcombe and H. White. Universal approximation of an unknown mapping and its derivatives using multilayer feedforward networks. *Neural Networks*, 3(5), 1990.

[151] T. Hur, L. Kim and D.K. Park. Quantum convolutional neural network for classical data classification. *Quantum Machine Intelligence*, 4(1), 2022.

[152] H.Y. Huang, M. Broughton, M. Mohseni, R. Babbush, S. Boixo, H. Neven and J.R. McClean. Power of data in quantum machine learning. *Nature Communications*, 12(2631), 2021.

[153] W. Huggins, P. Patel, K.B. Whaley and E.M. Stoudenmire. Towards quantum machine learning with tensor networks. *Quantum Science and Technology*, 4(2), 2019.

[154] IBM Quantum Roadmap: Expanding the IBM Quantum roadmap to anticipate the future of quantum-centric supercomputing. IBM, 2022.

[155] IBM: Updated Quantum Roadmap, 2024. https://www.ibm.com/quantum/blog/ibm-quantum-roadmap-2025, 2024.

[156] IonQ: Duke University and IonQ Develop New Quantum Computing Gate, Only Possible on IonQ and Duke Systems, February 2022. https://investors.ionq.com/news/news-details/2022/Duke-University-and-IonQ-Develop-New-Quantum-Computing-Gate-Only-Possible-on-IonQ-and-Duke-Systems/default.aspx, 2022.

[157] IonQ: Accelerated Roadmap and New Technical Milestones, 2024. https://investors.ionq.com/news/news-details/2024/IonQ-Unveils-Accelerated-Roadmap-and-New-Technical-Milestones-to-Propel-Commercial-Quantum-Advantage-Forward/default.aspx, 2024.

[158] IQM: IQM Quantum Computers achieves new technology milestones with 99.9% 2-qubit gate fidelity and 1 millisecond coherence time, 2024.

https://www.meetiqm.com/newsroom/press-releases/iqm-achieves-new-technology-milestones, 2024.

[159] E. Ising. Beitrag zur Theorie des Ferromagnetismus. *Zeitschrift für Physik*, 31(1), 1925.

[160] B. Jackson, J.D. Scargle, D. Barnes, S. Arabhi, A. Alt, P. Gioumousis, E. Gwin, P. Sangtrakulcharoen, L. Tan and T.T. Tsai. An algorithm for optimal partitioning of data on an interval. *IEEE Signal Processing Letters*, 12(2), 2005.

[161] S. Jansen, R. Seiler and M.-B. Ruskai. Bounds for the adiabatic approximation with applications to quantum computation. *Journal of Mathematical Physics*, 48(102111), 2007.

[162] M.R. Jerrum, L.G. Valiant and V. Vazirani. Random generation of combinatorial structures from a uniform distribution. *Theoretical Computer Science*, 43, 1986.

[163] M. Johnson, M.H.S. Amin, S. Gildert, T. Lanting, F. Hamze, N. Dickson, R. Harris, A.J. Berkley, J. Johansson, P. Bunyk, E.M. Chapple, C. Enderud, J.P. Hilton, K. Karimi, E. Ladizinsky, N. Ladizinsky, T. Oh, I. Perminov, C. Rich, M.C. Thom, E. Tolkacheva, C.J.S. Truncik, S. Uchaikin, J. Wang, B. Wilson and G. Rose. Quantum annealing with manufactured spins. *Nature*, 473(7346), 2011.

[164] T. Kadowaki and H. Nishimori. Quantum annealing in the transverse Ising model. *Physical Review E*, 58(5), 1998.

[165] A. Kandala, A. Mezzacapo, K. Temme, M. Takita, M. Brink, J.M. Chow and J.M. Gambetta. Hardware-efficient variational quantum eigensolver for small molecules and quantum magnets. *Nature*, 549, 2017.

[166] H. Karimi and G. Rosenberg. Boosting quantum annealer performance via sample persistence. *Quantum Information Processing*, 16(7), 2017.

[167] R. Karp. Reducibility among combinatorial problems. In *Complexity of Computer Computations*. Plenum Press, 1972.

[168] T. Kato. On the adiabatic theorem of quantum mechanics. *Journal of the Physical*

Society of Japan, 5(6), 1950.

[169] J. Katz and Y. Lindell. Introduction to Modern Cryptography. Chapman & Hall/CRC Cryptography and Network Security, 3rd Edition, 2020.

[170] P. Kaye, R. Laflamme, and M. Mosca. An Introduction to Quantum Computing. OUP, 2007.

[171] H. Kellerer, R. Mansini and M.G. Speranza. Selecting portfolios with fixed costs and minimum transaction lots. *Annals of Operations Research*, 99(1-4), 2000.

[172] H. Kellerer and U. Pferschy. Knapsack Problems. Springer, 2004.

[173] I. Kerenidis, J. Landman and A. Prakash. Quantum algorithms for deep convolutional neural networks. arXiv:1911.01117, 2019.

[174] D. Khachatryan. Variational quantum eigensolver. github:DavitKhach, 2020.

[175] A. Khoshaman, W. Vinci, B. Denis, E. Andriyash, H. Sadeghi and M.H. Amin. Quantum variational autoencoder. *Quantum Science and Technology*, 4(1), 2019.

[176] P. Kidger and T. Lyons. Universal approximation with deep narrow networks. *Conference on Learning Theory*, 2020.

[177] M. Kim, K. Kim, J. Hwang, E.-G. Moon and J. Ahn. Rydberg quantum wires for maximum independent set problems. *Nature Physics*, 18, 2022.

[178] M. Kjaergaard, M.E. Schwartz, J. Braumüller, P. Krantz, J. I-Jan Wang, S. Gustavsson and W.D. Oliver. Superconducting qubits: current state of play. *Annual Review of Condensed Matter Physics*, 11, 2020.

[179] E. Knill, R. Laflamme and G.J. Milburn. A scheme for efficient quantum computation with linear optics. *Nature*, 409, 2001.

[180] D. Knuth. The Art of Computer Programming, Volume 3 / Sorting and Searching. Addison-Wesley, 2nd Edition, 1998.

[181] D. Koller and N. Friedman. Probabilistic Graphical Models. MIT Press, 2009.

[182] I. Kolotouros and P. Wallden. An evolving objective function for improved variational quantum optimisation. *Physical Review Research*, 4(2), 2022.

[183] A. Kondratyev. Curve dynamics with artificial neural networks. *Risk*, 31(6), 2018.

[184] A. Kondratyev. Non-differentiable learning of quantum circuit Born machine with genetic algorithm. *Wilmott*, 2021(114), 2021.

[185] A. Kondratyev. Quantum machine learning. *Presentation at the Quant Insights Conference*, 2021.

[186] A. Kondratyev and G. Giorgidze. Evolutionary algos for optimising MVA. *Risk*, 30(12), 2017.

[187] A. Kondratyev and C. Schwarz. The market generator. *Risk*, 33(2), 2020.

[188] A. Kondratyev, C. Schwarz and B. Horvath. The data anonymiser. *Risk*, 33(8), 2020.

[189] A. Kondratyev and D. Venturelli. Beyond Markowitz with quantum annealing. *Risk*, 32(6), 2019.

[190] A. Koshiyama, N. Firoozye and P. Treleaven. Generative adversarial networks for financial trading strategies fine-tuning and combination. *Quantitative Finance*, 21(5), 2021.

[191] P. Krantz, M. Kjaergaard, F. Yan, T. P. Orlando, S. Gustavsson and W.D. Oliver. A quantum engineer's guide to superconducting qubits. *Applied Physics Reviews*, 6(2), 2019.

[192] A. Kratsios and L. Papon. Universal approximation theorems for differentiable geometric deep learning. *Journal of Machine Learning Research*, 23(196), 2022.

[193] A. Krizhevsky. Learning multiple layers of features from tiny images. University of Toronto, Technical Report, 2009.

[194] S. Kshatriya and P.K. Prasanna. Genetic algorithm-based portfolio optimization with higher moments in global stock markets. *Journal of Risk*, 20(4), 2018.

[195] D. Kudrow, K. Bier, Z. Deng, D. Franklin, Y. Tomita, K.R. Brown and F.T. Chong. Quantum rotations: a case study in static and dynamic machine-code generation for quantum computers. *International Symposium on Computer Architecture*, 2013.

[196] L. Lamata, U. Alvarez-Rodriguez, J.D. Martín-Guerrero, M. Sanz and E. Solano. Quantum autoencoders via quantum adders with genetic algorithms. *Quantum Science and Technology*, 4(1), 2018.

[197] L.D. Landau and E.M. Lifshitz. Quantum Mechanics. Non-Relativistic Theory. Pergamon Press, 1965.

[198] R. Landauer. Irreversibility and heat generation in the computing process. *IBM Journal of Research and Development*, 5(3), 1961.

[199] J. Landman. Quantum algorithms for unsupervised machine learning and neural networks. *International Conference on Learning Representations*, 2020.

[200] R.Y. Li, R. Di Felice, R. Rohs and D.A. Lidar. Quantum annealing versus classical machine learning applied to a simplified computational biology problem. *Quantum Information*, 4(14), 2018.

[201] Q. Li, H. Wu, W. Qian, X. Li, Q. Zhu and S. Yang. Portfolio optimization based on quantum HHL algorithm. *International Conference on Artificial Intelligence and Security*, 2022.

[202] N. Linden, A. Montanaro and C. Shao. Quantum vs. classical algorithms for solving the heat equation. *Communications in Mathematical Physics*, 2022.

[203] A. Lipton and A. Treccani. Blockchain and Distributed Ledgers: Mathematics, Technology, and Economics. World Scientific, 2022.

[204] J.-G. Liu and L. Wang. Differentiable learning of quantum circuit Born machine. *Physical Review A*, 98(062324), 2018.

[205] L. Liu, S. Yang and D. Wang. Particle swarm optimization with composite particles in dynamic environments. *IEEE Transactions on Systems, Man, and Cybernetics, Part B:*

Cybernetics, 40(6), 2010.

[206] S. Lloyd. Universal quantum simulators. *Science*, 273(5278), 1996.

[207] S. Lloyd and C. Weedbrook. Quantum generative adversarial learning. *Physical Review Letters*, 121(4), 2018.

[208] M. López de Prado. Advances in Financial Machine Learning. Wiley, 2018.

[209] M. López de Prado. A robust estimator of the efficient frontier. SSRN:3469961, 2019.

[210] R. Loredo. Learn Quantum Computing with Python and IBM Quantum Experience. Packt, 2020.

[211] J. Lu, and S. Yi. Autoencoding conditional GAN for portfolio allocation diversification. arXiv:2207.05701, 2022.

[212] A. Lucas. Ising formulations of many NP problems. *Frontiers in Physics*, 2(5), 2014.

[213] X.-Z. Luo, J.-G. Liu, P. Zhang and L. Wang. Yao.jl: Extensible, efficient framework for quantum algorithm design. *Quantum*, 4, 2020.

[214] M. Magris, M. Shabani and A. Iosifidis. Bayesian bilinear neural network for predicting the mid-price dynamics in limit-order book markets. *Journal of Forecasting*, 42(6), 2023.

[215] F. Mallet, F.R. Ong, A. Palacios-Laloy, F. Nguyen, P. Bertet, D. Vion and D. Esteve. Single-shot qubit readout in circuit quantum electrodynamics. *Nature Physics*, 5, 2009.

[216] S. Mandrà and H.G. Katzgraber. A deceptive step towards quantum speedup detection. *Quantum Science and Technology*, 3(4), 2018.

[217] H. Markowitz. Portfolio selection. *Journal of Finance*, 7(1), 1952.

[218] J. Marshall, D. Venturelli, I.Hen and E.G. Rieffel. The power of pausing: advancing understanding of thermalization in experimental quantum annealers. *Physical Review Applied*, 11(044083), 2019.

[219] J.M. Martinis. Superconducting phase qubits. *Quantum Information Processing*, 8, 2009.

[220] M. Marzec. Portfolio optimization: applications in quantum computing. In *Handbook of high-frequency trading and modeling in Finance*. John Wiley & Sons, 2016.

[221] K.-P. Marzlin and B.C. Sanders. Inconsistency in the application of the adiabatic theorem. *Physical Review Letters*, 93(160408), 2004.

[222] J.R. McClean, S. Boixo, V.N. Smelyanskiy, R. Babbush and H. Neven. Barren plateaus in quantum neural network training landscapes. *Nature Communications*, 9(4812), 2018.

[223] D. McClure and J. Gambetta. Quantum computation center opens. www.ibm.com/blogs/research/2019/09/quantum-computation-center, 2019.

[224] C. McGeoch. Adiabatic quantum computation and quantum annealing: theory and practice. Synthesis Lectures on Quantum Computing, Morgan & Claypool, 2014.

[225] C. McGeoch, P. Farre and K. Boothby. The D-Wave Advantage2 Prototype. D-Wave Technical Report, 2022.

[226] C. McGeoch and C. Wang. Experimental evaluation of an adiabiatic quantum system for combinatorial optimization. *Proceedings of the ACM International Conference on Computing Frontiers*, 2013.

[227] P.A. Mello. The von Neumann model of measurement in quantum mechanics. *American Institute of Physics Conference Proceedings*, 1575, 2014.

[228] K. Mitarai, M. Negoro, M. Kitagawa and K. Fujii. Quantum circuit learning. *Physical Review A*, 98(032309), 2018.

[229] M. Mitchell. An Introduction to Genetic Algorithms. MIT Press, 1998.

[230] N. Moll, P. Barkoutsos, L.S. Bishop, J.M. Chow, A. Cross, D.J. Egger, S. Filipp, A. Fuhrer, J.M. Gambetta, M. Ganzhorn, A. Kandala, A. Mezzacapo, P. Müller, W. Riess, G. Salis, J. Smolin, I. Tavernelli and K. Temme. Quantum optimization using varia-

tional algorithms on near-term quantum devices. *Quantum Science and Technology*, 3(030503), 2018.

[231] A. Montanaro. Quantum speedup of Monte Carlo methods. *Proceedings of the Royal Society A*, 471(2181), 2015.

[232] M. Morini. One more model risk when using Gaussian copula for risk management. SSRN:1520670, 2009.

[233] A. Mott, J. Job, J.R. Vlimant, D. Lidar and M. Spiropulu. Solving a Higgs optimization problem with quantum annealing for machine learning. *Nature*, 550, 2017.

[234] W.C. Myrvold. On the relation of the laws of thermodynamics to statistical mechanics. Preprint. http://philsci-archive.pitt.edu/id/eprint/19361, 2021.

[235] H. Naomichi and M. Suzuki. Finding exponential product formulas of higher orders. *Quantum annealing and other optimization methods*. Springer, 2005.

[236] J. Nemirovsky and Y. Sagi. Fast universal two-qubit gate for neutral fermionic atoms in optical tweezers. *Physical Review Research*, 3(013113), 2021.

[237] N.H. Nguyen, E.C. Behrman and J.E. Steck. Quantum learning with noise and decoherence: a robust quantum neural network. *Quantum Machine Intelligence*, 2(1), 2020.

[238] M.A. Nielsen and I.S. Chuang. Quantum Computation and Quantum Information. Cambridge University Press, 10th Anniversary Edition, 2010.

[239] S.M. Nikolskii. Approximation of Functions of Several Variables and Imbedding Theorems. Springer, 1975.

[240] J.L. O'Brien. Optical quantum computing. *Science*, 318(5856), 2007.

[241] B. Øksendal. Stochastic Differential Equations. Springer, 5th Edition, 2000.

[242] G. Ortiz, J.E. Gubernatis, E. Knill and R. Laflamme. Quantum algorithms for fermionic simulations. *Physical Review A*, 64(022319), 2001.

[243] R. Orús, S. Mugel and E. Lizaso. Quantum computing for finance: overview and prospects. *Reviews in Physics*, 4, 2019.

[244] C. Paar and J. Pelzl. Understanding Cryptography. Springer, 2010.

[245] M. Paini. Quantum tomography via group theory. arXiv:quant-ph/0002078, 2000.

[246] M. Paini. Quantum Finance: the road to business applications. *Presentation at the London Quantum Computing Meetup*. Quantum Finance, 2018.

[247] M. Paini, A. Kalev, D. Padilha and B. Ruck. Estimating expectation values using approximate quantum states. *Quantum*, 5, 2021.

[248] S. Park, C. Yun, J. Lee and J. Shin. Minimum width for universal approximation. *International Conference on Learning Representations*, 2021.

[249] F. Pedregosa, G. Varoquaux, A. Gramfort, V. Michel, B. Thirion, O. Grisel, M. Blondel, P. Prettenhofer, R. Weiss, V. Dubourg, J. Vanderplas, A. Passos, D. Cournapeau, M. Brucher, M. Perrot and É. Duchesnay. Scikit-learn: Machine learning in Python. *Journal of Machine Learning Research*, 12, 2011.

[250] A. Pérez-Salinas, A. Cervera-Lierta, E. Gil-Fuster and J.I. Latorre. Data re-uploading for a universal quantum classifier. *Quantum*, 4, 2020.

[251] A. Pérez-Salinas, D. López-Núñez, A. García-Sáez, P. Forn-Díaz and J.I. Latorre. One qubit as a universal approximant. *Physical Review A*, 104(1), 2021

[252] A. Peruzzo, J. McClean, P. Shadbolt, M.-H. Yung, X.-Q. Zhou, P. J. Love, A. Aspuru-Guzik and J.L. O'Brien. A variational eigenvalue solver on a photonic quantum processor. *Nature Communications*, 5213, 2014.

[253] E. Peters, J. Caldeira, A. Ho, S. Leichenauer, M. Mohseni, H. Neven, P. Spentzouris, D. Strain and G.N. Perdue. Machine learning of high dimensional data on a noisy quantum processor. *Quantum Information*, 7(1), 2021.

[254] R.C. Pfaffenberger and J.H. Patterson. Statistical Methods for Business and Economics. Irwin, 3rd Edition, 1987.

[255] F. Phillipson and H.S. Bhatia. Portfolio optimisation ssing the D-Wave quantum annealer. In *Computational Science – ICCS*. Lecture Notes in Computer Science, 12747, Springer, 2021.

[256] E.A. Poe. The Complete Tales and Poems of Edgar Allan Poe. Vintage, 1975.

[257] B. Pokharel, Z.G. Izquierdo, P.A. Lott, E. Strbac, K. Osiewalski, E. Papathanasiou, A. Kondratyev, D. Venturelli and E. Rieffel. Inter-generational comparison of D-Wave quantum annealers in solving hard scheduling problems. *Quantum Information Processing*, 22(364), 2023.

[258] R. Poli. Analysis of the publications on the applications of particle swarm optimisation. *Journal of Artificial Evolution and Applications*, 2008.

[259] J. Preskill. Quantum Computing in the NISQ era and beyond. *Quantum*, 2, 2018.

[260] Qiskit: An open-source framework for quantum computing. qiskit.org, 2019.

[261] Quantinuum: Quantum Roadmap. https://www.quantinuum.com/press-releases/quantinuum-unveils-accelerated-roadmap-to-achieve-universal-fault-tolerant-quantum-computing-by-2030, 2024.

[262] J.R. Quinlan. Australian Credit Approval Dataset. UCI Machine Learning Repository. http://archive.ics.uci.edu/ml. UC Irvine, 1987.

[263] J.R. Quinlan. Simplifying Decision Trees. *International Journal of Man-Machine Studies*, 27(3), 1987.

[264] S. Raschka and V. Mirjalili. Python Machine Learning. Packt, 3rd Edition, 2019.

[265] S.E. Rasmussen and N.T. Zinner. Simple implementation of high fidelity controlled-iSWAP gates and quantum circuit exponentiation of non-Hermitian gates. *Physical Review Research*, 2(033097), 2020.

[266] P. Rebentrost and S. Lloyd. Quantum computational finance: quantum algorithm for portfolio optimization. arXiv:1811.03975, 2018.

[267] Rigetti Quantum Roadmap: Rigetti Computing Reports First Quarter 2022 Financial

Results and Provides Business Update. https://investors.rigetti.com/node/7371/pdf, 2022.

[268] Rigetti. Novera QPU. https://www.rigetti.com/novera, 2024.

[269] J.J. Rissanen. Fisher information and stochastic complexity. *IEEE Transactions on Information Theory*, 42(1), 1996.

[270] R. Rivest, A. Shamir and L. Adleman. A Method for Obtaining Digital Signatures and Public-Key Cryptosystems. *Communications of the ACM*, 21(2), 1978.

[271] A. Robert, P.K. Barkoutsos, S. Woerner and I. Tavernelli. Resource-efficient quantum algorithm for protein folding. *Quantum Information*, 7(38), 2021.

[272] R. Robinett. Quantum Mechanics: Classical Results, Modern Systems, and Visualized Examples. Oxford University Press, 2nd Edition, 2006.

[273] T.F. Rønnow, Z. Wang, J. Job, S. Boixo, S.V. Isakov, D. Wecker, J.M. Martinis, D.A. Lidar and M. Troyer. Defining and detecting quantum speedup. *Science*, 345(6195), 2014.

[274] G. Rosenberg, P. Haghnegahdar, P. Goddard, P. Carr, K. Wu and M. López De Prado. Solving the optimal trading trajectory problem using a quantum annealer. *IEEE Journal of Selected Topics in Signal Processing*, 10(6), 2016.

[275] R. Salakhutdinov, A. Mnih and G. Hinton. Restricted Boltzmann machines for collaborative filtering. *Proceedings of the 24th International Conference on Machine Learning*, 2007.

[276] G.E. Santoro, R. Martonak, E. Tosatti and R. Car. Theory of quantum annealing of an Ising spin glass. *Science*, 295(2427), 2002.

[277] V. Schmitt. Design, fabrication and test of a four superconducting quantum-bit processor. PhD Thesis, Université Pierre et Marie Curie, 2015.

[278] A. Schrijver. Theory of Integer and Linear Programming. Wiley, 1998.

[279] M. Schuld, V. Bergholm, C. Gogolin, J. Izaac and N. Killoran. Evaluating analytic

gradients on quantum hardware. *Physical Review A*, 99(3), 2019.

[280] M. Schuld, A. Bocharov, K. Svore and N. Wiebe. Circuit-centric quantum classifiers. *Physical Review A*, 101(3), 2020.

[281] M. Schuld, R. Sweke and J.J. Meyer. Effect of data encoding on the expressive power of variational quantum-machine-learning models. *Physical Review A*, 103, 2021.

[282] F. Schwabl. Quantum Mechanics. Springer, 4th Edition, 2007.

[283] A. Selby. D-Wave: comment on comparison with classical computers, 2013.

[284] S. Sethi. Optimal Control Theory. Applications to Management Science and Economics. Springer, 2019.

[285] A. Sethia, R. Patel and P. Raut. Data augmentation using generative models for credit card fraud detection. *4th International Conference on Computing Communication and Automation*, 2018.

[286] A.D. Shapiro. King-Rook vs. King-Pawn Chess Dataset. UCI Machine Learning Repository. http://archive.ics.uci.edu/ml. UC Irvine, 1987.

[287] A.D. Shapiro. Structured Induction in Expert Systems. Addison-Wesley, 1987.

[288] R.R. Sharapov and A.V. Lapshin. Convergence of genetic algorithms. *Pattern Recognition and Image Analysis*, 16(3), 2006.

[289] P.W. Shor. Polynomial-time algorithms for prime factorization and discrete logarithms on a quantum computer. *SIAM Journal on Computing*, 26(5), 1997.

[290] S. Sim, P.D. Johnson and A. Aspuru-Guzik. Expressibility and entangling capability of parameterized quantum circuits for hybrid quantum-classical algorithms. *Advanced Quantum Technologies*, 2(12), 2019.

[291] B. Simon. Holonomy, the quantum adiabatic theorem, and Berry's phase. *Physical Review Letters*, 51(2167), 1983.

[292] H. Situ, Z. He, Y. Wang, L. Li and S. Zheng. Quantum generative adversarial network for generating discrete distribution. *Information Sciences*, 538, 2020.

[293] G.D. Smith. Numerical Solution of Partial Differential Equations: Finite Difference Methods. Oxford University Press, 1985.

[294] P. Smolensky. Information processing in dynamical systems: Foundations of harmony theory. Parallel Distributed Processing Explorations in the Microstructure of Cognition, 1: Foundations. MIT Press, 1986.

[295] K. Sörensen. Clustering in Financial Markets: A Network Theory Approach. Master of Science Thesis, Royal Institute of Technology, Stockholm, 2014.

[296] S. Sra. Metrics induced by Jensen-Shannon and related divergences on positive definite matrices. *Linear Algebra and its Applications*, 616, 2021.

[297] M. Steffen, W. van Dam, T. Hogg, G. Breyta and I. Chuang. Experimental implementation of an adiabatic quantum optimization algorithm. *Physical Review Letters*, 90(6), 2003.

[298] S.A. Stein, B. Baheri, D. Chen, Y. Mao, Q. Guan, A. Li, B. Fang and S. Xu. QuGAN: a Generative Adversarial Network through quantum states. arXiv:2010.09036, 2020.

[299] F.W. Strauch, P.R. Johnson, A.J. Dragt, C.J. Lobb, J.R. Anderson and F.C. Wellstood. Quantum logic gates for coupled superconducting phase qubits. *Physical Review Letters*, 91(167005), 2003.

[300] M. Suchara. Quantum algorithms and their applications. *CQE-Protiviti Design Thinking Workshop*, 2021.

[301] Y. Sung, L. Ding, J. Braumüller, A. Vepsäläinen, B. Kannan, M. Kjaergaard, A. Greene, G.O. Samach, C. McNally, D. Kim, A. Melville, B.M. Niedzielski, M.E. Schwartz, J.L. Yoder, T.P. Orlando, S. Gustavsson and W.D. Oliver. Realization of high-fidelity CZ and ZZ-free iSWAP gates with a tunable coupler. *Physical Review X*, 11(021058), 2021.

[302] R.S. Sutor. Dancing with Qubits, Packt, 2019.

[303] M. Suzuki. General decomposition theory of ordered exponentials. *Proceedings of Japan Academy*, 69(B), 1993.

[304] G.W. Taylor, G.E. Hinton and S.T. Roweis. Two distributed-state models for generating high-dimensional time series. *Journal of Machine Learning Research*, 12(28), 2011.

[305] M. Tegmark. Life 3.0. Being Human in the Age of Artificial Intelligence. Allen Lane, Penguin Random House, 2017.

[306] J.L. Ticknor. A Bayesian regularized artificial neural network for stock market forecasting. *Expert Systems with Applications*, 40(14), 2013.

[307] J. Tilly, H. Chen, S. Cao, D. Picozzi, K. Setia, Y. Li, E. Grant, L. Wossnig, I. Rungger, Ivan and G. Booth and J. Tennyson. The variational quantum eigensolver: a review of methods and best practices. *Physics Reports*,986, 2022.

[308] L. Tomawski, I. Mrózb and Z. Kukułac. From Thomson formula to resonant equivalent diagrams. *Acta Physica Polonica A*, 139(3), 2021.

[309] T. Toffoli. Reversible computing. *International Colloquium on Automata, Languages, and Programming*, 1980.

[310] T.T. Tran, M. Do, E.G. Rieffel, J. Frank, Z. Wang, B. O'Gorman, D. Venturelli and J.C. Beck. A hybrid quantum-classical approach to solving scheduling problems. *The 9th Annual Symposium on Combinatorial Search*, 2016.

[311] H.F. Trotter. On the product of semi-groups of operators. *Proceedings of the American Mathematical Society*, 10(4), 1959.

[312] C.A. Trugenberger. Probabilistic quantum memories. *Physical Review Letters*, 87(067901), 2001.

[313] H. Tuy. Minimax theorems revisited. *Acta Mathematica Vietnamica*, 29(3), 2004.

[314] F. Vaezi, S.J. Sadjadi and A. Makui. A portfolio selection model based on the knapsack problem under uncertainty. *PLoS ONE*, 14(5), 2019.

[315] J. van Apeldoorn and A. Gilyén. Improvements in quantum SDP-solving with applications. *International Colloquium on Automata, Languages, and Programming*, 2019.

[316] W. van Dam, M. Mosca and U. Vazirani. How powerful is adiabatic quantum computation? *Proceedings, Symposium on the Foundations of Computer Science*, 2001.

[317] L. Vandenberghe and S. Boyd. Semidefinite programming. *SIAM Review*, 38(1), 1996.

[318] V. Vapnik and A. Chervonenkis. On the uniform convergence of relative frequencies of events to their probabilities. *Theory of Probability and its Applications*, 16(2), 1971.

[319] D. Ventura and T. Martinez. Quantum associative memory. *Information Sciences*, 124(1), 2000.

[320] D. Venturelli and A. Kondratyev. Reverse quantum annealing approach to portfolio optimization problems. *Quantum Machine Intelligence*, 1(3), 2019.

[321] D. Venturelli, D.J.J. Marchand and G. Rojo. Quantum annealing implementation of job-shop scheduling. arXiv:1506.08479, 2015.

[322] J. von Neumann. Zur Theorie der Gesellschaftsspiele. *Mathematische Annalen*, 100: 295-320, 1928.

[323] X. Wang, Y. Du, Y. Luo and D. Tao. Towards understanding the power of quantum kernels in the NISQ era. *Quantum*, 5, 2021.

[324] S. Wang, E. Fontana, M. Cerezo, K. Sharma, A. Sone, L. Cincio and P.J. Coles. Noise-induced barren plateaus in variational quantum algorithms. *Nature Communications*, 12(1), 2021.

[325] S. Wei, ShiJie, Y. Chen, Z. Zhou and G. Long. A quantum convolutional neural network on NISQ devices. *AAPPS Bulletin*, 32(1), 2022.

[326] M. Wiese, B. Wood, A. Pachoud, R. Korn, H. Buehler, P. Murray and L. Bai. Multi-asset spot and option market simulation. arXiv:2112.06823, 2021.

[327] M. Wiese, R. Knobloch, R. Korn and P. Kretschmer. Quant GANs: deep generation of financial time series. *Quantitative Finance*, 20(9), 2020.

[328] J.W.J. Williams. Algorithm 232 — Heapsort. *Communications of the ACM*, 7(6), 1964.

[329] S. Wissman, B. Leggio and H.-P. Breuer. Detecting initial system-environment correlations: Performance of various distance measures for quantum states. *Physical Review A*, 88(2), 2013.

[330] J. Wurtz and P.J. Love. Counterdiabaticity and the quantum approximate optimization algorithm. *Quantum*, 6, 2022.

[331] N. Xu, J. Zhu, D. Lu, X. Zhou, X. Peng and J. Du. Quantum factorization of 143 on a dipolar-coupling nuclear magnetic resonance system. *Physical Review Letters*, 108(13), 2012.

[332] R. Yalovetzky, P. Minssen, D. Herman and M. Pistoia. NISQ-HHL: Portfolio optimization for near-term quantum hardware. arXiv:2110.15958, 2021.

[333] D. Yarotsky. Error bounds for approximations with deep ReLU networks. *Neural Networks*, 94, 2017.

[334] I.C. Yeh and C.H. Lien. Default of credit card clients dataset. UCI Machine Learning Repository. http://archive.ics.uci.edu/ml. UC Irvine, 2009.

[335] I.C. Yeh and C.H. Lien. The comparisons of data mining techniques for the predictive accuracy of probability of default of credit card clients. *Expert Systems with Applications*, 36(2), 2009.

[336] K. Yosida. Functional Analysis. Springer, 1965.

[337] T. Zhao, C. Sun, A. Cohen, J. Stokes and S. Veerapaneni. Quantum-inspired variational algorithms for partial differential equations: Application to financial derivative pricing. arXiv:2207.10838, 2022.

[338] X. Zhou, S. Li and Y. Feng. Quantum circuit transformation based on simulated annealing and heuristic search. *IEEE Transactions on Computer-Aided Design of Integrated Circuits and Systems*, 39(12), 2020.

[339] E. Zhu, S. Johri, D. Bacon, M. Esencan, J. Kim, M. Muir, Mark N. Murgai, J. Nguyen, N. Pisenti, A. Schouela, K. Sosnova and K. Wright. Generative quantum learning

of joint probability distribution functions. *Bulletin of the American Physical Society*, 2022.

[340] D. Zhu, N.M. Linke, M. Benedetti, K.A. Landsman, N.H. Nguyen, C.H. Alderete, A. Perdomo-Ortiz, N. Korda, A. Garfoot, C. Brecque, L. Egan, O. Perdomo and C. Monroe. Training of quantum circuits on a hybrid quantum computer. *Science Advances*, 5(10), 2019.

[341] C. Zoufal, A. Lucchi and S. Woerner. Quantum generative adversarial networks for learning and loading random distributions. *Quantum Information*, 5(103), 2019.

Index

‹packt›

Subscribe to our online digital library for full access to over 7,000 books and videos, as well as industry leading tools to help you plan your personal development and advance your career. For more information, please visit our website.

Why subscribe?

- Spend less time learning and more time coding with practical eBooks and Videos from over 4,000 industry professionals
- Improve your learning with Skill Plans built especially for you
- Get a free eBook or video every month
- Fully searchable for easy access to vital information
- Copy and paste, print, and bookmark content

Did you know that Packt offers eBook versions of every book published, with PDF and ePub files available? You can upgrade to the eBook version at www.packtpub.com and as a print book customer, you are entitled to a discount on the eBook copy. Get in touch with us at customercare@packtpub.com for more details.

At www.packtpub.com, you can also read a collection of free technical articles, sign up for a range of free newsletters, and receive exclusive discounts and offers on Packt books and eBooks.

Other Books You Might Enjoy

If you enjoyed this book, you may be interested in these other books by Packt:

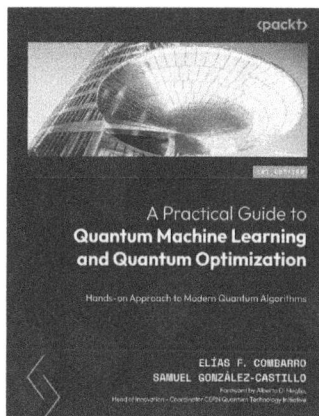

A Practical Guide to Quantum Machine Learning and Quantum Optimization

Elías F. Combarro, Samuel González-Castillo

ISBN: 978-1-80461-383-2

- Review the basics of quantum computing
- Gain a solid understanding of modern quantum algorithms
- Understand how to formulate optimization problems with QUBO
- Solve optimization problems with quantum annealing, QAOA, GAS, and VQE
- Find out how to create quantum machine learning models
- Explore how quantum support vector machines and quantum neural networks work using Qiskit and PennyLane
- Discover how to implement hybrid architectures using Qiskit and PennyLane and its PyTorch interface

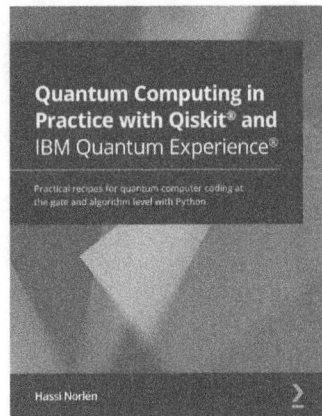

Quantum Computing in Practice with Qiskit® and IBM Quantum Experience®

Hassi Norlén

ISBN: 978-1-83882-844-8

- Visualize a qubit in Python and understand the concept of superposition
- Install a local Qiskit® simulator and connect to actual quantum hardware
- Compose quantum programs at the level of circuits using Qiskit® Terra
- Compare and contrast Noisy Intermediate-Scale Quantum computing (NISQ) and Universal Fault-Tolerant quantum computing using simulators and IBM Quantum® hardware
- Mitigate noise in quantum circuits and systems using Qiskit® Ignis
- Understand the difference between classical and quantum algorithms by implementing Grover's algorithm in Qiskit®

Packt is searching for authors like you

If you are interested in becoming an author for Packt, please visit `https://partners` `hips.packt.com/contributors/` and apply today. We have worked with thousands of developers and tech professionals, just like you, to help them share their insight with the global tech community. You can make a general application, apply for a specific hot topic that we are recruiting an author for, or submit your own idea.

Share Your Thoughts

Now you've finished *Quantum Machine Learning and Optimisation in Finance, Second Edition*, we'd love to hear your thoughts! Scan the QR code below to go straight to the Amazon review page for this book and share your feedback.

`https://packt.link/r/1836209614`

Your review is important to us and the tech community and will help us make sure we're delivering excellent quality content.

Download a free PDF copy of this book

Thanks for purchasing this book!

Do you like to read on the go but are unable to carry your print books everywhere? Is your eBook purchase not compatible with the device of your choice?

Don't worry, now with every Packt book you get a DRM-free PDF version of that book at no cost.

Read anywhere, any place, on any device. Search, copy, and paste code from your favorite technical books directly into your application.

The perks don't stop there, you can get exclusive access to discounts, newsletters, and great free content in your inbox daily.

Follow these simple steps to get the benefits:

1. Scan the QR code or visit the link below:

https://packt.link/free-ebook/9781836209614

2. Submit your proof of purchase
3. That's it! We'll send your free PDF and other benefits to your email directly